Plant Anatomy

Plant Anatomy

— *Authors* —

T. Pullaiah

Department of Botany,
Sri Krishnadevaraya University
Ananatapur

K.C. Naidu

K. Lakshminarayana

B. Hanumantha Rao

Department of Botany,
Andhra University
Visakhapatnam

P. Suresh Narayana

Department of Botany,
S.S.B.N. Junior College,
Ananatapur

2022

Regency Publications
A Division of
Astral International Pvt. Ltd.
New Delhi – 110 002

ISBN: 9789354616372 (HB)

Published by : **Regency Publications**
A Division of
Astral International Pvt. Ltd.
– ISO 9001:2015 Certified Company –
4736/23, Ansari Road, Darya Ganj
New Delhi-110 002
Ph. 011-43549197, 23278134
E-mail: info@astralint.com
Website: www.astralint.com

Preface

The first four authors of this book earlier wrote a book in 2007 entitled *Plant Development*. This book was written taking into consideration of UGC syllabus of M.Sc. Botany/Plant Sciences. Although this book was intended to the students, Publishers published this as Library edition with a little higher cost. Although the book was received well the sales was sluggish for this higher cost. Further much of the information in this book was on Plant Anatomy. Hence, teachers and students of Botany requested us to revise the book and retitle it as Plant Anatomy. The result of all this is the present book. This is the revised and enlarged edition of our earlier book *Plant development*. We are bringing this book both as library edition and also as student edition so that it is affordable. We co-opted fifth author P. Suresh Narayana for revising this book.

We thank all the teachers and students for giving valuable inputs for revision of this book. All the chapters are thoroughly revised and updated. Quite a good amount of information on the modern aspects are given. We request the readers of this book to bring to our notice errors and omission if any. We thank Anil Mittal and others of Regency Publication for their continued cooperation in bringing out this edition in a presentable form within a short time.

T. Pullaiah

K.C. Naidu

K. Lakshminarayana

B. Hanumantha Rao

P. Suresh Narayana

Contents

The Authors

Dr. T. Pullaiah is a former Professor at the Department of Botany at Sri Krishnadevaraya University in Andhra Pradesh, India, where he has taught for more than 35 years. He has held several positions at the university, including Dean, Faculty of Biosciences, Head of the Department of Botany, Head of the Department of Biotechnology, and Member of Academic Senate. He was President of Indian Botanical Society (2014), President of the Indian Association for Angiosperm Taxonomy (2013) and Fellow of Andhra Pradesh Akademi of Sciences. He was awarded the Panchanan Maheshwari Gold Medal, the Prof. P.C. Trivedi Medal, the Dr. G. Panigrahi Memorial Lecture award of the Indian Botanical Society and Prof. Y.D. Tyagi Gold Medal of the Indian Association for Angiosperm Taxonomy, and a Best Teacher Award from Government of Andhra Pradesh. Under his guidance 54 students obtained their doctoral degrees. He has authored 52 books, edited 23 books, and published over 330 research papers, including reviews and book chapters. His books include *Advances in Cell and Molecular Diagnostics* (published by Elsevier), *Camptothecin and Camptothecin producing Plants* (Elsevier) *Ethnobotany of India* (5 volumes published by Apple Academic Press), *Global Biodiversity* (4 volumes, Apple Academic Press), *Red Sanders: Silviculture and Conservation* (Springer), *Genetically Modified Crops* (2 volumes, Springer), *Monograph on Brachystelma and Ceropegia in India* (CRC Press), *Flora of Andhra Pradesh* (5 volumes), *Flora of Eastern Ghats* (4 volumes), *Flora of Telangana* (3 volumes), *Encyclopaedia of World Medicinal Plants* (7 volumes, 2nd edition), and *Encyclopaedia of Herbal Antioxidants* (3 volumes). He was also a member of Species Survival Commission of the International Union for Conservation of Nature (IUCN). Professor Pullaiah received his PhD from Andhra University, India, attended Moscow State University, Russia, and worked as Post-Doctoral Fellow during 1976-1978.

Prof. K. Chandrasekhar Naidu, is a retired Professor of Botany, Andhra University, Visakhapatnam. He obtained his Ph.D degree in Botany from Andhra University in the year 1982. He has more than 30 years of Teaching and research experience in Andhra University. Chandrasekhar Naidu authored 2 books and published about 80 Research Papers in National and International Journals. He got three funded Research Projects that were sponsored by DST, UGC, New Delhi and Calyx Therapeutics Inc, California, USA. He was also a former Executive Counsel Member of JNTUA, Anantapur, A.P. He is a good Philanthropist and has founded an engineering College-Vemu Institute of Technology and running it very successfully for the past 14 years.

Prof. K. Lakshminarayana took his M.Sc. (with distinction) and Ph.D. degrees from Andhra University in 1974 and 1978 respectively. He is recipient of the National Associateship of UGC in 1985. He has been in the teaching faculty from 1979 to 2012. He has completed five research projects and has published about 100 research papers in several National and International journals of repute. He is in the editorial board of Flora and Fauna. His research works are cited in several international standard books. He is an active member of various academic and research associations and actively participated in several conferences/seminars and symposia. Under his guidance 10 students obtained their Ph. D degrees and 15 students their M.Phil. degrees.

Prof. B. Hanumantha Rao obtained his M.Sc. (Gold Medalist) and Ph.D. Degrees from Andhra University. He is a former Professor, Head and Chairman Board of Studies in the Department of Botany, Andhra University, Visakhapatnam, with many years of Teaching and Research experience. Prof. Rao published more than 80 Research Papers in Journals of National and International repute. Several students obtained their Ph.D. and M.Phil. Degrees under his guidance. He is the Editor of some Books in Telugu and English for Under graduate and Postgraduate students. He is associated with several National and International Professional Bodies.

P. Suresh Narayana obtained his M. Sc. degree (1971) in Botany from Sri Venkateswara University. Later, he joined the Department of Botany, S. S. B. N. Junior College, Anantapur (1971) as a Junior Lecturer and served the institution for more than 37 years. He rose to become the Principal of the same institution (2006). He took active interest in popularizing Botany among College and High School students. Many a time he participated in District Science Fairs and Children's Science Congresses as an adjudicator and encouraged the students to develop the scientific temper. He participated in several National Symposia in Botany, Environmental Sciences and Biotechnology. He gave many radio talks on plants, environment and biotechnology over All India Radio. He actively participates in the social activities among the elite of Anantapur. He is a member of a few scientific societies of India. He also organized a couple of symposia in Plant Sciences and Biotechnology, Quiz programmes and elocution and essay writing competitions. During his tenure as Principal, he established the NSS unit in the college and conducted many programmes such as '*Clean and Green*', '*Planting*' and '*Environmental Awareness*'. He conducted many botanical tours and accompanied students for collections. His hobbies include reading, correspondence, photography and collection of books and stamps. He is also an Executive Member of a few local educational institutions and organizations. His other publications are *Eminent Indian Botanists – Past and Present – (Biographies and Contributions)*, *Text Book of Pharmacognosy* (3 volumes), *Research Methodology in Plant Sciences* (revised 2nd edition), *Research Methodology in Zoology* and *Botany for Competitive Exams.* He retired from active service in June 2008.

Chapter 1

Seed Germination and Seedling Growth

Introduction

Seed is the representative of the next generation of a plant and is an embryonic plant enclosed in a protective covering, the seed coat. The seed formation is a part of the process of reproduction in all the seed plants (spermatophytes – both the Gymnosperms and the Angiosperms). Seed is the product of a ripened ovule (after fertilization by the pollen and some growth within the mother plant). The embryo is developed from the zygote and the seed coat from the integuments of the ovule. The seeds have been an important development in the reproduction and success of the spermatophytes to get established in their environments. The term – seed – also has a general meaning that antedates the above – anything that can be sown (for example, seed potatoes, sunflower seeds, *etc.*). The seeds serve many functions such as: the nourishment of the embryo – dispersal to a new location – dormancy under unfavourable conditions *etc.*

Seed Germination

The seeds are the connection between the past and the future. The seed contains the genetic wisdom of the past and the potential for its perpetuation in the future. The natural packaging of the genetic information in a seed is remarkable in itself a package to protect the genotype for longer periods of time when properly stored in germplasm collections. The timing of seed germination dictates a seedling's seasonal exposure to potentially lethal environmental factors, and thus has strong fitness consequences. The seeds remain in a state of dormancy induced by desiccation and the hormone abscissic acid until the conditions for growth become favourable. The germination in a particular situation and season is determined by the interaction between the dormancy releasing factors which influence the termination of dormancy or the initiation of germination and seedling growth in many plant species (such as phytohormones, light, temperature, water, nutrients, moisture or mechanical cues).

The identification of the time of physiological maturity has been a controversial subject among different authors, and three concepts are dominant - identified as maximum seed dry matter accumulation – reached when there is no further significant increase in the dry weight of the seed – and occurs when the seeds reach maximum dry weight, germination and vigour. The first concept was proposed by Shaw and Loomis (1950), as the stage in which the seed possesses maximum dry weight and yield. The physiological maturity has been termed a "relative maturity" (Aldrich, 1943), "morphological maturity" (Anderson, 1955), "mass maturity" (Ellis and Pieta Filho, 1992), while others prefer to use the terms like "harvest maturity", "agronomic maturity" and "time of harvest" to identify the same stage. According to Hartmann *et al.,* 3 conditions are required for germination to be initiated. They are: the seed must be viable; the seed dormancy must be overcome; and the seed must be subjected to appropriate environmental conditions such as water, temperature, oxygen, and light.

Seed germination is a critically important junction in the life cycle of the plant and the decision made by an imbibing seed to initiate germination can be considered as a critical regulatory step in the development of the plant. The function of the seed is to establish a new plant. To fulfil its function, seed has to germinate to give rise to the seedling and eventually matures into an adult plant. Germination is a process by which a seed embryo develops into a seedling, and this involves the reactivation of the metabolic pathways that lead to the growth and the emergence of the radicle (seed root) and the plumule (shoot). The emergence of the seedling above the soil is the next phase, called the seed establishment.

Seed Germination - Stages

Germination involves three distinct stages, namely:

Stage-1: imbibition of water to hydrate the seed and the activation of metabolism *i.e.,* increased respiration and protein synthesis:

Stage-2: the digestion of the stored food (the starch into sugars in the cotyledons or the endosperm) and the translocation to the embryo (when the sugars move to the embryo for its growth): and

Stage-3 - cell division and growth - resulting in the development of the seedling.

The stages 2 and 3 are also referred to as the "lag phase" – and emergence of the radicle. The germination commences with the uptake of water by the dry seed (a process called imbibition = soak up). The embryo imbibes water which causes it to swell, thus splitting the seed coat. The rate of imbibition is dependent on the permeability of the seed coat, amount of water in the environment and the area of contact the seed has to the source of water. Germination (the sprouting of a seed, a spore, or any other reproductive body) occurs usually after a period of dormancy. The absorption of water, the passage of time, chilling, warming, availability of oxygen, and the exposure of light may all operate in the process of initiation.

During the process of germination, water is absorbed by the embryo and it results in the rehydration and expansion of the cells. Shortly after water uptake, the respiratory rate increases, and many metabolic processes resume. The process of germination involves considerable amount of the metabolic activities. These events are associated with the structural changes in the organelles in the cells of the embryo. The enzymes necessary for the initiation of metabolic activity are present within the dry seed. The resumption of respiratory activity commences. Then, the events including the synthesis of nucleic acids, proteins, metabolization of the storage reserves are associated with growth of the seedling are continued until seedling becomes autotrophic.

Seedling Emergence

The emergence of the seedling begins with the emergence of the primary root (radicle) from the seed, though in some species, the shoot (plumule) emerges first as in coconut. Within a short time, the cell division begins in the young root and the shoot, and thereafter growth and organogenesis associated with combination of increasing cell number and enlargement of individual cells. Until the seedling becomes nutritionally self-supporting, it depends upon the reserves provided by the parent (stored in the endosperm, in residual tissues of the ovule, in the body of the embryo and in the cotyledons). As the reserve materials are partly in insoluble form – as starch grains, protein granules, lipid droplets, *etc.*, the early metabolism is concerned with mobilizing and translocation of these materials, and their products to the active areas. The reserves outside the embryo are digested by the embryo, and also by the special cells of the endosperm.

The environmental factors also play an important part in determining the orientation of the seedling for establishment as a rooted plant. The response to gravity is important. The radicle grows downward into the soil and is said to be positively geotropic, while the young shoot is negatively geotropic as it moves away from the soil. The shoot rises by the extension of either the hypocotyl (the cotyledons come out) or the epicotyl (the cotyledons remain in the soil). Similarly, the light affects both the orientation and its form, and the adaptive responses are governed by the reactions in which the light-sensitive pigment, phytochrome plays an important role. The shoot and the root may be positively phototropic or negatively phototropic.

Germination Rate and Germination Capacity

The germination rate describes how many seeds of a particular plant species, variety and seed lot are likely to germinate over a given period. This is a measure of germination time course and is usually expressed as a percentage (for example, an 85 per cent germination rate indicates that about 85 out of 100 seeds will germinate under proper conditions over the germination period given). This is determined by the genetic composition of the seed, morphological features of the seed and the environmental factors. This is useful for calculating the number of seeds needed for a given area or desired number of plants. This is the reciprocal of time taken for the process of germination to complete starting from the time of sowing. The number of seeds able to complete germination in a population is referred to as the germination capacity.

Metabolism and Nucleic Acids

During germination and early seedling growth, both cell division and cell elongation take place. At the same time both protein synthesis and nucleic acids synthesis take place. Later various substances including nucleic acids move out of the cotyledons and are transferred to other parts of the growing embryo. It decreases in their amount in the storage tissue and increases in the growing embryo. As a result of it, an increase in the weight of the hypocotyls and loss in weight in the cotyledons has been observed. In this process although both DNA and RNA increase in amount, they do not increase at the same rate.

DNA synthesis occurs before or after cell division at the beginning or at the end of cell cycle. There is a change in the amount of phosphorylated protein in the nucleus. Some such proteins disappear and others are synthesized. Protein kinases seem to have a regulatory role at some stage of the cell cycle. The amount of polysomes is very low or entirely absent in dry seeds. Studies on cotton seed revealed that the ribosome formation and RNA synthesis occurs during the early stages of germination. the cotton seeds also contain long-lived mRNA and the *de novo* synthesis of a protease. Another thing is that the mRNA was formed during embryogenesis, but its translation is blocked at this stage. In some cases, ribosomes must be synthesized, while in other cases sufficient ribosomes are present in the embryo for polysome formation.

The long-lived mRNA may not be present in all the seeds. Such mRNA is seen in the ribosomal fraction of dry seed embryos. A part of the long-lived mRNA lack polyA and during germination a new polyA RNA is formed. In addition to long-lived mRNA, new mRNA is formed rapidly after the onset of germination. During germination, in addition to mRNA formation, other forms of RNA, namely r-RNA and t-RNA are formed.

In addition to the presence of long-lived mRNA, the dry seeds such as peas, rye, wheat *etc.*, contain ribonucleoproteins, which are capable of sustaining protein synthesis activity. Only a few of the enzymes involved in RNA metabolism have been studied. RNA polymerase activity increases prior to RNA synthesis and the enzymes required for RNA synthesis are present in the mature wheat seed.

Proteins and their Mobilization

Storage Proteins

Plants accumulate protein reserves in developing seeds. The major amount of these reserves consist of specific storage proteins like globulin, which predominate in dicotyledonous seeds, or polyamines, which are the major storage proteins of most cereals. Proteins are generated mainly during seed production and stored either in the cotyledons or in the endosperm, which serve as nitrogen sources for the developing embryo during germination. They are deposited in membrane-bound compartments, the protein bodies (PB) with diameter from 0.1 to 2.4 μM, and are generated directly from the endoplasmic reticulum (ER), as in maize or from the protein storage vacuoles (PSV) as in majority of species. The average protein content of cereal grains is 10-15 per cent of their dry weight and that of Fabaceae seeds 20-25 per cent.

The proteins can show the following properties.

- The proteins have no enzyme activities.
- They serve as nitrogen source for the germinating seeds.
- They occur normally in an aggregated stage within a membrane (Protein bodies, aleuron grains)
- They are often built from a number of different polypeptide chains.

Protein bodies may contain only one type of storage protein. In certain legumes, for example, some protein bodies contain only albumin or vicilin or legumins, while others contain vicilin and legumin.

According to Osborne's classification, seed proteins can be divided into four classes on the basis of their solubility, (i) albumins – soluble in water and dilute buffers of neutral pHs, (ii) globulins – soluble in salt solutions, (iii) glutenlins – soluble in dilute acid or alkali solutions,(iv) prolamins- soluble in aqueous alcohols (70-90 per cent). Characteristically, storage proteins are oligomeric, and are made up of two or more subunits. These subunits are made up of a number of polypeptide chains. The number of genes encoding the different proteins varies considerably among the different prolamins.

Legume seeds are the second most important protein sources, on a world basis, after cereals. The major storage proteins are globulin, which account for up to 70 per cent of the total seed nitrogen. The globulins consists of two major families of proteins which differ in molecular mass and which sediment during ultracentrifugation with sedimentation coefficients (S values) of approximately 7 (average 7-8) and 11 (average 11-13). These are the 7 S (called the vicilin group) and the 11 S (the legumin group) proteins.

Hydrolysis of storage protein (polypeptides) to their constituent amino acids requires a class of enzymes called proteinases, some of which effect total hydrolysis whereas others produce small polypeptides that must be degraded further by peptidases. Cotyledons and endosperm of dry and germinating seeds contain a variety of proteolytic enzymes some of which are present already in the dry seeds while others appear during germination. The proteolytic enzymes can be divided into proteinases and peptidases depending on the size of the molecule. The peptidases can be divided into endopeptidases and exopeptidases. Exopeptidases can be divided into carboxypeptidases and aminopeptidases. The liberated amino acids are utilized for protein synthesis in other tissues.

Protein Mobilization

The mobilization of storage proteins is one of the most important post-germinate events in the development of seedling. During germination period, the storage proteins are degraded by a variety of proteases, which convert the insoluble storage proteins into soluble peptides and free amino acids. These are mobilized to the embryonic axis to support and also provide energy by oxidation of the carbon skeleton after deamination. The protein reserve mobilization can be studied in both cereals and Dicotyledons.

In Cereals

In cereals, the proteases responsible for the degradation of storage proteins are synthesized in aleurone layer and the synthesis of the enzymes is upregulated by gibberellins and downregulated by ABA (Abscisic acid). Reserve proteins are stored in two regions of cereal grain: (a) in the aleurone grains of the aleurone layer, and (b) in the protein bodies of the starchy endosperm. The role of the embryo or embryonic axis in mobilization of storage proteins is minor and less understood. Proteinases are produced in the aleurone layer, in some species under the control of gibberellins, and these may be synthesized *de novo*. One or more of these enzymes hydrolyze the proteins within the aleurone grains, and the resulting amino acids are utilized for the synthesis of proteins (*e.g.*, α-amylase) in this tissue.

Proteinases, effective in breaking down the protein reserves within the starchy endosperm, may come from two sources: the aleurone layer or the starchy endosperm itself. Endopeptidases that are released from the aleurone layer may play a central role in endosperm protein hydrolysis in barley. The performed proteinases that are present within the endosperm of dry grains may be involved in protein mobilization. In addition to their action in degrading reserve proteins, proteinases in the starchy endosperm activate and release enzymes to dissolute the cell walls by hydrolyzing links between glucan and protein components.

Proteolytic activity within the starchy endosperm results in the production of amino acids, dipeptides and a number of small oligopeptides. The embryo rapidly absorbs these soluble products via the scutellum, for its growth.

In Dicotyledons

Storage protein mobilization has been predominantly investigated in seeds of cultivated plants such as grain legumes, rape, sunflower *etc.*, that during domestication, underwent selection for increased protein accumulation in cotyledon and other storage tissue. In cotyledons storage protein mobilization only becomes measurable after the radical has protruded and germination is finished. At this stage, inactive precursors of proteinases are synthesized on the rER, from where they are transported through the endomembrane system into PB. Here enzyme activation takes place and protein breakdown is initiated.

The first group of proteinases to increase in activity following germination, as a result of *de novo* synthesis, have been called "proteinase(s) A', which are usually SH-dependent acid endopeptidases. These enzymes act on the insoluble native 11 S (legumin) and 7S (vicillin) proteins, resulting in the following consecutive series of events.

a) Short-chain peptides are cleaved from the native proteins, which results in a marked increase in their susceptibility to proteinases.
b) Further proteolysis leads to an increase in their negative charge, perhaps resulting from the removal of short basic peptides.
c) An increased solubilisation of the proteins resulting from the initial proteolysis.
d) A stepwise decrease in the size of the acidic subunits of the 11 S proteins that are cleaved into polypeptides of intermediate size (10-18kDa, from the native 20-40 kDa).

The native storage protein, here a disulphide linked insoluble legumin, is initially trimmed by proteinase A (endopeptidase) activity (A) to render it more soluble as small oligopeptides are released. Further, proteinase A activity and that of protein B endopeptidases (A and B) result in hydrolysis of protein to amino acids and small peptides which are transported into cytosol. Hydrolysis of peptides in the protein bodies is aided by carboxypeptidases (C). The released oligopeptides are degraded further by aminopeptidases and peptidases within the cytosol to yield amino acids.

The cellular changes that accompany proteolysis have been studied most thoroughly in the cotyledons of mung beans (*Vigna radiata*), in which the major storage protein is vicilin that comprises some 70-80 per cent of the total, and the enzyme responsible for its hydrolysis is an endopeptidase, vicilin peptidohydrolase; with some participation by a carboxy peptidase. The ER is the site of vicilin peptidohydrolase synthesis and associated with the synthesis of this enzyme the ER undergoes various modifications. Initially the vicilin is cleaved from 50- to 63-kDa components to 20-30-kDa components, and these are then hydrolyzed more slowly.

Then, the protein bodies come to contain hydrolytic enzymes other than the peptidohydrolase. During protein hydrolysis a *de novo*-synthesized ribonuclease is inserted into the protein body (after synthesis on ER and transport via vesicles). These organelles have other hydrolase activities also: α-mannosidase and glucosaminidase (for the hydrolase of mannose and glucosamine residues from vicilin, a glycoprotein), acid phosphatase, phosphodiesterase, and phospholipase D (for digestion of membrane phospholipids).

As protein digestion proceeds, the emptying storage vesicles fuse to form a large vacuole containing hydrolytic enzymes, analogous in function to the lysosome. Digestion of cell contents by the enzymes of the vacuole is achieved when and autophasic process internalizes vesicles. Therefore, disintegration of the cotyledons is achieved by cellular autolysis.

Mobilization of Food Reserves

The mobilization of stored reserves in the storage organs commences after radical elongation. Then, the reserve food that is present within the storage organs of the seed are mobilized and are converted into soluble forms and transported to the sites where they are required (growing organs) for the support of energy. During germination, the enzymes hydrolyze the stored products such as carbohydrates, lipids, phosphates, proteins *etc.*

Stored Carbohydrate Catabolism

The most common reserve carbohydrate in seeds is starch. Other stored carbohydrates like hemicelluloses (galactomannans), are to be found as cell wall components of the endosperm of legumes. There are two catabolic pathways of starch: one hydrolytic and the other phospholytic.

The amylase and amylopectin in the native starch grain are first hydrolyzed by α-amylase, which breaks the α (1→4) glycosidic links between the glucose residues. The released oligosaccharides are further hydrolyzed by α-amylase until glucose and maltose are produced:

Amylose → Glucose + maltose (in presence of α-amylase)

Amylopectin → Glucose + maltose + limit dextrin (in presence of α-amylase)

Limit dextrin → Glucose oligomers (in presence of α- Isoamylase Debranching (R-) enzyme) (Limit dextrinase) amylase → Maltose + Glucose (in presence of glucose α-Glucosidase)

The β-amylase cannot hydrolyze native starch grains; rather it cleaves away successive maltose units. Amylopectin also cannot be completely hydrolyzed but yields maltose and limit dextrin. Further, maltose is hydrolyzed into glucose by maltase.

Amylose → Maltose (in presence of β-amylase)

Amylopectin → Maltose + limit dextrin (in presence of β-amylase)

Maltose → Glucose (in presence of α-Glucosidase or Maltase)

Mobilization of Stored Carbohydrate Reserve in Cereals

It is of two types: The embryo reserves and the endosperm reserves.

Embryo Reserves

Low-molecular weight sugars are stored in lower quantities within the embryo. These probably provide an early source of food substrate during germination and early seedling growth, until the stored food materials available from the endosperm. Sucrose and raffinose are present in the ratio of about 3:1 in dry seeds of barley, and both decline their levels after imbibition.

Endosperm Reserves

The starch in the endosperm is degraded to glucose by the action of four enzymes: α-amylase, β-amylase, the debranching enzyme(s), and α-glucosidase. While mobilizing starch and carbohydrates, cell walls of endosperm are also hydrolyzed by these enzymes.

The α-amylase occurs in the region of the scutellum and aleurone layers. In barley, the α-amylase is released into the starchy endosperm from the scutellum. Later, the enzyme is also synthesized within the aleurone layer and it hydrolyze the starch reserve. Although α-amylase from the scutellum is important at early stages of mobilization, in barley, maize, wheat *etc.*, later most hydrolysis is affected by isoenzymes from the aleurone layer.

β-amylase is involved in the digestion of cereal starch. This enzyme is not *de novo* synthesized. Also, it is not released from either the scutellum or the aleurone layer. But, it is present in the starchy endosperm of the mature, dry and un-germinated grains. It becomes active after initial digestion of the starch grains by α-amylase from scutellum and aleurone layer.

Amylosis of the amylase compartment of starch can be achieved by α- and β-amylases, but digestion of amylopectin ceases with the production of limit dextrins. For further degradation, debranching enzymes are needed probably, the limit dextrinase.

The major product of α- and β-amylosis is maltose that is hydrolyzed by α-glucosidase (maltase), an enzyme present in low amounts in the embryo and aleurone layer of un-germinated barley. The enzyme increases in activity at both the sites after germination. Hydrolysis of the endosperm cell walls accompanies the digestion of the starch reserves.

Mobilization of Stored Food Reserves in Dicots

Large amounts of proteins, lipids, and starch are accumulated in seeds of dicots during seed development and maturation. In some plants, hemicelluloses, especially galactomannans and mannans are deposited in the cell walls of storage organs. These reserves are mobilized, and the products are used as building blocks for macromolecules and as fuel for energy during seedling growth.

Lipids (triacylglycerols) are hydrolyzed by lipases to yield fatty acids and glycerol. Fatty acids are used for membrane synthesis and for gluconeogenesis via β-oxidation and the glyoxylate cycle (Figure 1.1).

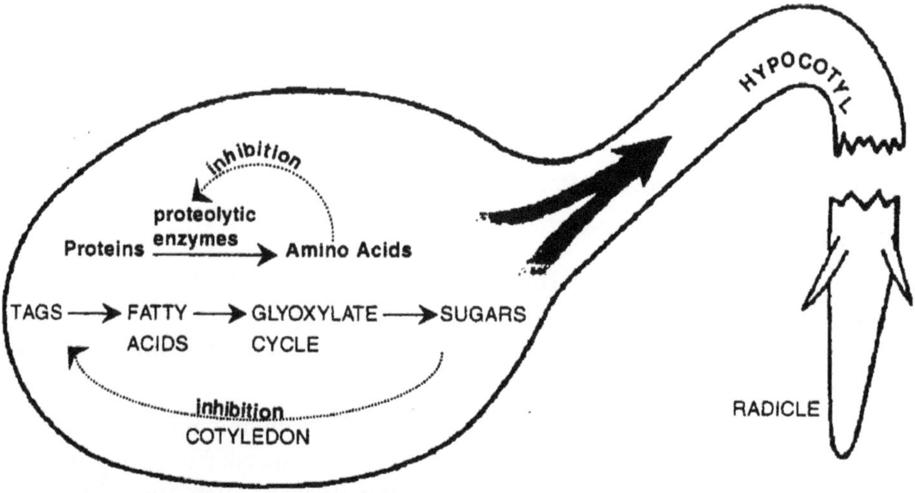

Figure 1.1: Mobilization of Reserve Proteins and Triacylglycerols (TAGs) in Cotyledons of Cucumber (after Bewley and Black, 1994).

Several proteases (*e.g.*, cysteine endopeptidases, aspartic proteases, other endo- and exo-peptidases) and α-amylases have also been isolated from a number of plants. Some of these enzymes are stored in the dry seed whereas in others are synthesized *de novo* from amino acids released by the hydrolysis of stored proteins.

The digestion of galactomannans in cell walls requires the action of three enzymes: an endo-β-mannanase, α-galactosidase and β-mannosidase. It is believed that GAs and/or cytokinins from the embryonic axis move to the endosperm where they bring about the *de novo* synthesis of endo- β-mannanase and α-galactosidase. The oligomannans and galactose residues are released by the action of these two enzymes. Then, they move from the endosperm into the cotyledons where the oligomannans are further broken down into mannose residues by the enzyme β-mannosidase. The products of hydrolysis, galactose and mannose residues are converted into sucrose and utilized for seedling growth.

In legumes, galactomannan deposits in the cell walls of endosperm of extensively that the cells die (*e.g.*, fenugreek). An aleurone-like layer surrounding the dead tissue is synthesized and secretes the three wall hydrolases. In other endospermic legumes (*e.g.*, carob), where the deposits may not be so heavy, the endosperm cells are metabolically active after germination and are the source of hydrolases.

Stored Triacylglycerol Catabolism

Triglycerols are the storage lipid in seeds. Their initial hydrolysis (lipolysis) is by lipases, enzymes that catalyze the three-stage hydrolytic cleavage of the fatty acid ester bonds in triacylglycerols (TAGs) to yield glycerol and free fatty acids (FFA).

Triacylglycerol → Diacylglycerol + FFA → Monoacylglycerol+2FFA

Monoacylglycerol → Glycerol + 3 FFA

Glycerol enters the glycolytic pathway after its phosphorylation and oxidation to the triose phosphates (dihydroxyacetone phosphate - glyceraldehyde-3-phosphate), which are then condensed by aldolase in the reversal of glycolysis to yield hexose units. Alternatively, the triose phosphates may be converted to pyruvate and then oxidized through the citric acid cycle.

The FFA released by lipase may be utilized in oxidation reaction to yield compounds containing fewer carbon atoms. The predominant oxidation pathway is β-oxidation (and glyoxylate cycle).

Catabolism of tricycloglycerol reserves within the storage tissues of germinated seeds involves three distinct organelles found within oil-containing cells. These are, (a) the TAG-storing oil body (hydrolysis of TAGs to FA and glycerol commences within the oil bodies), (b) the glyoxysome (within the glyoxysome the fatty acids are oxidized and synthesis of succinate occurs via the glyoxylate cycle), and (c) the mitochondrion (the succinate is converted to malate or oxaloacetate within the mitochondrion). The malate or oxaloacetate is processed further in the cytosol to give rise to sucrose.

Utilization of the Liberated Amino Acids

The amides *i.e.*, asparagines and glutamine are the forms of amino acids transferred to the growing organs of seedling. Later, these amino acids further metabolize.

Aspartate +glutamine → Asparagine + glutamate + ATP AMP + PPi (in presence of Asparagine synthase)

Glutamine itself is formed from glutamate as follows:

Glutamate + NH_3 + ATP → Glutamine + ADP + Pi (in presence of Glutamine synthase)

Glutamine + α-ketoglutarate → 2 Glutamate + NAD (P) (in presence of GOGAT)

The amino acids that are released from the storage protein by hydrolysis in Castor are aspartate, glutamate, alanine, glycine, and serine and are converted to sucrose and transported as the sugar.

Tropisms

In botany, Cholodny-Went model (1927) is an early model which described tropism in emerging shoots of monocots, including the tendencies for the stalk to grow towards light (phototropism) and the root to grow downward (gravitropism) and in both cases, the directional growth is considered to be due to asymmetrical distribution of auxin, a plant growth hormone. A tropism (from Greek – meaning "*a turning*") is a biological phenomenon indicating the growth or turning movement of a biological organism, usually a plant, in response to an environmental stimulus.

The tropisms are directional movement responses (either growth or turning movement) that occur in response to a directional external stimulus. The plant responses to environmental stimuli including unilateral light (or the colours of light), gravity and water, *etc.*, which involve altered growth (Tropisms) may be positive involving growth towards the stimulus, or negative (Aphototropism or Skototropism in vines), away from the stimulus.

Different Types of Tropisms

One of the most commonly observed tropic responses in plants is phototropism (plant stems grow towards unidirectional source of light – this can be best observed in plants kept near the window where the stem tends to lean towards the window where the light is usually stronger than inside the room while the roots grow away from light). Phototropism is dependent on blue light photoreceptors, called phototropins. Another common tropic response is gravitropism or geotropism (the plant will grow in such a way that it stays oriented relative to the source of gravity or the gravitational force – the Earth. The roots are positively geotropic while the shoots are negatively geotropic or apogeotropic). This is dependent on the presence of amyloplasts (starch-filled plastids). *Heliotropism* is a type of phototropism in which stems and flowers follow the path of the sun from east to west as it moves across the sky (Sunflower heads - this makes the plants more attractive to pollinators which may be positive or negative or apheliotropic). Thigmotropism (Heptotropism) refers to the plant growth in response to touch or contact with a solid object (a tactile stimulus). The positive thigmotropism is demonstrated by climbing plants or vines, which have tendrils (a thread- like appendage used for twining around the solid structures – which is a modified leaf, stem or petiole). The roots exhibit negative thigmotropism as they grow away from an object in the ground. Hydrotropism refers to the growth in response to water or moisture (water gradient) which is important in plant as for protection against drought through positive hydrotropism (pro-hydrotropism) and against over-saturation of water through negative hydrotropism. Thermotropism is the growth or movement in response to heat or temperature changes, while chemotropism is the growth in response to chemicals. The roots are highly chemotropic and respond either positively or negatively. Similarly, pollination is an example of positive chemotropism.

Other types known in plants and bacteria are: Magnetotropism is response to the magnetic field either positively or negatively as a source of direction towards the poles; Garbotropism is the attraction to garbage; Aerotropism is growth of plants towards or away from a source of oxygen; Electrotropism is the movement or growth in response to an electric field; Exotropism is the continuation of growth outward (in the previously established direction); Hygrotropism is the movement or growth in response to moisture or humidity; Orthotropism is the movement or growth in the same line of action as the stimulus; Plagiotropism is the movement or growth at an angle to a line of stimulus like gravity or light; Selenotropism is the motion of plant parts in response to the direction of the moon; Traumatotropism is the response to wound lesion; Galvanotropism is the response to electric current; and Diatropism is the movement at right angles to the direction of the stimulus; *etc.*

Tropisms involve a receptor to sense the stimulus involved and a mechanism to transmit that stimulus to the region of the plant, in which the altered growth will take place.

Tropisms have attracted a lot of interest since Charles Darwin first studied them in the 1870s. Much of the work has concentrated on the responses of the coleoptiles (a sheath that covers and protects the primary leaf, and shows rapid growth) of grasses.

Photoreception

When a coleoptile is exposed to unilateral light for a short period (1-2 hrs), it bends towards the light. The phototropically active light has to be absorbed by blue light, β-carotene and riboflavin. Among various wavelengths of lights, the blue light is the most effective in causing phototropic curvature. In a seedling, the photoreceptor pigment may be located in the tip of coleoptile.

Mechanism

Auxin was thought to be a transmitting signal in phototropism. The bending of the shoot or root is due to differential growth on the two sides of unilaterally exposed organs. In shoot tissues, the growth on the darkened or shaded side is more than on the illuminated side. In the phototropic responsive roots, the situation is reverse; the growth rate is retarded on the shaded side. The unequal distribution of auxins was thought to result from a combination of three different mechanisms. (i) Photodestruction of auxins on the illuminated side. (ii) Auxin causes increased growth on the dark side of the shoot or coleoptile stimulating acidification of the cell wall, resulting in loosening of the wall matrix. (iii). The turgor pressure against the loosened wall results in elongation.

Geotropism (Gravitropism)

The perception of gravity requires the presence of an object that responds and shows movement to the gravitational field. Such objects are called statoliths. Starch-filled plastids (amyloplasts) move against endoplasmic reticulum (ER) at the cortex of specialized cells termed statocytes (Figure 1.2). The statocytes are in the root cap or in a layer, the starch sheath (endodermis) adjacent to the vascular tissue. The removal of the root cap results in a loss of sensitivity to gravity in primary roots. The *scr* (*scarecrow*) mutant of *Arabidopsis*, which has no layer of starch-containing cells around its vascular tissue, does not respond to gravity.

Mechanism

Auxin is thought to transmit the signal from the statocytes to the growth zone. In roots, high levels of auxin on the lower side inhibit growth, while lower levels on the upper side permit it; in shoots, it is vice-versa. The transport and unequal distribution of auxin is caused by the changing permeability of cells due to the migration of amyloplasts. Calcium has also been suggested to be involved in graviresponse.

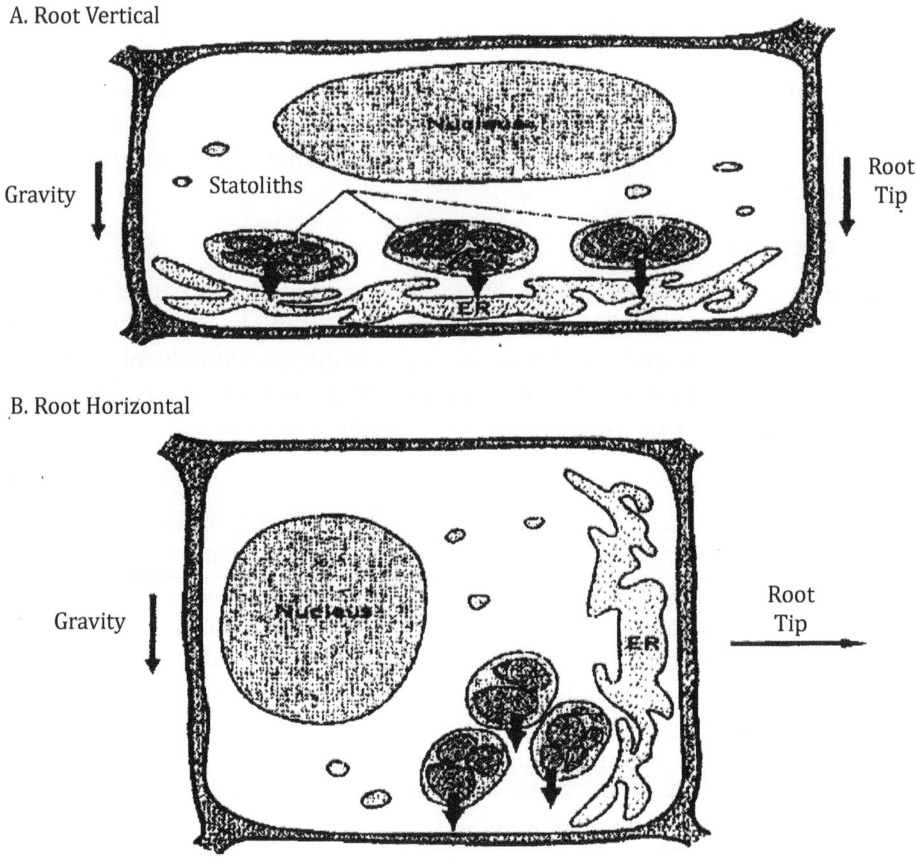

Figure 1.2: Schematic Representation of the Graviperception Model. (A) When the root is oriented vertically starch-laden statoliths exert pressure uniformly on the endoplasmic reticulum; (B) In the horizontal position the statoliths are redistributed in the cell in response to gravity, exerting unequal pressure on the ER.

Thigmotropism

The reaction or growth of plant part to the stimulus of touch is called thigmotropism or heptotropism. Thigmotropic curvature is apparent in the stems of climbers and in tendrils. The entire stem of some vines grows in coils after coiling around the supporting object, as is seen in many members of Convolvulaceae. When the stems come in contact with the support, they respond positively by bending towards the object. But some plants such as grapes, peas, *Vitis*, cucurbits *etc.*, have tendrils, which coil around an object to get support and to climb.

Mechanism

The thigmotropic response in tendrils and stems is also due to unequal growth of the organ. Although no direct involvement of auxins has been demonstrated, the hormone might be playing some role as it affects, tendril coiling. The movement of salts and electrolytes is probably involved in thigmotropic responses. In pea tendrils, rapid changes in ATP and inorganic phosphate takes place upon touch stimulus. Perhaps, rapid changes in the permeability of membrane takes place, which besides, facilitating the movement of these metabolites, also accelerate the water uptake and cell growth.

Hydrotropism

Response of plant organs to water content is hydrotropism. It can be seen in roots of some species, which show a directional growth towards the region of high water content in the soil. However stems, roots of seedlings of many species do not show any response to the soil water content. But positive hydrotropism has been observed in the growth of pollen tubes.

Chemotropism

Movement by part of a plant in response to a chemical stimulus is called chemotropism. The response by the plant is termed 'positive' if the growth is towards the stimulus or 'negative' if growth is away from the stimulus. Fertilization of flowers by pollen is achieved because the ovary releases chemicals that produce a positive chemotropic response from the developing pollen tube.

Hormonal Control of Seedling Growth

Development is quantitative changes that involve increase in tissue and differentiation of these tissues into various organs of the plant body. The formation of root, stem and leaves *etc.*, is known as morphogenesis. The plant development from seedling to adult plants up to seed production is a complex process which involves various phases like: (i) cell division, (ii) cell elongation, and (iii) cell differentiation *etc.* The different phytohormones regulate or influence a range of various morphogenic, cellular and physiological processes.

Hormonal Control

Hormones are naturally occurring organic substances that, at low concentrations, influence physiological processes such as growth, differentiation and development. Several growth regulators are collectively called the plant hormones or phytohormones. This group includes auxin, cytokinin, gibberellin, abscissic acid (ABA) and ethylene. Many plant hormones are transported from one place in the seedling to another, thus coordinating growth throughout the plant, while others act in the tissue in which they are produced. The hormone must first bind to protein receptor, either on the cell surface or inside the cell. This activates a signal transduction pathway which amplifies the signal and to changes in enzyme activities, gene expression and other physiological responses. Plant scientists investigated these phenomena by analyzing the levels of biochemical forms of the hormones present in relation to differences in development. Recently the use of mutant and bioengineered plants in which growth or development is abnormal has enabled us to start understanding how hormones work. This research has been coupled with the isolation of the genes and proteins that are needed for the normal functioning of the hormone.

Biosynthesis of Hormones

Auxin (indole acetic acid, IAA) is synthesized from indoleglycerophosphate, the precursor to the amino acid tryptophan, and in some plants, from tryptophan itself. Gibberellin is synthesized via the isoprenoid pathway. Cytokinins are synthesized by the attachment of an isopentenyl side chain to adenosine phosphate. Abscicic acid is synthesized via carotenoid molecules. Ethylene is derived from methionine via non-protein amino acid ACC (1-amino-cyclopropane-1-carboxylic acid).

Transport of Hormones

Most hormones simply travel along with the contents of the xylem or phloem by both diffusion and bulk transport. Auxin is special as it is transported primarily in the cells of the vascular cambium initials. Auxin enters the cell from the cell wall above as an un-ionized molecule that can cross the cell membrane. At the neutral pH inside the cell it becomes ionized, preventing its outward diffusion through the cell membrane. Outward transport is on special carrier proteins located only at the base of the cell, so movement is downward (in roots, the situation is reversed).

Signal Transduction

A hormone is a molecule, which must bind to a receptor protein to have its effect. The auxin binding protein is located in the endoplasmic reticulum (ER), from which it also migrates to the cell membrane. After detection, the signal from the hormone molecule has to be translated into action. There are various steps in this process. Usually the hormone triggers the phosphorylation of an activator protein, which then binds to the regulatory region of a gene, thus turning on gene transcription. This gene may produce the final product, or may itself produce a gene regulator. Prior to the phosphorylation of the regulatory protein may include an interaction with a membrane G-protein that in turn releases other factors and the opening of calcium channels in the membrane permitting an increase in the cytoplasmic level of calcium.

Role of Auxins in Cell elongation

Indole acetic acid (IAA) is the most widely distributed, naturally occurring auxin in vascular plants. It regulates various processes such as embryo development, induction of cell division, coleoptiles and stem elongation, apical dominance, induction of rooting, vascular tissue differentiation at early stage of seedling development, tropic movements *etc.* Cell elongation is a vital part of growth. Auxin causes cell elongation within ten minutes by making the cell walls more extensible. This occurs through a series of steps. Auxin stimulates the pumping of hydrogen ions out of the protoplast via proton pumps driven by ATP, so acidifying the wall, this activates an enzyme called expansin, which is activated by acid conditions (about pH 4.5). Expansin breaks the hydrogen bonds between the cellulose microfibrils of the wall and the other sugar chain molecules that cross-link the microfibrils; the cell walls are made more extensible; and the cell then elongates because of the turgor pressure inside the cell. Auxin commonly inhibits root elongation, but high auxin concentrations will promote the initiation of secondary, or branch roots.

Role of Gibberellins in α-amylase Production and Stem elongation

Two main types of GAs are recognized on the basis of their structure. They have 20 carbon atoms, the C_{20}-GAs and C_{19}-GAs. Biologically active gibberellins in plants are C_{19} compounds. The action of gibberellins in the cereal grain germination is the best understood hormonal mechanism in plants. In barley, first gibberellins move from the embryo to the aleurone layer where the synthesis of α-amylase and protease enzymes takes place. The protease converts an inactive β-amylase to the active form. α- and β-amylase together digest starch to glucose, which is mobilized to the growing embryo for its metabolic activity.

It can be proved by *in vitro* studies. Barley grains may be transected to produce two half-seeds, with one half-seed containing the embryo and one without (Figure 1.3). The embryo containing half-seed will secrete α-amylase, digest the starchy endosperm and germinate. The embryo-less half-seed neither can germinate nor produce elevated levels of α-amylase. But if embryo-less half-seed is treated with gibberellic acid it will produce high level of α-amylase. Gibberellin is needed for cell and stem elongation.

Cytokinins

The most widespread naturally occurring cytokinin in higher plants is zeatin. Zeatin and other naturally occurring cytokinins are commonly found with a ribose (the riboside) or ribose-phosphate (the ribotide) at the 9th position. In addition to stimulating cell division, cytokinins also influence shoot and root differentiation in tissue culture.

Abscisic Acid (ABA) and Stomatal Closure

Abscisic acid is a single compound. ABA promotes stomatal closure by causing the exit of potassium ions from the guard cells. K^+ is the main solute causing turgor in the guard cells and opening the stoma. ABA binds to cell-surface receptor on the surface of the guard cells. This causes a calcium ion influx and an increase in the level of inositol triphosphate. The Ca^{2+} brings influx and an increase in the level of inositol triphosphate. The Ca^{2+} brings about a membrane depolarization, triggering the outward K^+ ion channels to open.

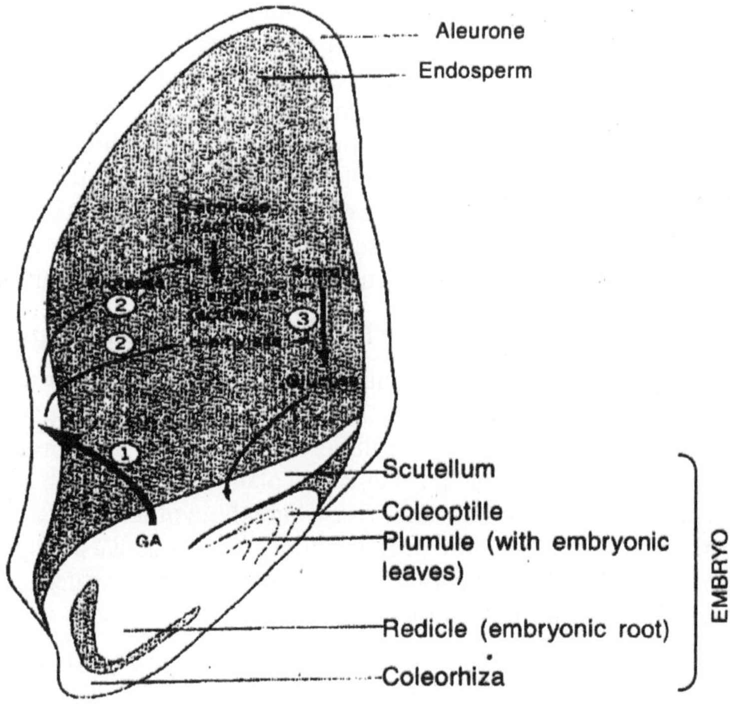

Figure 1.3: A Schematic Illustrating Gibberellins-Induced Release of Enzymes and Carbohydrate Mobilization during Germination of Barley (*Hordeum vulgare*) Seed.

Ethylene and Seedling Stem Growth

Ethylene is synthesized primarily in response to stress and produced in large amounts by tissues undergoing senescence. Exposure of *Arabidopsis* seedlings to ethylene usually causes stunted growth. Ethylene's receptor interacts with a protein that blocks an ion channel. In the presence of ethylene, the receptor causes the protein to unblock the channel. The entry of (unknown) ions then activates other second messengers. Activation results in the synthesis of a transcription factor, triggering the synthesis of specific enzymes that can cause stunted growth.

Gene Expression in Seed Germination

Seed germination begins with water absorption (imbibition). Then the seed food reserves are mobilized. Next the embryo grows by utilizing the mobilized food materials. Mobilization of food reserves varies depending on the nature of food reserves of the seed. The process has been studied in detail in barley.

Cereal seeds contain a large reserve of starch that is broken down to sugars that supply the developing embryo. The process has been studied in barley. A barley seed has an embryo, an endosperm containing storage reserves, and a layer surrounding the endosperm called the aleurone (Figure 1.4). If a dormant seed imbibes water simultaneously the growing embryo produces gibberellic acid. The newly produced gibberellic acid diffuses to the aleurone layer, resulting in the synthesis of new proteins, including starch degrading enzymes such as α-amylase, which break down the starch of the endosperm. Gibberellic acid greatly enhances the rate of transcription of the α-amylase gene. Gibberellic acid response elements (GARE) are present at the transcription site of the gene. Later they aggregate and form gibberellic acid response complex (GARC).

Seed + Water → Imbibed seeds → Embryo produces gibberellin → Gibberellin diffuses to aleurone layer → Gibberellin responsive genes get activated → α-amylase produced → this diffuses to endosperm → Starch reserves mobilized → which diffuse to embryo.

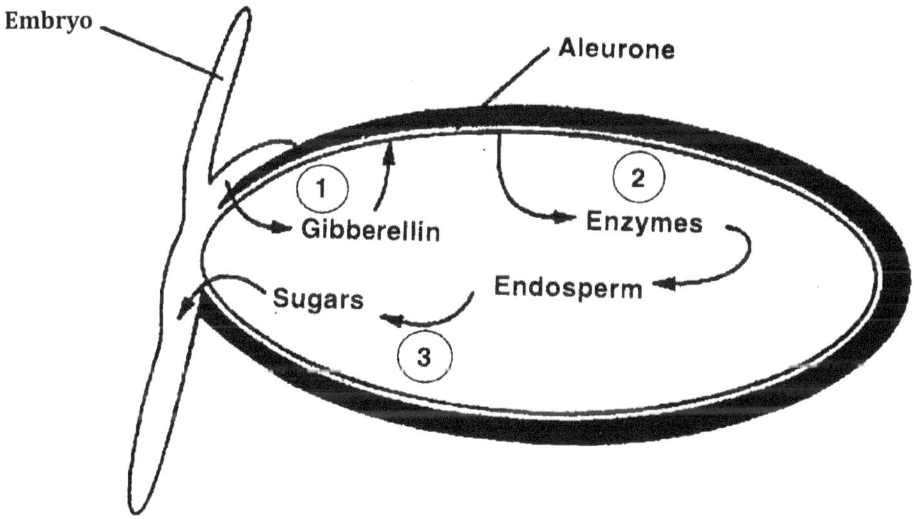

Figure 1.4: Parthway in Mobilization of Storage Reserves in Barley Grain.

The gibberellic acid response complex (GARC) is activated by the binding of transcription factors. Proteins are produced in response to gibberellic acid that binds to DNA. Of these GAMYB, is one of the transcription factors (MYBs) known to be involved in development and is upregulated by gibberellic acid well before α-amylase gene expression starts.

During the post germination stages, the shoot can photosynthesize and supply the nutrients required for growth. In grass, production of a fully functional photosynthetic apparatus occurs in three phases. In the first phase, the coleoptiles and primary leaf contain etioplasts (precursors of chloroplasts), but that cannot photosynthesize. In the second phase plastids go on dividing and increase in abundance. Genes are activated which generate the photosynthetic apparatus and enzymes of the Calvin cycle. This involves activity of genes of both the plastid genome (plastome) and nuclear genome and the efficiency of the photosynthetic system increases as more components are added. In the third phase, the activity of the photosynthetic apparatus is maintained and repaired, if the damage occurred.

Chapter 2

Cell, Cytoskeleton and Ergastic Substances

Introduction and History

The cells are the basic, structural, functional and fundamental (biological) units of life and the smallest independent component that is responsible for all the life processes with which the living organisms are constructed. The study of cells is called cell biology, cellular biology or cytology. Robert Hooke (1665) was the first biologist who discovered the cells using simple lenses and gave the term '*Cellulae*' (from Latin: `*cella*` meaning `*a small room*`) for the small cavities (minuscule structures that reminded him of small rooms) surrounded by walls in a piece of bottle cork under the microscope. Later, he observed similar cells in other plant tissues and that they also contained '*Juice*' (Matzke, 1943). Later, Anton von Leeuwenhoek observed the cells (protista, sperms and bacteria) under compound microscope with a higher magnification (lenses) and noted the motility as a result of which he concluded that the entities are alive (and named them as `*animalcules*`). Robert Brown (1883), a Scottish Botanist, provided the first insights into the structure of the cell and he was able to describe the nucleus present in the cells of orchids. Two important theories deal with the relation of the entire organism and the single cell, and they are – cell theory and organismal theory. According to 'cell theory', proposed by Schleiden (a botanist) and Schwann (a zoologist) the organism (plant or animal) consists of an enormous number of cells, each of which plays a role in determining the nature of the organism. According to this theory, - all organisms are composed of one or more cells – the cell is the smallest unit having the properties of life – and the continuity of life arises directly from growth and division of single cells. Rudolf Virchow described that the `cells always come from pre-existing cells` (in Latin – *Omne cellula e cellula*) The second theory namely 'organismal theory' contradicts the cell theory, gives less importance to the individual cells and gives more stress on unity of the protoplasmic mass of the entire organism. According to this theory, the organism as a unit, determines the nature of the cells. Both these theories are important.

Cell - Size

In biology, the cell is the basic, membrane-bound unit provided with the fundamental molecules of life and of which all the living things are made of. A single cell is considered to be a complete organism (such as a bacterium or yeast). The cells in multicellular organisms, acquire the specialized functions as they mature, and cooperate with other to become the building blocks of life (as in human beings and animals). The smallest known cells are tiny bacteria called mycoplasmas (spheres as small as 0.2 μm (about 0.000039 inch) with a total mass of 10^{-14} gram – equal to that of 8,000,000,000 hydrogen atoms. The human cells have a mass of 400,000 times larger than the mass of a single mycoplasma (about 20 μm across). About 10,000 human cells are required to cover the head of a pin and the human body is composed of more than 30,000,000,000,000 cells.

Cell - Structure and Components

The liquid substance within the cell was discovered by Hanstein (1880) who used the term 'protoplasm' to indicate the unit of cell substance found in a single cell. The term 'cell' in plants includes the protoplast together with the wall. Until very long, the cell wall was considered as a non-living excretion of the living cell matter. But recently, more and more evidence has been found that organic unity exists between the protoplast and the wall, and both of them together for a single biological unit. Kolliler (1862) introduced the term `cytoplasm` (the liquid substance or the protoplasm of the cell in between the wall and the nucleus).

Structurally, all the cells have 3 basic parts, namely: A plasma membrane, that separates each cell from the environment, permits the flow of molecules across the membrane and contains the receptors that can affect the cell's activities – a nucleus, DNA containing region that occupies a portion of the interior – and the cytoplasm, which contains the membrane -bound compartments (absent in bacteria), particles, filaments which are all bathed in a semi-fluid substance. Prokaryotic and eukaryotic cells are present in all plants and animals except bacteria and blue-green algae. The eukaryotic cells are defined by their possession of a membrane-bound nucleus while the prokaryotic cells go without it (bacteria). The fluid mosaic model of the cell membrane considers that the fluid portion of the membrane is made of phospholipids (these molecules are composed of a hydrophilic head and two hydrophobic tails). The molecules are surrounded by water and their fatty acid tails cluster together as a result of which a bi-layer results. The heads are at the outer faces of a 2-layer sheet, and this is the structural foundation for all the cell membranes. The membrane proteins serve as transport proteins, receptor proteins, recognition proteins, and adhesion proteins.

More and more knowledge has been accumulated on the structure and biology of the cell after the discovery of the electron microscope. The protoplast consists of membranous and non-membranous components. When observed under an electron microscope, the membranes are seen as two dark lines each about 2.5 nm thick, separated by a lightly stained line of 3.5 mm thick. This type of membrane consists of a bio-molecular lipid layer covered on each side with a layer of protein.

Cell - Components

The main components of plant cell are: Cell wall, Cytoplasm, Nucleus and Cyto-membrane system.

1. Cell wall

The cell wall is the stiff, rigid structure that surrounds the cell membrane to support/protect the cell outside of the plasma membrane. Microscopic pores allow water and solute passage to and from underlying membrane. In plants, the bundles of cellulose strands form the primary cell wall more pliable than the secondary cell wall which is more rigid. The plasmodesmata between the adjacent walls connect the cytoplasm of the neighbouring cells.

2. Cytoplasm

The cytoplasm is the clear, thick, jelly-like substance inside the cell membrane that contains the other organelles, and is forced into a very narrow zone between the central vacuole and the plasma membrane

3. Nucleus

The nucleus controls access to DNA and also contains the code for protein assembly, and permits easier packing of DNA during cell division. The DNA makes it easier to sort out hereditary instructions during cell division. It also contains nuclear membrane (discontinuous and porous membrane, serving as a protective envelope), nucleolus (distinct and small), chromosomes (thread-like structures that carry the hereditary characters and in the form of a network), and nucleoplasm (fluid content of the nucleus).

4. Cyto-membrane System

The cytoplasm (cyto-membrane system) contains many, independent organelle such as: Endoplasmic reticulum, Golgi apparatus, Mitochondria, Plastids, Central Vacuole, Micro-bodies, Ribosomes, Lysosomes, Peroxisomes, Glyoxysomes, and Sphaerosomes.

a. Endoplasmic Reticulum

ER is a collection of inter-connected tubes and flattened sacs that begin at the nucleus and spread out through the cytoplasm. Two types of ER can be distinguished: Rough ER [with the ribosomes attached] and Smooth ER [has no ribosomes]. ER modifies newly formed polypeptide chains and is also involved in lipid synthesis.

b. Golgi Apparatus

Also called as Golgi body – within which the proteins and the lipids undergo final processing, sorting and packaging. This modifies, sorts out and ships proteins. It also plays a role in the synthesis of lipids for secretion or internal use.

c. Mitochondria

These are the efficient factories of ATP production and respiratory organelle. The outer membrane is smooth, The inner membrane gets infolded into cristae that are associated with energy transformation. These have their own DNA and ribosomes and they are independent units in the cell

d. Plastids

These are of three types – Chloroplasts, Chromoplasts and Amyloplasts.

Chloroplasts – These green plastids are oval or discoid, bounded by a double membrane and are critical to photosynthesis. Grana region contains pigments and enzymes to trap energy from sunlight to form ATP, while stroma is the region where sugars are formed in the fluid. Chlorophyll gives the green colour to the plastid.

Chromoplasts - These are the coloured plastids which contain carotenoids and xanthophylls and impart red - to - yellow-colours, and

Amyloplasts – These are colourless plastids or *Leucoplasts* (white) which are without any pigments and store starch grains as in tuber of potato.

e. Central Vacuole

These are present in all mature plant cells, central in position, and occupy 50-90 per cent, store amino acids, sugars, ions and wastes. This is excretory in function),

f. Micro-bodies

These are also called as Cytosomes, and are usually vesicles which are spherical in shape. These are visible only under EMS, and are surrounded by a single phospholipid bilayer membrane. They contain a matrix including the enzymes and other proteins. The genetic material is absent and they do not perpetuate.

g. Ribosomes

These are independent entities which are rich in RNA and take part in protein synthesis. These are prefabricated within the nucleolus before they come out into the cytoplasm.

h. Lysosomes

These are the vesicles that bud out from Golgi bodies, they carry powerful enzymes that can digest the contents of other vesicles, worn-out cell parts, bacteria or other foreign particles. These are the suicidal bags of the cells. They help in the disposal of cells when they are infected, by the digestive enzymes or lysozymes contained within them.

i. Peroxisomes

These are the vesicles that contain the enzymes which break down the fatty acids and amino acids, and release hydrogen peroxide which is degraded by other enzymes. These are discovered and named by C. de Duve.

j. Glyoxysomes

These are the specialized peroxysomes in plants, and molds which convert the stored lipids into carbohydrates. These also possess the key enzymes of the glyoxylate cycle.

k. Sphaerosomes

These are also called as Oleosomes, the small cell organelles which take part in storage and synthesis of lipids. These were discovered by Pemer (1953) and these are small, spherical and refractile. These arise from ER and are surrounded by a single but half unit membrane.

The cell generally contains a single nucleus, but in some cells like the sieve elements of the phloem. The nucleus is absent. There are also some cells like laticifers with numerous nuclei in them. Such a structure is called a coenocyte. The coenocytes aroused a lot of interest in phylogenetic and ontogenetic studies.

The Cytoskeleton

The cytoskeleton is involved in the maintenance of cell shape. The cytoskeleton is a complex, dynamic network of interlinking protein filaments (fibers, threads and lattices) present in the cytoplasm of all cells including bacteria and archaea. It extends from the cell nucleus to the cell membrane and is composed of similar proteins in different organisms. It gives the cells their internal organization, overall shape and capacity to move. It is mainly composed of 3 main components, namely, microfilaments, intermediate filaments and microtubules all assembled from protein subunits in eukaryotes. All these are capable of rapid growth or disassembly dependent on the requirement of the cell. The cytoskeleton performs many functions, and its primary function is to give the shape to the cell as well as the mechanical resistance to deformation and through the association with extracellular connective tissue and other cells it gives stability to all the tissues. It also contracts to deform the cell and thus allows the cells to migrate. It is also involved in cell signalling pathways and also in the uptake of the extracellular material (by endocytosis), the segregation of chromosomes during division, the cytokinesis, and also in intracellular transport as well as a template for cell wall construction. It can also form special structures such as flagella, cilia, *etc.* It also plays an important role in cell mobility and the colloidal changes in the cytoplasm. The microtubules and a variety of microfilaments and micro-trabeculae in the cytoplasmic matrix form a kind of dynamic and spongy cytoskeleton. It provides a framework for soluble proteins, enzymes and ribosomes.

History

The name "cytoskeleton" was coined many years ago by Nikolai K. Koltsov (1903), who proposed that the shape of the cells was determined by a network of microtubules but it was abandoned because many structures observed in cytoplasm with light microscope were considered to be fixation artifacts. The existence of an organized fibrous array in the protoplasm was postulated by Koltzoff in 1928. He opined that "each cell is a system of liquid components and rigid skeletons, which generate the shape, and even though we rarely see the skeletal fibrils in living and fixed cells, that only means that these fibrils are very thin or that they are not distinguished by their refractive index from the surrounding solution". Thus, he conceived of a cytoskeleton which determines both the shape of the cell and the changes in its form. The concept of a protein mosaic that dynamically coordinated the cytoplasmic biochemistry was later proposed by Rudolph Peters (1929), and the term – cytosquelette (in French) was first introduced by Paul Wintrebert, a French Embryologist (1931). The cytoskeleton was thought to be an uninteresting gel-like substance when it was first introduced that helped the organelles to stay in a place within the cell. But, the Electron microscope have confirmed the assumption and revealed that the cytoplasm of most eukaryotic cells contains the cytoskeletal basic fabric formed of microtubules, microfilaments and intermediate filaments. These are related to the various forms of cell motility. The main proteins that are present in the cytoskeleton are tubulin, actin, myosin, tropomyosin *etc.* Initially, it was thought that the cytoskeleton was exclusive to eukaryotes, but, in 1992, it was discovered in prokaryotes as well, and this came after the realization that the bacteria possess the proteins that are homologous to tubulin and actin – the main components of eukaryotic skeleton.

i) Microtubules

These are the structures commonly present in the cytoplasm of eukaryotic cells. These are characterized by their tubular appearance and their uniform properties in the different cell types. De Robertis and Franchi (1953) first observed these tubular structures in the axoplasm extruded from myelinated fibre. These are elongated, unbranched, hollow cylinders (with the lumen diameter of approximately 15 nm) of 20-30 (commonly 23) nm in their diameter and indefinite length, which commonly comprise 13 proto-filaments that are polymers of alpha- and beta-tubulin. These are very dynamic binding GTP for polymerization, and are commonly organized by the centrosome. Microtubules have been observed in a variety of animal cells. Cytoplasmc microtubules are uniform in size and are straight. In a cross section, it shows an annular configuration with dense wall of about 6 nm thick and a light centre. Each microtubule is surrounded by a zone of low electron density. All the microtubules studied showed almost the same morphological characteristics but differ in other properties. For example, the microtubules of cilia and flagella are more resistant to various treatments. Cytoplasmic microtubules normally disappear if stored at 0°c or after treatment with colchicines.

They form the centrioles in 9-triplet sets (star-shaped), and they form cilia and flagella in 9-doublets oriented about 2 additional microtubules (wheel-shaped). The latter formation is often referred to as a 9+2 arrangement, where in each doublet is connected to another by the protein, dynein. The cilia are of two types, motile and non-motile cilia, and these are chorat and more numerous than the flagella. The motile cilia have a rhythmic wave or beat motion, compared to the non-motile cilia which receive the sensory information, and processing signals from the other cells or fluids surrounding it. In addition, the microtubules also control the beating of the cilia and flagella. The dynein arms attached to the microtubules also function as the molecular motors. The motion of the cilia and the flagella is created by the microtubules sliding past one another and requires ATP. In addition, the microtubules also play key role in intracellular transport (associated with dyneins and kinesins, transport the organelles like mitochondria and vesicles). Cross section through a cilium shows the 9+2 arrangement of the microtubules, the axoneme of the cilia and flagella, and the mitotic spindle – and in the synthesis of the cell wall in plants. Stuart Hameroff and Roger Penrose have also proposed that the microtubules also function in consciousness.

Tubulin: Main Protein of Microtubules

Microtubules are composed of protein subunits which are similar though they are found in different cell types. The tubulin is the principal protein of cilia and flagella and is also the protein of cytoplasmic microtubules, and is a dimer of 110,000 to 120,000 daltons. Two different monomers – Tubulin A and Tubulin B are present in flagella. Mostly, tubulin is a heterodimer having two monomers of different kind. The amino acid composition of tubulin from different sources shows little variation. Some enzymatic activities like proteinkinase activity have been reported for tubulin.

Microtubules can be assembled *in vitro*. For this, presence of GTP (Guanosine triphosphate) and high concentrations of tubulin, as well as some associated proteins are necessary. The polymerization is both temperature and pressure sensitive. When microtubules are disassembled, a mixture in the form of rings is produced. These rings are composed of tubulin dimers and associated proteins. The role of these rings in microtubule assembly is questionable because they are not observed *in vivo* and appear only *in vitro*. The assembly-disassembly process is a dynamic equilibrium (Figure 2.1).

The assembly of tubulin in the formation of microtubules is a specifically oriented and programmed process. Centrioles, basal bodies and centromeres are sites of orientation for this assembly. Calcium and calmodulin may be regulating factors in the *in vivo* polymerization of tubulin. The level of polymerized tubulin is high at interphase and metaphase (*i.e.*, spindle microtubules) and lower at prophase and anaphase. Assembly involves the hydrolysis of GTP to GDP. The polarity of the assembly can be demonstrated, under the Electron microscope, by the addition of tubulin.

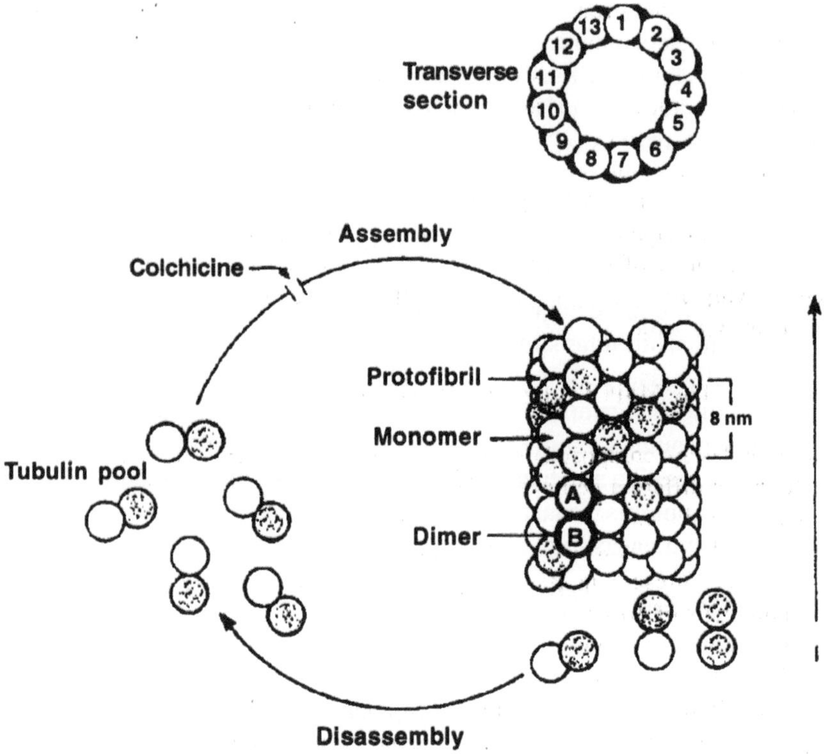

Figure 2.1: Diagram of a Microtubule and the Process of Assembly and Disassembly of Tubulin. *Top*, transverse section of a microtubule showing 13 protofibrils. *Bottom*, longitudinal view showing the helicoidal protofibrils made of tubulin dimers and the polarization of the microtubule. Observe that the microtubule is being disassembled at the bottom while being simultaneously assembled at the top. Colchicine, by blocking the assembly process, produces depolymerization of the microtubules.

Cell transformation (*i.e.*, cancer) produces a disorganization of microtubules. Several functions have been attributed to microtubules - (1) Cell motility in primitive forms. (2) Microtubules play a mechanical function. The shape of the cell and cell processes is dependent on microtubules. This function is apparent during cell differentiation of certain placodes, in the nerve cells and in spermatogenesis. (3) The polarity and directional gliding of cultured cells depend on microtubules. (4) Microtubules are also associated with transport of molecules, granules, and vesicles within the cell. (5) They play a role in the contraction of the spindle and movement of chromosomes and centrioles as well as in ciliary and flagellar motion. (6) A possible role in sensory transduction also has been postulated.

ii) Microfilaments (Actin filaments)

Thin microfilaments ranging between 5-7 nm in width represent the active or motile part of the cytoskeleton and appear to play the major role in cyclosis amoeboid motion. With high voltage Electron microscopy, a three-dimensional view of microfilaments has been obtained (*i.e.*, an image of microtrabecular lattice). These microfilaments are sensitive to cytochalasin-B, an alkaloid that also impairs many cell activities such as the beat of heart cells, cell migration, cytokinesis, endocytosis and exocytosis. Applied to the alga *Nitella*, cytochalasin-B produces a rapid halt in cyclosis. It is generally assumed that the cytochalasin-B-sensitive microfilaments are the contractile machinery of non-muscle cells.

The microfilaments are composed of linear polymers of G-actin proteins, and these generate force when the plus (growing) end of the filaments pushes against a barrier (cell membrane). These also act as tracks for the movement of myosin molecules which affix to the microfilaments and walk along with them. The major protein of the microfilaments is actin. The G-actin monomer combines to form a polymer which continues to form the actin filament, and these subunits intertwine into F-actin chains. The myosin along the F-actin filaments generates the contractile force in actomyosin fibres (both in muscle and non-muscle types). The actin structures are controlled by the Rho family of small GTP-binding proteins for contractile actomyosin filaments (stress fibres), like Rac lamellipodia and Cdc42 for all filopodia. The functions of these microfilaments include: muscle contraction, cell movement, intracellular transport (trafficking), and maintenance of eukaryotic cell shape, cytokinesis and cytoplasmic streaming.

The contractile proteins such as actin and myosin, as well as tropomyosin, tropoins-e, a-actinin and other protein found in muscle, represented in the microfilaments. These proteins can be localized with specific antibodies. Actin filaments are made of globular actin molecules (G-actin). In the presence of myosin they become 'decorated' as can be seen with the electron microscope.

Actin is one of the most abundant proteins in many cells. The globular (G-actin) fibrillar transition is the basis of the classical *sol-gel* transition in the cytoplasm of moving cells. Myosin is found in amoebae, blood platelets and slime molds, but in a much lower concentration than actin. It contains a Ca^{2+} activated ATPase as is found in muscle myosin. Identification of myosin filaments is more difficult.

iii) Intermediate Filaments

These are a part of the cytoskeleton of many cells of eukaryotes, averaging from 10 nanometers in their diameter, and are more stable (strongly bound) and heterogeneous constituents than the microfilaments. These maintain the cell shape by bearing the tension, and also organize the internal tri-dimensional structure of the cell, anchor the cell organelles and serve as the structural components of the nuclear lamina, and also participate in cell-cell and cell-matrix junctions. In addition to the microtubules and the microfilaments, in the cytoskeleton of many cells, there is a group of 10 nm filaments called intermediate filaments, because they range in thickness between the microtubules and the microfilaments. These are highly heterogeneous in their biochemical properties. By their morphology and localization, these filaments can be grouped into four main types, namely neuron filaments, glial filaments, keratin filaments and a more heterogeneous type, which is composed of several proteins.

The intermediate filaments also provide the necessary protection for the organs against metabolic, oxidative and chemical stress. These are most commonly called as the supporting system (or scaffolding) for the cell and the nucleus and also play a role in some functions of the cell. In spite of their heterogeneity, intermediate filaments show a similar structural organization. When seen with X-ray diffraction, all members of this family contain considerable amount of d-helix. When examined under the EM after shadow casting, these show a periodic structure with a length of 21 nm. Intermediate filaments are insoluble in physiological solutions but dissolve at low and high pH. After dissolution, they can be reconstituted again to form intermediate filaments.

Intermediate filaments form networks that interconnect the nucleus with the cell surface. In fibroblasts, intermediate filaments attach to the nuclear envelop and form a case around the nucleus; at the plasma membrane, they appear associated with microfilaments at the adhesion plaques. The intermediate filaments that form the nuclear case are continuous with other intermediate filaments that form the nuclear case in the cytoplasm. Intermediate filaments that radiate through the cytoplasm are formed by protein subunits of a multigene class that are expressed differentially in various cell types. However, the different intermediated filament proteins are organized in a similar way forming tetrameric complexes.

The different intermediate filaments are: intermediate filaments made of vimentin are present in the mesenchyma – the filaments made of keratin (keratin filaments or tonofilaments) are present in the epithelial cells and contain alpha-keratins – the neuro-filaments are associated with the neural cells, and are present in axons, dendrites and perikarya of the neurons – cells made of lamin give structural support to the nuclear envelope – the glial filaments contain a very acidic protein and these are present in astrocytes - and heterogenous filaments have a similar morphology and localization but contain a variety of proteins, in particular, the cells made of desmin have an important role in the structural and mechanical support of the muscle cells, where they constitute a cytoskeleton that holds the myofibrils within the muscle cell. This cytoskeleton also contains vimentin and synemin. Vimentin filaments are wavy and in many cells tend to form perinuclear caps. Intermediate filaments are resistant to colchicines and cytochalasin-B and are sensitive to proteolysis. They are involved in mechanical function contributing to the determination of the cell shape and to the integration of cellular compartments. During mitosis of some epithelial cells, intermediate filaments undergo disintegration and restoration.

iv) Septins

These refer to a group of highly conserved GTP binding proteins in eukaryotes. Different types of septins form the protein complexes with each other, and can assemble to filaments as well as the rings. Hence, these can be considered as a part of the cytoskeleton. In the cells, the function includes to serve as a localized attachment site for other proteins, and to prevent diffusion of certain molecules from one compartment of the cell to another. In the cells of yeast, they help in building of scaffolding which provides structural support during the cell division and compartmentalize parts of the cell. The recent research findings in human cells suggest that the septins build cages around the bacterial pathogens, immobilizing the harmful microbes and prevent them from invading other cells.

v) Spectrin

This is a cytoskeletal protein that lines the intracellular side of the plasma membrane in the eukaryotic cells, which forms pentagonal or hexagonal arrangements, forming a scaffolding and also playing an important role in the maintenance of the plasma membrane integrity as well as the structure of the cytoskeleton.

vi) Yeast Cytoskeleton

In the cells of budding yeast, an important model organism, actin forms the cortical patches, actin cables and a cytokinetic ring and the cap. The cortical patches represent the discrete actin bodies on the membrane that are vital for endocytosis, especially for the recycling of glucan synthase important for cell wall synthesis. The actin cables are the bundles of actin filaments involved in the transport of the vesicles towards the cap (with a number of different proteins to polarize the growth of the cell) and also in the positioning of the mitochondria. The cytokinetic ring forms and constricts around the site of cell division.

vii) Prokaryotic Cytoskeleton

The cell wall was believed to be the deciding factor for the cell shapes of many bacteria, including the rods and spirals, prior to the work of Jones *et al.* (2001). The cytoskeleton was once thought to be a feature of only eukaryotic cells (Erickson, before 1992), but homologues of all the major proteins of eukaryotic cytoskeleton have been found in prokaryotes. But, research in the early 1990s suggested that the bacteria and archaea had the homologues of actin and tubulin, and these were the basis of microtubules and microfilaments of eukaryotes. Though the evolutionary relationship is distant, the similarity of the 3-dimensional structure of the proteins and similar functions provide strong evidence that both the cytoskeletons are truly homologous. FtsZ, a protein already known as a key player in the bacterial cytokinesis was independently discovered, and had the tubulin signature sequence that is present in alpha-, beta- and gamma-tubulins. Even then, some structures in the bacterial cytoskeleton have yet to be identified.

viii) FtsZ

This was the first protein of the prokaryotic cytoskeleton to be identified, and like tubulin, this forms the filaments in the presence of GTP (guanosine triphosphate), but these filaments do not group into tubules. This is the first protein to move to the site of division during cell division, and is very essential for recruiting other proteins that synthesize the new cell wall between the dividing cells.

ix) MreB and ParM

The prokaryotic actin-like proteins (like MreB) are involved in the maintenance of the shape of the cells. All non-spherical bacteria have the genes that encode actin-like proteins, and these form a helical network beneath the cell membrane which guides the proteins involved in the biosynthesis of the cell wall. Some plasmids encode a separate system which involves an actin-like protein, ParM. The filaments of this protein exhibit dynamic instability, and help in partition of the plasmid DNA into the dividing daughter cells by a mechanism which is analogous to that used by the microtubules during eukaryotic mitosis.

x) Crescentin

Crescentin is the third protein in the bacterium, *Caulobacter crescentus* which is related to the intermediate filaments of the eukaryotic cells. This is involved in maintaining the cell shape (helical and vibrioid forms of bacteria) but the mechanism is unclear to date. But, the curvature can be described by the displacement of crescentic filaments, after the disruption of peptidoglycan filaments.

xi) Cytoplasmic Streaming

Also known as *cyclosis*, this refers to the active movement of the cell contents along the components of the cytoskeleton. Mainly seen in the plants, all the cell types use this process for the transport of waste, nutrients, and organelles to other parts of the cell. The cells of plants and algae are larger than many other cells, and cytoplasmic streaming is important in these cells, because the cell's extra volume requires streaming in order to move the organelles throughout the entire cell. The organelles move along the microfilaments in the cytoskeleton driven by myosin motors binding and pushing along actin filament bundles.

Ergastic Substances

These are also called as "deutoplasmic" substances or "paraplasmic" substances and are non-living substances (non-protoplasmic materials) either organic or inorganic products of metabolism classified as reserve material, excretory material, pigments and mineral contents, that include crystals, oil drops, gums, tannins, resins and other compounds that help in defence, maintenance of the cellular structure or a substance of storage. The living protoplasm of a cell is sometimes called the "bioplasm" which is distinct from these substances. If present in the cells, the cell becomes different from other cells and is called an idioblast. The ergastic substances include carbohydrates such as cellulose and starch, and in majority these are formed by photosynthesis and stored in the tissues of the plant (such as the vacuoles or in the cell wall). Plants synthesize and accumulate a variety of compounds in them. Some of these compounds have high nutritional value like starches, proteins and fats but some were considered to be 'waste products'. During recent times, there is a tendency to avoid/dismissing compounds as mere waste products, with no value. Harborne (1982) opined that many of these ergastic substances which were once believed as 'waste products' are not really 'waste', but they are useful to the plants in many ways. These ergastic substances make the plants either less nutritious or bitter and unpalatable and deter the forage animals from eating them. A few of these ergastic substancs are crystalline in nature and effectively inhibit the insects to chew the plant parts or lay eggs on them.

The chemical composition of the ergastic substances seen in the histological/ anatomical preparations was not known quite a long time and similarly the advantage they confer to the plants. But, with the advent of several new techniques recently, it is possible to analyze those small amounts microscopically and know their chemical nature and properties. The ergastic substances can be broadly classified as:

Ergastic Subtances

- Food products
  - Non-nitrogenous: Starch, Inulin, Hemicellulose, Cellulose, Sugars
  - Nitrogenous: Proteins, Amino - compounds
  - Fats and fatty oils
- Secretory products
  - Enzymes, Pigments, Nectar
- Waste products
  - Non-nitrogenous: Tannins, Mineral crystals, Latex, Essential oils, Gums
  - Nitrogenous: alkaloids

Food Products

These are the substances manufactured by the protoplasm of the cells by using simple inorganic substances like CO_2, H_2O and stored in the cells as food materials. The food is partly used to make up new protoplasm and is partly broken down to provide the energy required for performing the normal activities and the remainder is stored as reserve food material in the protoplasm of the cells. The reserve food materials can be classified as (a) the carbohydrates, (b) nitrogenous products and (c) fats and fatty oils.

Carbohydrates

These are the products more or less directly derived from carbon dioxide and water during 'photosynthesis'. Some of the carbohydrates are insoluble while some are soluble in water. Cellulose and starch are the main ergastic substances of the plant cells; the former is the chief component of the cell wall while the latter occurs as a reserve material in the protoplasm. The most important insoluble product is the starch and soluble products are inulin and sugar.

Starch

This is an insoluble carbohydrate which is composed of long chain of glucose molecules. It appears in the form of grains which are formed in the chloroplasts. Later, the starch is broken down into sugars which then move into the storage tissues where it is re-synthesized in the amyloplasts. Starch grains are abundantly found in all parenchymatous cells of rhizomes, endosperm, corms, tubers and cotyledons of seeds *etc.* They are also present in the latex tubes of stems and roots (tuberous roots, underground stems, cortex of stems, endodermis), seeds of cereals, grains of rice, wheat and maize, fruits like banana *etc.* (Figure 2.2). The assimilatory starch is a temporary product synthesized by the chloroplasts, and this is broken down to sugar and is transported to the storage organs where the storage starch is synthesized by the amyloplasts.

The starch appears in the form of grains that stain bluish-black with iodine. Starch grains vary in their shape and size and is used as an identification feature of a taxon, *i.e.*, oval starch grains in potato, flat grains in wheat, polygonal in maize, spherical in pulses, dumb - bell or rod-shaped in the latex of Euphorbias. Starch grains of rice are the smallest and those of *Canna* are the largest. The variation in size ranges from 5-100μ. The starch grains arise exclusively in plastids, especially leucoplasts and amyloplasts.

Starch grains show carbohydrates deposited in layers around a point at the centre termed 'hilum', which gives the grain a striated appearance. The number of striations deposited one per day denotes the number of growth in days. The hilum is situated centrally in *Triticum durum* (*concentric*) or may be eccentric (the layers are deposited on one end) as in *Solanum tuberosum*. The number of layers in the starch grains of cereals corresponds to the number of days during which it grows. The commonest type of starch grain is concentric. If the starch grain possesses a single hilum, it is known as simple. Sometimes, 2, 3 or many grains are arranged in groups and have many hila, known as compound grains, *e.g.*, potato, rice, oats, *etc.* Starch grains turn to blue or black when aqueous Iodine solution is added and appears shining or luminous under a microscope except for dark cross.

The starch grains are often classified into simple grains, semi-compound (half-compound) grains, and compound grains. In simple grains, each grain is single, and has a solitary hilum, while in compound gains, two or more grains are aggregated together (as in *Oryza sativa, Avena sativa* and *Ipomoea batatas*) with separate

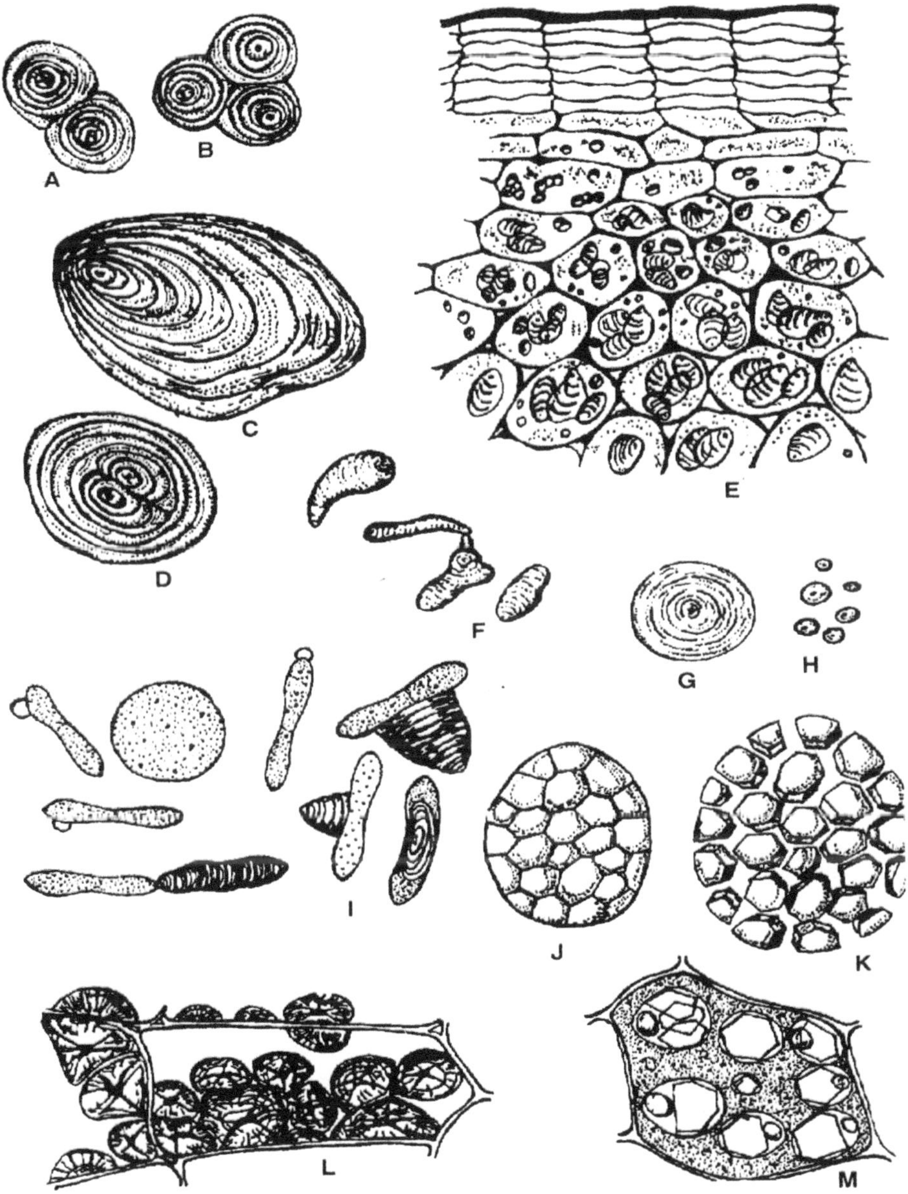

Figure 2.2: (A—D) Potato Starch Grains; (A, B) Compound starch grains; (C) Simple starch grain; (D) Half-compound starch grain; (E) Cross-section of the outer portion of a potato tuber; (F) Banana starch grain; (G, H) Starch grains of *Triticum durum*; (I) Stages in development of starch grains in chloroplasts of *Phaius maculata*; (J) Compound starch grain of *Avena*; (K) As in J but disintegrating; (L) Sphaerocrystals of inulin in cells of *Dahlia* tuber, precipitated when alcohol was added; (M) Aleurone grains in an endosperm cell of *Ricinus communis* from a section of material embedded in dilute glycerine.

hila. In Potato, the grains are half-compound, and consist of two hila with their lines of stratification aggregated together around which several common layers of carbohydrates are present. The grains of starch vary in their shape, size, and position of the hilum, solitary or in aggregates or in general appearance. The variation in morphology of the starch grains makes it possible to identify the plant species.

Inulin

This is a soluble carbohydrate of polysaccharide type. This on hydrolysis yields only fructose. It is found in the cell sap of roots of many members of Asteraceae. It is common in tuberous roots of *Dahlia* and *Helianthus tuberosus*. It is also found in Campanulaceae and Monocots. Inulin can be obtained in the form of precipitate by keeping of the roots of *Dahlia* in alcohol for 6-7 days. It normally appears in the form of spherical, star-shaped or wheel-shaped crystals. The crystals that are incompletely formed, can be seen as fan-shaped structures.

Hemicelulose

The food is stored in the form of hemicelluloses in some seeds. This type of storage is rather rare when compared to starch or sugar. Hemicellulose (reserve cellulose) is found in the seeds of certain palms.

Cellulose

This is an insoluble structural carbohydrate having a similar formula like that of starch, *i.e.*, $C_5H_{10}O_{5,}$ and is the major constituent of the cell wall of higher plants. The building blocks of cellulose are the glucose units (approximately 15,000 molecules of glucose residues) that are held together by beta-linkages However, the arrangement of the atoms in the molecule is quite different in starch and cellulose, hence they have different properties. In plants, the cellulose is made from sugars and serves as a building material in the formation of a cell wall.

Pectin

This is a structural polysaccharide that is mainly present in the middle lamella and the primary walls of Dicots. The pectins are rich in galacturonic acid, rhamnose, arabinose and galactose.

Sugars

The common sugars (monosaccharides - hexoses with 6-carbons) that are found abundantly are glucose, fructose and sucrose. Glucose and fructose had a similar formula, *i.e.*, $C_6H_{12}O_6$. They are composed of same kind of atoms, in the same proportion, but the arrangement in the molecule is different. These are monosaccharides and are the simplest of the natural sugar. Sucrose is a disaccharide and has twice as many carbon atoms and the formula is $C_{12}H_{22}O_{11}$. Sucrose is known as the "ordinary sugar" and is obtained from sugarcane or sugar beets. The sugars are the simplest carbohydrates and are quite soluble in water. Glucose and fructose have relatively simple molecules. They are good molecules for the building up of other substances or for furnishing energy. In sugarcane or sugar beet, sucrose is found as reserve food.

The oligosaccharides contain 2-10 units of monosaccharide; those that contain 2 units are called as disaccharides (sucrose or cane sugar) and those contain 3 units are called trisaccharides (raffinose). Sucrose is the "table sugar" from sugarcane. Similarly, maltose yields 2 sucrose molecules upon hydrolysis, while raffinose yields glucose, fructose and galactose on hydrolysis. The polysaccharides are made up of more than 10 units of monosaccharides and these are of two types, namely, "storage polysaccharides" and "structural polysaccharides" (inulin, starch, cellulose and pectin *etc.*).

Nitrogenous Products

Proteins and amino compounds are the two important nitrogenous compounds.

Proteins

The proteins are the most important compounds found in plants. They constitute the active matter of protoplasm as many chemical phenomena of the life processes are associated with them. The proteins are composed of amino acids which are linked together by peptide bonds. The proteins are highly complex organic nitrogenous substances consisting of carbon, hydrogen, oxygen and nitrogen. In some complex proteins, sulphur and phosphorus are also present. For example, protein, zein from Indian corn has the formula $C_{736}H_{116}N_{184}O_{208}S_3$, while Gliadin from wheat has the formula $C_{685}H_{1068}N_{196}O_{211}S_5$. In plants, the proteins occur as inactive, ergastic bodies – in an amorphous or crystalline (or crystalloid) form. A well known amorphous ergastic protein is gluten (wheat). These are found frequently as solid granules. Proteins are formed due to rearrangement of atoms of carbohydrates with the addition of nitrogen and sulphur commonly and sometimes phosphorus also. A protein molecule is made up of 100s or 1000s of amino acid molecules joined together into one or more chains which are variously folded. 20-24 different kinds of amino acids are commonly found in proteins. Most of these amino acids generally occur in any protein molecule. They are arranged in a chain sequence which is exactly the same in all molecules of a given kind of protein. The different arrangements of amino acid molecule are practically infinite. Every species have protein molecules peculiar to itself. The proportion of different amino acids in different proteins varies. Some proteins lack amino acids found in other proteins. Animal proteins are better as human food than plant proteins because the amino acids of animal proteins are more like human proteins. Some plant proteins lack some of the amino acids which are essential for the building of human proteins.

Storage protein is in the form of definitive bodies called the aleurone grains. These may be amorphous or crystalline. The former protein is found in the aleurone layer (the outermost layer of the endosperm) of *Triticum*. The latter possesses both the crystalline and colloidal properties, and as such these are called crystalloid proteins.

A common form of protein found in the endosperm of castor seed is known as aleurone grains. Each aleurone grain is a solid, ovate or rounded body which usually encloses a crystal like body in it known as crystalloid and a globule like body known as globoid. The crystalloid is proteinaceous in nature and occupies the major portion of aleurone grain, whereas the globoid is double phosphate of calcium and magnesium which occupies the narrower part of the grain. The aleurone grains vary in their shape and size in different plants. When they are found in oily seed of castor, they are larger in size, while when found with starch, they are very minute in size. Cuboidal crystalloid proteins are observed in the peripheral parenchyma cells of the tuber of *Solanum tuberosum*; and also in the fruit of *Capsicum annum*. Aleurone grains are bounded by a proteinous membrane, and this membrane may enclose the amorphous protein, globoids and crystalloids (as in *Myristica fragrans*) or the crystals of calcium oxalate (as in Umbelliferae). The grain of *Ricinus communis* contains one globoid and one crystalloid.

Amino Acid Compounds

These are simple nitrogenous food materials. They are found in the form of amino acids and amines which occur in the cell sap. These are found abundantly in the growing apices of the plants and less frequently in storage tissues.

Fats and Fatty Oils

The fats and oils are the glycerides of fatty acids, and the oils are liquids while the fats are solid at normal temperature. These provide more calories and as such they are valuable reserve food materials. Fats (lipids) and oils are widely distributed in the tissues of plants. These are abundant in the cells of endosperm or the perisperm of the seeds, fruits, spores, embryos and the meristematic cells and also in the differentiated tissues of the vegetative body of the plants. They also occur in the cytoplasm of the cell sap as dispersed or aggregated masses of larger size. The essential or volatile oils occur in all tissues (in conifers), in the petals of roses, in the fruit skins of oranges and other *Citrus* fruits and in a special layer (the second layer) of *Elettaria cardamom* seeds. The essential oils are generally produced by specialized secretory tissues. The substances like waxes, suberin, and cutin (other fatty substances) occur as protective layers in or on the cell wall (occur on the external cell walls of epidermal cells). These stain a reddish colour when treated with Sudan III or IV. The fats and fatty oils in plants are composed of glycerine and organic acids. They occur in the form of minute globules in the protoplasm. Special types of fats and fatty oils are composed of carbon, hydrogen and oxygen. They are characterized by a small proportion of oxygen as can be seen from the formula of common fats, *e.g.*, Palmitin [$C_{15}H_{98}O_6$], Olein [$C_{57}H_{104}O_6$] *etc.* Because of the very small percentage of oxygen contained in fats, the oxidation of fats produces large quantities of energy. Fats are insoluble in water but soluble in ether, chloroform and petroleum. Fats may be solid or liquid (fatty oils) according to temperature. Fats and oils may be produced by elaioplasts or spherosomes. Their distribution in plants is restricted than that of starch. Lipids occur as spherical droplets and can be easily detected. Lipids readily absorb stains like Sudan III and Sudan IV. Rarely lipids also occur in crystalline state, *e.g.*, needle-shaped fat crystals in the endosperm of the palm, *Elaeis*.

Secretory Products

Besides food materials, several other products are also secreted by the protoplasm which is not useful as nutritional products, but they may help or accelerate the various reactions in the cells. Though these are concerned with nutrition, these are valuable ergastic substances, and carry out specific roles in the life of the plants. They are: Enzymes, Pigments (or Colouring matter) and Nectar.

Enzymes

The enzymes are considered as "biological catalysts" which catalyze all the chemical reactions that occur in a plant cell. There are many metabolic pathways associated with different metabolism in a plant cell – such as carbohydrate metabolism, protein metabolism, fat metabolism *etc.* The enzymes associated with a plant cell catalyze each and every step of the chemical reactions (*e.g.* Zymase, Lipase, Protease, Invertase, Diastase *etc.*). These are soluble nitrogenous substances secreted by the protoplasm. They are digestive in function and convert the insoluble substances into soluble substances and complex compounds into simpler compounds (*e.g.*, the enzyme diastase converts the starch into sugar). Because of the action of this enzyme, an insoluble substance is converted into soluble one. The enzyme lipase breaks fats into glycerine and fatty acids. Papain converts proteins into amino acids.

Pigments or Colouring matter

The petals and fruits of a plant possess various colours which is due to the presence of the pigments in the cell. The leaves and the aerial organs possess chlorophyll which imparts the green colour for the process of photosynthesis. The pigments other than green are carotenoids and xanthophylls which impart orange to yellow colour of many flowers and fruits. Many other bright colours are due to the presence of red, blue, violet and pink produced by the anthocyanin-like pigments in the flowers and fruits. As such, the colours attract the agents (birds, insects, honey bees, and other larger animals) to carry out the pollination and dispersal of the fruits and the seeds that are vital in the life of the plant. The plant pigments are usually found in the plastids and in the vacuole. The green colour of the leaf is due to chlorophyll which is found in the chloroplasts. In the same plastids, carotenoids – the yellow to red pigments are also found, but they are masked by the chlorophyll. The carotenoids become noticeable when there is little or no chlorophyll as is the case in chromoplasts. Chromoplasts are also developed from proplastids. Another group of pigments, flavonoids (anthocyanins and flavones or flavonols) are generally present in vacuoles. They are water soluble and give various colours to many flowers and fruits. Anthocyanins give red, pink, lilac and blue colour. This depends on the pH (acid solution-orange red to lilac, basic pH 7.0, blue).

Nectar

Nectar is another useful secretion of the protoplasm produced and secreted by special glands, the nectarines (or nectaries). According to their position, they are called floral nectarines if they occur on the flower, and extra-floral nectarines, if they occur on the vegetative parts. The former occur on different positions of the flower while the latter occur on the stems, the leaves, the stipules, *etc.* The secretory tissue of the nectar is found on the epidermis or beneath it (a few sub-epidermal layers). The nectar is covered by a cuticle. The sugar of the nectarines (both floral and extrafloral) is derived from the phloem. The vascular tissue is found near the nectar. The nectar is excreted either through the cell wall and the ruptured cuticle or sometimes through stomata. The nectar is mainly composed of glucose, fructose and sucrose. The nectar attracts many insects by its flowers to obtain their nectar, and the insects during their visit carry the pollen from one flower to another effecting the pollination (the flowers are entomogenous).

Osmophors

The fragrance of the flowers is generally produced by volatile substances distributed throughout the epidermal region of the perianth parts. In some plants, the fragrance originates in special glands known as 'osmophors'. In Asclepiadaceae, Aristolochiaceae, Araceae, Burmanniaceae and Orchidaceae, various floral parts are differentiated as osmophors and they assume the form of flaps, cilia or brushe. The prolongation of the spadix of Araceae and the insect attracting tissues in the flowers of Orchidaceae are osmophors. The osmophors have secretory tissue which is several layers in depth.

Waste Products

These are the excretions of the plant cells. These products are stored in the dead cells. They are formed as a result of metabolic activities of protoplasm and hence are known as metabolic wastes of the plants. There are two types of metabolic wastes. They are: Non-nitrogenous waste products and nitrogenous waste products.

Non-Nitrogenous Waste Products

These are of various types like tannins, mineral crystals, latex, essential oils, gums, resins and organic acid. They are found in the cytoplasm of cell sap.

Tannins

These are heterogenous group of non-nitrogenous phenol derivatives which are present in the cell sap. In microscopical sections, they appear as granular masses which are coloured yellow, red or brown. Tannins are found in different parts of the plant like leaves, periderm, heartwood, vascular tissues, unripe fruits, seed coats *etc.* These are found in individual cells (idioblasts) or in special organs called 'tannin sacs' within the cell. The tannins are also found in the vacuole, or as droplets in the cytoplasm. Sometimes, for instance in the cork tissue, they penetrate into the cell walls. These are believed to protect the plants against dehydration, rotting and damage by the animals. The tannins can deter the insects feeding on the plants by interacting with and damaging the proteins of the gut. These can prevent the pathogenic attack. Tannins, as substances associated with starch metabolism, are useful in the transport of sugars and also act as antioxidants. Tannins are commercially useful in medicine, the tanning industry, *i.e.*, tanning of animal skins to obtain leather (in leather industry) and also in dyeing industry.

Crystals

These refer to the waste deposits within a cell, and are composed of either Calcium carbonate or Calcium oxalate which is more common. The cells that contain the crystals may easily be distinguished from other cells by their shape and size (termed as *idioblast*). The crystals originate from within the vacuoles, and they may remain single or as aggregates; and sometimes, a large crystal filling up the entire cavity. These lie loose in the cells but some are found to be suspended from the cell wall into the cavity (cystolith).

Mineral Crystals

The animals eliminate the excess of the inorganic materials – while the plants mostly deposit the material in the tissues. Such mineral salts are mainly those of calcium and anhydrides of silica. Various types of mineral crystals occur in plant cells. They occur in the cell cavity or in the cell walls. The crystals may lie loose in the cells or suspended into the cell cavity or in the cell walls. The crystals vary in shape and size. The crystals consist of calcium carbonate or calcium oxalate or silica.

Calcium Carbonate Crystals

These are commonly known as "cystoliths". They are present in the epidermal cells of leaves, found in the parenchyma, trichomes (as in *Humulus lupulus*) or hairs of many flowering plants like Moraceae, Urticaceae, Acanthaceae, Cucurbitaceae *etc.* A cell may contain solitary or more than one crystal. The leaf of *Momordica* contains a double cystolith. The main body of a cystolith is a cellular extension of the cell wall in which the calcium carbonate is deposited in the form of granules. The epidermal cell containing the cystolith is called a "lithocyst". In the leaf of *Ficus elastica*, the lithocyst contains a solitary cystolith which is much larger than the epidermal cell. The stalk of the lithocyst is made up of cellulose that develops and projects into the cell cavity, and calcium carbonate deposits around the stalk and the fully developed cystolith appears as grape-like cluster with a stalk that is attached to the cell wall from where it hangs into the lumen. Later, the epidermal cells are destined to become the lithocyst can be differentiated by their larger nucleus and denser cytoplasm. These cells do not divide but only enlarge and become lithocyst where the nucleus remains functional. Sometimes, double cystoliths are also found. These cytoliths project into the protoplasts in large, specialized cells. They vary in shape in different genera and families.

Calcium Oxalate Crystals

The leaves and other organs of many plants contain conspicuous crystals of calcium oxalate. The forms of these crystals are very diverse. These appear in definite forms such as – prismatic (rectangular), or pyramidal, either as single or twin prisms (*e.g.* leaves of *Citrus* and *Begonia etc.*)

Raphides

These are thin, elongated, needle-shaped (or acicular) crystalline form of calcium oxalate that is tapered at both the ends to a tip which is aggregated in bundles (in an idioblast which is larger than the neighbouring cells) within a plant cell. Because of their needle-like shape, and the larger numbers in the tissues, they make the leaves unpalatable to herbivores (*Dieffenbachia* and Taro). These are long, needle like crystals lying parallel to each other in a bundle, which sometimes found in a special sac like cell. When these sacs are mechanically injured, these raphides are released with jerk through a hole. The raphides are commonly found in the petals of *Impatiens, Alocasia*, leaves of *Colocasia,* and leaves of *Pistia etc.* The cells that contain the raphides may also contain mucilage. The rosettes in Umbelliferae and the aleurone grains in the seeds contain rosette crystals (large with equal length) which deposit around a centre along the radius in all directions to form a rosette. These raphides are very irritating and afford some protection to the plants from the foraging animals. The raphides can be destroyed by boiling. Hence, the food plants containing them are not irritating when cooked.

Druses, Rosette Crystals or Cluster Crystals

The druse is also known as the sphere crystal or "sphaeraphide". The druse is more or less spheroidal in their shape. Many prism-like or pyramidal crystals aggregate to form a "druse". Some projecting points are observed all over the surface, in these crystal aggregates. This is one of the commonest crystal and had the rosette shape, hence the name rosette crystal. These are quite common in the leaves of *Datura, Eucalyptus, Nerium,* and the petiole of *Carica, etc.*

Prismatic Crystals

These are single calcium oxalate crystals found in various plants. They may be square, prismatic, rhomboidal or pyramidal in shape.

Sand Crystals

These are commonly found in masses of microspheroidal crystals packed into a cell. These are found in the leaves and roots of *Atropa belladonna*. Silica is deposited mostly in the cell walls and sometimes in the lumen of the cell. Members of the family Poaceae are the best examples of a plant group with silica both in the cell walls and the lumen. The term "silica body" is used for the silica deposits of Cyperaceae and Poaceae.

Stegmata

These occur in the leaves and stem as longitudinal files of silica cells lining the fibre bundles or vascular bundle sheaths. These are generally restricted to epidermis or cells adjacent to vascular tissues only. Stegmata are valuable characteristic character of the members of the group Arecaceae. The distribution of these bodies in Orchidales is given for 130 species representing 105 genera. The silica bodies are of two types, namely, spherical and conical. The absence of stegmata and presence of stegmata with conical bodies are the two most common characteristics in the order. The stegmata containing the spherical bodies are the characteristics of Vandeae and Eriinae, Podochilineae and Dendrobiinae within Lepidendreae.

Latex

This is the milky or watery juice (*fluid*) found in the special long, branching tubes known as latex tubes (laticifers). The neighbouring latex tubes get connected to form a network. The latex oozes out when these tubes are cut or injured. Rubber, opium, chewing gum or the valuable substances derived from the coagulated latex. The latex usually consists of carbohydrates, starch grains, proteins, oils, tannins, gums, resins, alkaloids, salts, enzymes, organic acids, resins, rubber and also some poisonous substances. The latex may be colourless or coloured, milky-white, or yellow to orange.

Best known latex is the one obtained from rubber yielding plants. The latex of *Hevea brasiliensis* consists of 40-50 per cent of the rubber (and produces about 2240 kg rubber per hectare annually). Rubber particles, suspended in latex vary in size and shape. When the latex is released from the plants, the particles clump together, *i.e.*, the latex coagulates. This property is utilized in the commercial separation of rubber from latex. The latex may be transparent or clear (*Morus, Nerium*) or milky (*Asclepias, Calotropis, Euphorbia, Ficus etc.*), yellow-brown in *Cannabis*, yellow or orange in some members of Papaveraceae. The latex occurs in the families of Asclepiadaceae, Apocynaceae, Moraceae and Euphorbiaceae *etc.* Papain (the proteolytic enzyme), is present in the latex of *Carica papaya*. The opium (an alkaloid) is rich in the latex of *Papaver somniferum*.

Essential Oils or Volatile Oils

The odoriferous organs of a plant contain the essential or volatile oils which are produced is special cells or glands. These generally occur in the flowers, leaves and fruits *etc.* The best examples are: oil of peppermint from *Mentha piperita*, clove oil from *Eugenia caryophyllata*, eucalyptus oil from *Eucalyptus globulata etc.* Though these are excrtetory products, they have biological significance to attract the vectors to bring about pollination and dispersal of the fruits and seeds. These are volatile, very odoriferous and frequently occur in oil glands as in *Eucalyptus*. The orange peel contains large, oil glands which are formed by the breaking down of the cells containing the oil. The disintegration of the cells brings the oil into large cavity of the gland. Oil glands are also found in the petals of rose, jasmine and other flowers.

Gums

The gums are exuded from the stems (and/or bark) of many plants in the form of thick juice. The cellulose of the cell wall by its decomposition forms the gums. *Acacia* species like *A. senegal* (produces the gum Arabic of commerce, and this is used in medicine, confectionary and sizing and finishing materials in the textile industry), *A. modesta* and *A. arabica* produce the best gum. The gums are soluble in water and insoluble in alcohol. The gums swell up in water and form viscous mass. These are related to peptic compounds. They appear in plants mainly as a result of physiological disturbances which induce break down of cell walls and contents. Usually the gums are exuded naturally but sometimes, these are produced in response to the injuries to heal up the wounds in plants.

Resins

These are frequently found in the resin ducts of conifers. The resins may be solid, brittle or liquid in their nature and are complex in composition. These are produced in special glands or secretory cells present around the resin ducts. Resin ducts are formed either by the separation or disintegration of cells. The turpentine is obtained by cutting of resin ducts of pine trees. Various types of varnish and other resins are obtained from other conifers. Resins are insoluble in water, but soluble in alcohol. The well known mounting fluid used in microscopy, canada balsam is a resin obtained from the conifer, *Abies balsamea*.

Resin Ducts

These are long passages surrounded by glandular cells. These occur in stems and also in other parts of plants. In *Pinus*, resin ducts develop in the axial and ray system. Resin ducts develop as schizogenous intercellular spaces by the separation of parenchyma cells from each other. Some parenchyma cells form the lining or epithelium over the ducts and excrete resin. In *Pinus*, the epithelial cells are thin-walled, active for several years and produce abundant resin. In *Abies* and *Tsuga*, the epithelial cells have thick, lignified walls and most of them die during the year of origin. They produce little resin. Eventually, the resin duct may become closed by the enlarging epithelial cells. There is distinction between normal resin ducts and those that are traumatic (trauma = wound), *i.e.*, arise in response to injury. Normal ducts arise singly whereas the traumatic ducts are cyst like and occur in tangential series.

Organic Acids

The different types of organic acids are found in plants and some of them remain concentrated in the leaves and fruits. These organic acids are produced during different types of metabolism. Several vegetables and fruit juices are re-combined with peculiar bases and alkaloids. Several organic acids are found in various plants *viz.*, citric acid in citrus fruits, tartaric acid in grapes and tamarind, malic acid in the fruits of apple, gallic acid in mango seeds, oxalic acid in *Rumex*,

Oxalis, Nepenthes etc., remain concentrated in the leaves and fruits, and are the products of different types of metabolism.

Nitrogenous Waste Products

Excretory Materials

These are the ergastic substances that are the by-products of plant metabolism which are of no use to the plants. Such waste material remains stored in the dead cells. In plants there is no proper excretory system; as such they get rid of the excretory or the waste material through falling leaves, fruits, seeds and bark *etc.*

Glycosides

These are composed of glucose and some other substances (for example, sinigrin is a sulphur-containing glycoside present in *Brassica nigra* and digitonin is present in *Digitalis purpurea*).

Alkaloids

These are a group of nitrogen containing basic organic compounds present in plants of a few families of Dicots like Solanaceae, Papaveraceae, Apocynaceae *etc.* They are thought to be the end products of nitrogen metabolism (or nitrogenous wastes). These are found in the roots, the leaves, the bark, the seeds and other parts. These are important because of their poisonous and medicinal properties, *e.g.*, atropine (found in *Atropa belladonna* of Solanaceae), cocaine, morphine (from the fruits of *Papaver somniferum*), nicotine (from leaves of *Nicotiana tabacum*), quinine (occurs in the bark of *Cinchona*), strychnine (from *Strychnos nux-vomica*) *etc.*

Chapter 3

Cell Wall

A cell wall in plants is a structural layer around some types of cells, outside the cell membrane. It is tough, flexible and sometimes rigid. It provides structural support and protection to the cell and also acts as a filter. The cell walls are present in most of the prokaryotes (except molecule Bacteria), in algae, in fungi, and in eukaryotes which include the plants, but are absent in animals. The major function of the cell wall is to act as pressure vessels, to prevent the over-expansion of the cell when water enters into the cells.

The composition of the cell wall is variable between the species and depends on the types of the cells as well as the development stage of the cell. The primary cell wall of the land plants is made of polysaccharides – cellulose, hemicelluloses and pectin. Often, polymers like lignin, suberin or cutin are embedded in plant cell walls. In algae, the cell walls are made of glycoproteins and polysaccharides like carrageenan and agar which are absent in land plants. In bacteria, the cell wall is composed of peptidoglycan. The cell walls of archaea are formed of glycoprotein S-layers, pseudo-peptidoglycan or polysaccharides. The cell walls of fungi are made of N-acetylglucosamine polymer chitin. The diatoms have a cell wall composed of biogenic silica.

Cell Wall – History

The cell wall in plant cells was first observed and named (as a "wall") by Robert Hooke (1665). Karl Rudolph and J. H. F. Link (1804) proved that the cells possessed independent cell walls. Earlier, it was thought that the cells had shared cell walls and that the fluid passed between them. In the 19th century, Hugo von Mohl (1853 and 1858) advocated that the cell wall grows by apposition (or "lamination theory" by Edward Strasburger – 1882, 1889). Later, Carl Nageli (1858, 1862 and

1863) believed that the growth of the cell wall in thickness and the in the area was due to intussusception (Julius Wiesner – 1886). Ernst Munch (1930) coined the term – living symplast from the dead plant region that included the cell wall. By 1980s, some replaced the term cell wall (used for the plants) with a more precise term – *extracellular matrix* – used for the animal cells, while others preferred the older term for both.

Cell Wall – Properties

The cell walls give the required rigidity and strength and also offer protection against mechanical stress. In multicellular organisms, they build and hold a definite shape, and also limit the entry of large molecules which may be toxic to the cell. They also help to create stable osmotic environments by preventing osmotic lysis and help to retain water by the cells. During the cell cycle and growth conditions, the composition, properties and form may change.

Cell Wall - Rigidity

The cell wall in most cells, is flexible (it will bend rather than holding a fixed shape) but has a considerable tensile strength. Hydraulic (turgor) pressure creates this rigidity along with its structure (the rigidity results from inflation due to passive uptake of water). The flexibility is seen when the plants wilt (the stems and leaves begin to droop) or bend in water currents (as in sea weeds). A secondary wall in plants is a thicker, additional layer of cellulose which increases the rigidity of the wall. The additional layers may be formed by the lignin (in the walls of xylem) or suberin (in the walls of the cork). These are not only rigid but also water-proof and thus make the secondary wall stiff. The cells of the wood and the cells of the bark (in trees) have the secondary walls, while other plant parts like the petioles may also acquire reinforcement to resist the strain of the physical forces.

Cell Wall – Permeability

In most of the plant cells, the primary cell wall is freely permeable to small molecules (small proteins of the size of 30-60 kDa), and the pH is an important factor which governs the transport of the molecules through the cell walls.

Plant cell possess cell wall and this feature distinguishes them from animal cells. Cell wall is the non-living component of the cell and is synthesized by protoplasm. The synthesis of cell wall is controlled by Golgi bodies. The cell wall is protective in function.

The cell wall was discovered in 17th century even before the recognition of the protoplast. Since then, several studies have been made by using various methods – chemical, physical and morphological. These investigations made use of the advances made in organic chemistry, X-rays, the use of the Light, Polarizing and Electron Microscopes which helped in the better understanding of the origin and structure of the cell wall.

The cell wall is in contact with the protoplast and grows outside of it. Only in the pollen grains the outermost wall – exine develops from the contribution of both, *i.e.*, by the protoplast within and the tapetum which surrounds the pollen grains.

The cell walls of the plant cells have sufficient tensile strength to withstand the internal osmotic pressures of several times atmospheric pressure that result from the difference in the solute concentration between the cell interior and external solutions. The cell walls vary from 0.1 to several micro-meters in thickness.

The Cell Wall - Layers

The cell walls vary in thickness basing on their age and type of cells. Generally, the younger cells have thinner walls than the older ones. In mature thick walled cells, a concentric layering is usually distinct in the cell wall. These layers differ from one another in their physical and chemical nature. Up to three principal strata (or layers) are found in the cell wall of plant cells - (i) the Middle Lamella, (ii) the Primary cell wall, and (iii) the Secondary cell Wall.

1. The Middle Lamella

It is the cementing layer which binds the two adjacent cells in a tissue. It is therefore, present between two primary walls of neighbouring cells. The middle lamella is a layer rich in pectins and this is the outermost layer that forms an interface between the adjacent plant cells and glues them together. It is an amorphous substance, inactive in terms of polarized light (isotropic). In supporting tissues, it also fills intercellular spaces. The middle lamella consists of mainly pectic substances and usually impregnated with calcium and magnesium. Pectic substances are a mixture of polygalacturon (D-galacturonic acid) and polysaccharides (neutral sugars). The enzyme pectinase and chemical reagents which solubilise pectins, disintegrate the tissue into individual cells. The procedure is called maceration. The formation of this layer starts in the telophase of cell division and is formed by deposition.

2. The Primary Cell Wall

The growing plant cells are generally surrounded by a polysaccharide-rich primary wall. This is a part of the apoplast which is largely self-contiguous and contains everything that is located between the plasma membrane and the cuticle. The symplast is another unique feature which exists because tube-like structures (plasmodesmata) connect the cytoplasm of different cells.

The primary cell wall is in general a thin, flexible and extensible layer formed by the cell during its growth. The primary cell walls are composed predominantly of polysaccharides along with lesser amounts of structural glycoproteins (hydroxy-proline-rich extensins), phenolic esters (ferulic acid and coumaric acids), ionically and covalently bound minerals (calcium and boron) and enzymes. The walls also contain proteins (expansins) that are believed to have a role in the regulation of cell wall expansion. Lignin composed of highly cross-linked phenolic molecules is a major component of the secondary walls. The primary wall is the first layer of individual cell, synthesized by protoplasm. It is 1-3 μm thick, bordering the middle lamella. The primary wall consists of major polysaccharides such as: hemicelluloses (up to 50 per cent - branched polysaccharides structurally homologous to celluloses with a backbone composed of 1,4-linked β-D-hexosyl residues), cellulose (up to 25 per cent - composed of 1,4-linked β-D-glucose residues) and smaller amounts of pectic substances (pectin – a family of complex polysaccharides – 3 important classes are: homogalacturonans, rhamnogalacturonans and substituted galactorunans) which contain 1,4-linked α-D-galacturonic acid, and also hemi-cellulosic polysaccharides (xyloglucan [predominant in primary cell walls], glucuronoxylans, arabinoxylan, glucomannan, arabinogalactans, galactomannan, found both in primary and secondary cell walls) and forms the matrix of wall in which cellulose microfibrils are embedded. The primary cell wall is elastic and undergoes extension with the growth of the cell wall. In many roots, fleshy stems, fruits and leaves the cells contain only the primary wall and the middle lamella. The primary wall is not uniformly thickened but it may be alternatively thickened or thinned. The changes that occur in this wall are reversible. The walls may lose a thickening previously acquired and the chemical substances may be removed or replaced by others. For example, cambial walls show seasonal changes in thickness. The primary walls of the endosperm cells are digested during germination. The primary wall is optically active (anisotropic).

Primary Cell Wall - Functions

Some of the functions of the primary cell wall include: to give the structural and mechanical support – to maintain and determine the shape of the cell – to resist the internal turgour of the cell – to control the rate and direction of the growth – to be responsible for the architecture of the plant and its form ultimately – to regulate the diffusion of the material through the apoplast – to store the carbohydrates so that the seeds may be metabolized – to protect against the pathogens, dehydration and other environmental factors – to be the source of the biologically active signalling molecules and also to control the cell-to-cell-interactions.

3 The Secondary Cell Wall

All the plants form two types of cell wall that differ in their composition and function. The primary walls provide mechanical strength and also expand to allow the cell to grow and divide, while the secondary wall which is much thicker and stronger accounts for most of the carbohydrate in biomass and is deposited once the cell has ceased to grow. The secondary walls of xylem fibres, tracheids and sclereids are further strengthened by lignin. Both the walls contain cellulose, hemicelluloses and pectin in different proportions. Pectin and hemicellulose are present in equal amounts in dicot primary walls while hemicellulose is more abundant in grasses (switchgrass).

The secondary cell wall is a thick layer that is formed insides the primary cell wall after the cell is fully grown. It is formed next to the primary wall towards protoplasm. It appears usually after the cell ceases to grow. Hence, surface growth is normally absent in it. It is about 5-10 μm thick. It is very strongly anisotropic and is characteristic of highly specialized cells which undergo irreversible changes in their development. Rarely, the active cells with living protoplasts like xylem ray and xylem parenchyma cells also have secondary walls. The secondary wall is laid down over primary wall except over pit membrane. The secondary wall is more massive than primary wall and in most cells it forms the major part.

The secondary wall is not found in all cell types. Some cells like the conducting cells (in xylem) possess a secondary wall containing lignin which gives strength to the wall and also make it water-proof. In majority of tracheids and fibres, three layers can be observed in secondary wall; the outer layer (S_1), the central layer (S_2) and the inner layer (S_3). Of these layers, the central layer is usually the thickest. Sometimes, a new term - tertiary wall (or terminal layer) is applied to the inner layer (S_3). The tertiary wall (S_3) also differs from the other layers of the secondary wall in its chemical composition. The S_1 layer is covered by the noncellulosic film often with lumps called warts and known as "warty layer". This may be formed by the remains of a disorganized protoplast, *e.g.*, conifer tracheids and fibers and vessels of many Dicotyledons. In some cells, the number of layers may be more than three. Some authors use the term tertiary wall for the innermost lamella (tertiary lamella) that differs with secondary wall in some properties. It is differentiated into two strata – "membranogenic stratum" and a "warty stratum". The secondary wall is composed of cellulose, hemicelluloses and lignin. Because of high cellulosic content, it is strongly anisotropic. Hemicellulose is less in the secondary wall when compared to the primary wall. The secondary wall provides mechanical strength to the plant body and is usually associated with non-living tissues at maturity. It is also characteristically found in cells which undergo irreversible changes in their development.

Cell Wall - Formation

During the telophase, of the mitosis, phragmoplast widens and becomes barrel-shaped. At the same time, on the equatorial plane, the cell plate, *i.e.*, the first partition between the new protoplasts begins to form inside the phragmoplast. In the area where cell plate forms, the microtubules of the phragmoplast disappear, but successively are regenerated at the circumference of the cell plate. With the enlargement of the cell plate, the microtubules approach the wall of the cell. With electron microscope it has been shown that the cell plate formation is initiated by the concentration and fusion of large number of vesicles derived by the Golgi bodies. ER vesicles also may be involved in this. These microtubules of the phragmoplast seem to be involved in directing the vesicles towards the equatorial region. Individual plasmatic linkages remaining in the gaps of the cell plate form the plasmodesmata.

The middle lamella is first laid down, formed from the cell plate during cytokinesis, and the primary cell wall is then deposited inside the middle lamella.

The primary cell wall can be defined as composed of cellulose micofibrils aligned at all angles. The cellulose microfibrils are produced at the plasma membrane by the cellulose synthase complex, which is made of a hexameric rosette that contains 3 cellulose synthase catalytic subunits for each of the 6 units. The microfibrils are held together by the hydrogen bonds to provide a high tensile strength. The cells are held together and share the gelatinous membrane, the middle lamella, which contains magnesium and calcium pectate (salts of the pectic acids). The cells interact through plasmodesmata, which are the inter-connecting channels of cytoplasm that connect to the protoplasts of the adjacent cells across the cell wall.

The actual structure of the cell wall is not clearly defined and several models exist – the covalently-linked cross model, the tether model, the diffuse layer model and the stratified layer model are some of them.

The Covalently-Linked Cross Model

This model was proposed by Peter Albersheim and his colleagues (1973) who proposed that the wall matrix polymers (xyloglucan, pectin and glycoprotein) are covalently - linked to one another. The binding of xyloglucan to cellulose microfibrils results in a non-covalently cross-linked hemicellulose network which gives the tensile strength to the wall. The model has been questioned because of the lack of the evidence for the existence of covalent linkages between xyloglucon, pectin and glycoprotein.

The Tether Model

The xyloglucan molecules are hydrogen bonded to and cross-link with cellulose microfibrils. The cellulose-xyloglucan network is enmeshed in a non-covalently cross-linked pectic network.

The Diffuse Layer Model

The xyloglucan molecules are hydrogen-bonded to the surface of the cellulose microfibrils but do not directly cross-link them. The tightly-bound xyloglucan is surrounded by a layer of less-tightly bound polysaccharides, and the cellulose and xyloglucan are embedded in a pectic matrix.

The Stratified Layer Model

The xyloglucan molecules are hydrogen-bonded to and cross-link, with cellulose microfibrils. The cellulose-xyloglucan lamellae are separated by the strata of pectic polysaccharides.

On both sides of the middle lamella, thin lamellae are laid down by the daughter protoplasts. Formation of these lamellae is the starting point in the development of the new walls of the daughter cells. The walls consist of cellulose microfibrils and non-cellulose matrix. The matrix consists of pectic substances and hemicelluloses. This wall matrix is secreted by the Golgi vesicles. In some cases ER also contributes in the production of matrix.

Along the line of contact of the new wall and the wall of the mother cell, the new and old middle lamellae are separated by the primary wall of the mother cell. In the primary wall of the mother cell, a cavity which is triangular in cross section develops along the line of contact of the new and old walls. This cavity continues to enlarge till it reaches the middle lamella of the mother cell and so connection is made between the new and old middle lamellae. If the cavity continues to grow and the intercellular substance does not fill it, thus an intercellular space lined with inter cellular substance is formed.

In some plants and cell types, after a maximum size (and point) of development has been reached, a secondary wall is constructed between the plasma membrane and the primary cell wall. The cellulose microfibrils are aligned in parallel layers, the orientation changing slightly with each additional layer so that the structure becomes helicoidal, unlike the primary wall. The cells with the secondary walls can be rigid (as in the gritty sclereid cells in pear and quince fruits). The cell to cell communication is possible through the pits in the secondary cell wall that allow the plasmodesmata to connect the cells through the secondary cell walls.

Cell Wall - Composition

In the growing (primary) plant cell wall, the major carbohydrates are cellulose, hemicellulose and pectin. The cellulose microfibrils are linked via the hemicellulosic tethers to form the cellulose-hemicellulose network, which is embedded in the pectin matrix. Xyloglucan is the most common hemicellulose in the primary cell wall. The xyloglucan and pectin are reduced in abundance in grass cell walls, and are partially replaced by glucuron-arabinoxylon, another type of hemicellulose. The primary cell walls grow by acid growth mediated by expansins (the extra-cellular proteins activated by the acidic conditions that modify the hydrogen bonds between the pectin and the cellulose). This increases the extensibility of the cell wall. The outer part (of the primary cell wall) of the epidermis is usually impregnated with cutin and wax to form a permeability barrier (called the plant cuticle).

The secondary cell walls contain a wide range of additional compounds that modify their mechanical properties and permeability, and the major polymers that make up the wood (secondary walls) include: 35-50 per cent of cellulose; 20-35 per cent of a hemicellulose, called xylan; 10-25 per cent of a complex phenolic polymer, called lignin, which penetrate the spaces in the cell wall between the cellulose, hemicelluloses, and pectin components, to drive out water and strengthens the wall. In addition, 1-5 per cent of structural proteins are found in most plant cell walls – and these are classified as hydroxyproline-rich glycoproteins (HRGPs), arabinogalactan proteins (AGPs), glycine-rich protein (GRPs), and proline-rich proteins (PRPs). The glycoprotein is a characteristic, highly repetitive protein sequence, and most are glycosylated and contain hydroxyproline (Hyp) to become cross-linked in the cell wall. These are concentrated in specialised cells and in the corners of the cells. The cell walls of the epidermis may contain cutin; the Casparian strips in the endodermis of the roots and the cork cells of the plant, bark contain suberin. Both of these are polyesters which function as permeability barriers in the movement of water. The composition of the carbohydrates and proteins varies between the plants and the cell type and age. The cell walls of the plant also contain many enzymes (hydrolases, esterases, peroxidises and trans-glycolases) which cut, trim and cross-link wall polymers. The secondary walls (in grasses) also contain microscopic silica crystals which strengthen the walls and protect it from herbivores. In some plant tissues, the cell walls also function as storage deposits for the carbohydrates (which are broken down and resorbed to supply the metabolic and growth needs of the plant) *e.g.*, glucans and other polysaccharides in the endosperm walls of cereals and *Nasturtium* which are digested during the germination of the seeds, and the simple sugars thus formed help in the nourishment of the growing embryo.

Fine Structure of Cell Wall

Usually cell wall components are: cellulose, hemicellulose, pectin, lignin, suberin, cutin, proteins *etc.* The proportions of these substances differ from species to species. Of these, the architecture of the cell wall is mainly based on cellulose. The cell wall is composed of microfibrin, 0.5 µm in width, which are visible under the light microscope. The macrofibril consists of a bundle of microfibrils, each of which is about 250Å in diameter. The microfibril is visible only under the Electron microscope. The microfibril in turn consists of a bundle of micelle or elementary fibrils, each about 100Å in diameter. The micelle contain about 100 cellulose is a polysaccharide with linear chains of glucose ($C_6H_{10}O_5$). In the cell wall, cellulose forms the framework interpenetrated by the matrix representing the non cellulosic substances (*e.g.*, lignin).

Micelles are composed of approximately 100 individual cellulose chains with cross-sectional area of about 3000Å. The molecules of cellulose, however, consist of long chains of about 3000-6000 glucose units (hexose) and in the cell wall these units are associated parallel to one another to form microfibris. Microfibrils in turn consist of smaller sub-units called micelles. About 20 micelles with a cross-sectional area of about 62,500Å form a microfibril.

The spaces among the micellar strands constitute intermicellar spaces. Thus, the microfibrillar system of cellulose is interpreted by a system of intermicellar space (capillaries) of various sizes. These capillaries are filled with non-cellulosic substances and water. Though these capillaries are filled with these materials, other substances can still pass through them, thus rendering the cell wall porous.

In the cell wall, orientation of microfibrils is in various ways. In the primary wall the microfibrils are randomly oriented transverse to the long axis in the first instance subsequently becoming longitudinally inclined as the primary wall starts growing. In the secondary wall, separation of three S layers is mainly based on the orientation of the microfibrils. The helix of microfibrils is lax and nearly horizontal in S1 and S3 and steep in S_2 which is the thickest. The primary wall differs from the secondary wall by possessing random orientation of the microfibrils. The chemical composition of these layers also differs.

Lignification

Lignin (Latin *lignum* 'wood') is a polyphenolic polymer deposited directly in the cell wall of specialized cells. It is not only restricted to plant woody tissues but represents an integral feature ensuring the proper cellular function of many other cell types in different tissues/organs of the plant. Mechanically weaker than cellulose, lignin nevertheless adds a significant reinforcement to any cell wall, providing an additional tensile strength of 25–75 MPa and a Young's modulus of 2·5–3·7 GPa. Lignification of the cell wall is a complex process which occurs exclusively in higher plants and its exclusive function is to strengthen the vascular body in higher plants. The process involves the deposition of ill-defined phenolic polymers (the lignins) in the extracellular polysaccharide matrix. The polymers arise from the oxidative coupling of 3 cinnamyl alcohols in a non-random reaction, in which the polysaccharides appear to influence the freedom of cinnamyl alcohol of radicals, giving rise to a highly orchestrated process. The lignin monomers, called monolignols, are the non-methoxylated *p*-coumaryl alcohol, the monomethoxylated coniferyl alcohol and the dimethoxylated sinapyl alcohol which respectively form H- (hydroxyphenyl), G- (guaicyl) and S- (syringyl) units in the lignin polymer. In the secondary wall lignin gets deposited and the cell becomes dead. The impregnation of cell wall with lignin is called "lignification". Consequently cells become hard, thick-walled and dead (*e.g.*, sclerenchyma, xylem and phloem). The process of lignification ranges from full autonomy to complete operation depending on the type of the cell. The different roles of lignin for the function of each specific plant cell are clearly illustrated by the multiple phenotypic defects exhibited by the knock-out mutants in lignin synthesis. The range of the phenotypic effects observed include the altered xylem sap transport, the loss of mechanical support, the reduced seed protection and dispersion and/or increased pest and diseases susceptibility. The lignified cell types in plants include: tracheary elements (TEs), cells of sclerenchyma, cells of endodermis, cells of the seed coat, siliques cells *etc.* Lignifications is controlled by the cell and the enzymes involved in the biosynthesis of monolignols synthesised at polysomes and are released into the cytosol during the formation of the secondary wall.

Suberization

Suberin is a waterproofing waxy substance found in higher plants. It is the main constituent of cork, which is the secondary cell wall. Suberization is the process of formation and deposition of waxy substance. Suberin, a complex polyester biopolymer, is lipophilic, and composed of long chain fatty acids called suberin acids, and glycerol. Suberins and lignins are considered covalently linked to lipids and carbohydrates, respectively, and lignin is covalently linked to suberin, and to a lesser extent, to cutin. Suberin is a major constituent of cork, and is named after the cork oak, *Quercus suber*. Its main function is as a barrier to movement of water and solutes. Cell walls of cork cells and casparian strips of endodermis get deposited with a layer of suberin by a process known as suberization. Suberin is found in the phellem layer of the periderm (or cork). This is outermost layer of the bark. The cells in this layer are dead and abundant in suberin, preventing water loss from the tissues below. Suberin can also be found in various other plant structures. For example, they are present in the lenticels on the stems of many plants and the net structure in the rind of a netted melon is composed of suberised cells. The knowledge gaps and ambiguities associated with wound healing and wound-induced suberization are of great importance because of the critical roles that rapid suberization plays in minimizing food-quality deterioration, shrinkage, defects, and perhaps the most easily and widely recognized problem of infection by a wide range of bacterial and fungal pathogens.

Special Structure of Cell Walls

Primary Pit Fields

The primary walls of young cells stretch and increase in surface area and thickness as the cell grow. However, certain areas of the wall remains thin and these portions are termed as "primary pitfields" or "primary pits" or "primordial pits". A characteristic feature of these primary pit fields is the presence of thin prosoplasmic strands known as 'plasmodesmata'. The plasmodesmata connect the protoplasts of the neighbouring cells and are present in all living cells of higher plants. The living protoplasts are united to form single unit because of the plasmodesmata. The interconnected protoplast is called 'synplast' and the compartment external to the protoplast the 'apoplast'. The apoplast consists of cell walls, intercellular spaces and the lumen of non-living cells such as tracheids and fibres. Plasmodesmata are easily seen in the endosperm of seeds like *Phoenix*, *Aesculus* and *Diospyros* and in the cotyledons of some plants. On plasmolysis, the protoplast withdraws from the wall in all places except where plasmodesmata are present. Plasmodesmata occur in groups sometimes more or less evenly distributed over the entire walls. In a variety of cells, branched plasmodesmata are observed, for example between companion cells and sieve elements *etc.* The branches depart from median companion cells and sieve elements *etc.* There is a possibility that such branches are primary plasmodesmata which become joined by the dissolution of a median portion of the common wall of the neighbouring cells. Plasmodesmata play an important role in the transport of materials and the relay of stimuli. Viruses can pass from one cell to another via the plasmodesmata.

Pits

Certain portion of the cell wall remains thin even after the secondary wall is formed and they consist of only primary wall material. These areas which are of variable shape are called pits. Pits can develop over the primary pit fields and then one or more pits may develop in the pit field or pits may also develop on the primary wall devoid of pit fields. The pits are areas through which substances pass from cell to cell. The concentration of plasmodesmata in living cells in the region of the pit membranes is an additional proof of the pit being a channel of exchange. Each pit has a complementary pit exactly opposite to it in the wall of the neighbouring cell. Such pits form a morphological or functional unit called "pit cavity". The membrane representing the primary wall and middle lamella that separates the two pit cavities of the pit pair is called the "pit membrane" or "closing membrane". The opening of the pit on the inner side of the cell wall, *i.e.*, on that side facing the lumen of the cell is called the pit aperture.

Two principal types of pits are recognized: Simple pits and Bordered pits. The main characteristic of bordered pit is that the secondary wall develops over the pit cavity to form an overarching roof with a narrow pore in its center (Figure 3.1). In a simple pit no such development of the secondary wall is present. If the two pairs of pits are simple they are called – a simple pit pair (Figure 3.1), if the two pits are bordered pit pair, if one pit is simple and the other bordered – a half bordered pit (when a bordered pit is opposed by a simple pit), if the pit has no complementary pit in the adjacent cell or if it is opposite to (or/and opens into) an intercellular space, it is termed a "blind pit". In some cases, the branched pits formed by the fusion of many pits due to an increase in the cell wall thickness, they are called as "ramified pits". Sometimes, two or more pits are found opposite to one large pit – such an arrangement is called unilateral compound pitting. Sometimes, the pits are provided with minute outgrowth from the wall surface of the pit chamber, called "vestured pits". Simple pits are found in parenchyma cells with thickened walls, libriform fibres and sclereids. Bordered pits are present in tracheary elements and fibre tracheids.

The bordered pit is more complicated than the simple pit in its structure (Figure 3.1) and variously shaped. In the bordered pit, the cavity that is formed by the over arching of the secondary wall is called pit chamber and the opening in the secondary wall that faces the cell lumen is called the pit aperture. If the secondary wall is very thick a canal, the pit canal, is formed between the cell lumen and the pit chamber. In the pit canal, two openings are there – that facing the cell lumen is termed, the inner aperture and that nearest the pit chamber, the outer aperture. In some bordered pits, the pit membrane is thickened in its central portion which is disc-shaped and is termed 'the torus'. The diameter of the torus is wider than that of the pit pair, water can easily pass from one tracheid to another. In a pit pair where the torus is in a lateral position, *i.e.*, present against one of the pit apertures, the passage of water is prevented. Most of the tori in late wood and all of them in heartwood are always in a lateral position. In this condition, the pit is called 'aspirated'. The presence of a torus is characteristic of the bordered pits of Gnetales, *Ginkgo* and most of the Coniferales. Tori occur rarely in Ophioglossales and Angiosperms. Bordered pits found in tracheary elements vary in shape and arrangement.

Patterns of Bordered Pits

The bordered pits in vessel walls of Angiosperms show four main types of arrangement in their distribution: (1) Scalariform pitting (2) Opposite pitting and (3) Alternate pitting and (4) Sieve pitting.

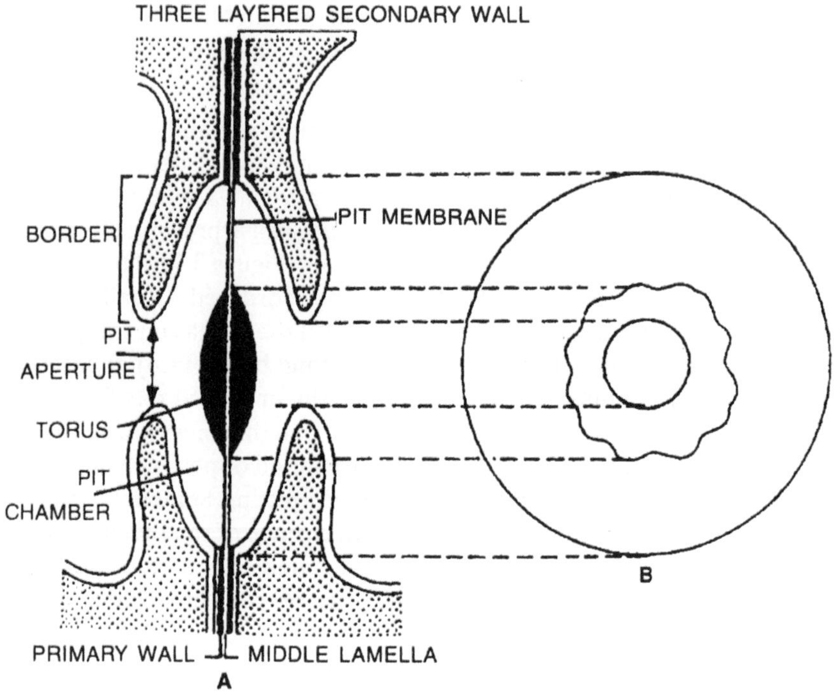

Figure 3.1: Diagrams of Bordered Pit Pairs. (A) Sectional view; (B) Front (surface) view.

If the pits are elongated or linear and are arranged in a ladder-like series, it is called scalariform pitting. When the pits are crowded, and arranged in horizontal rows or in pairs, they are so closely placed and the outline becomes rectangular in their surface view, and are called opposite pitting. When the pits are arranged in diagonal rows, the arrangement is called the alternate pitting. In this case, the pits are closely placed and their outline becomes hexagonal. When the small pits are seen in clusters, they look like sieves, and this is called sieve pitting.

Chemical Nature of the Cell Wall

The most common component in plant cell wall is carbohydrate cellulose. It is associated with other compound carbohydrates like non-cellulosic polysaccharides hemicellulose and pectic compounds. The fatty compounds cutin, suberin and waxes also occur in varying amounts in some cells. Various other compounds and mineral substances are also present. Water is common constituent of cell walls.

Cellulose

It is hydrophilic, crystalline compound with chemical formula $(C_6H_{10}O_5)^n$. It is a polysaccharide and closely related to starch. Its molecules are chains or ribbon like structures with 100 or more of the glucose residues held together to O_2 bridges.

Non-cellulosic Polysaccharides

The galactans, the xylans and the arabans which are closely allied to cellulose are commonly found in the cell walls.

Pectic Substances

These are derivatives of polygalacturonic acid and occur in 3 general types: Propectin, pectin and pectic acid. Pectic compounds are amorphous, colloidal substances, plastic and highly hydrophilic. The pectic substances constitute the inter-cellular substances that binds together the wall of the individual cells.

Gums and Mucilages

These are also compound carbohydrates of the cell walls related to pectic compounds and possess the property of swelling in water. Gums exude as a result of physiological and pathological disturbances that induce breakdown of walls and cell contents where as the mucilage occurs in some gelatinous type of cell walls.

Lignins

This is an organic compound of high carbon content, distinct from the carbohydrates. It is found in all the three wall layers, *i.e.*, primary wall, middle lamella and secondary wall. The impregnation of lignin starts in the intercellular lamella and spreads to primary and secondary walls. The process of conversion of cellulose into lignin is called lignification.

Cutin, Suberin and Waxes

The cutin and suberin are fatty substances which are not meltable and are insoluble in fat solvents. Cutin and suberin are closely related, highly polymerized compounds consisting of fatty acids. Suberin occurs along with cellulose in the cork cells of periderm, endodermis and exodermis of many plants. The phenomenon of impregnation of walls with suberin is known as "suberization". Usually cutin forms a continuous layer, the cuticle, on the surface of epidermis. The process of impregnation of walls with cutin is called cutinisation and the formation of cuticle is "cuticularization".

Waxes are also fatty substances which are meltable and are easily extracted with fat solvents. Waxes are associated with cutin and suberin and are deposited on the surface of the cuticle in various forms. The deposition of wax is responsible for glaucous condition of many fruits, leaves and stems. The deposition of these fatty substances checks the transpiration and protect the leaves from the leaching effects. The hard varnish like cuticle protects against the penetration of potential parasites and also against mechanical injuries.

Mineral Substances

Mineral substances like silica, calcium carbonate, calcium oxalate and several organic compounds like tannins, resins, volatile oils also may impregnate the cell walls. Silica or sand particles are common in the walls of grasses and horse tails (*Equisetum* sps.). Calcium carbonate is deposited in the form of cystoliths in *Ficus* sps. and others. Organic compounds are frequently deposited in the xylem walls when it changes from sapwood to heart wood.

Intercellular Substances

Usually the meristematic or young cells do not possess any intercellular spaces among them, but as they become older, their walls split at certain places giving rise to small spaces called intercellular spaces, and usually these are filled up with air or water. In general, there are two types of inter-cellular spaces, *viz.*, schizogenous intercellular spaces and lysigenous intercellular spaces.

Schizogenous Intercellular Spaces

The most common intercellular spaces result from the separation of the cell walls from each other along more or less extended areas of their contact. The spaces are most commonly formed due to the dissolution of the intercellular substance partly, as a result of which the space becomes bigger in size and this is called as schizogenous cavity. The ordinary intercellular spaces and schizogenous cavities form and an intercommunicating system of long, intercellular canals which facilitates the diffusion of gases and liquids from one part of the body to the other. The resin ducts of Coniferales and the secretory ducts in Asteraceae and Apiaceae are typical examples. The cells lining the cavity are secretory (glandular and epithelial) in nature and release their products in the intercellular duct canal.

Lysigenous Intercellular Spaces

These spaces arise through the dissolution of entire cells and hence known as Lysigenous intercellular spaces. These cavities of intercellular spaces store up water, gases and essential oils. The common examples are found in water plants and many monocots. The secretory cavities in *Eucalyptus, Citrus* and *Gossypium* are good examples.

Schizo-lysigenous Cavities

These occur when a storage space initiates schizogenously, but the glandular epithelial cells lining the space undergo autolysis during later development further enlarging the storing cavity.

Chapter 4

Meristems

In marked contrast to animals, the plants, vascular plants in particular, exhibit the open system of growth comprising the formation of organs or the parts of organs and various types of tissues throughout the entire period of their life. In the vascular plants, such a type of distinctive growth is governed by or restricted to the existence of a region bearing perpetually dividing cells and the region is confined mostly to the apices of the plant body, *i.e.*, the shoot and the root. Such a region with perpetually dividing cells is called a "meristem". The term was first used in 1858 by Carl Wilhelm von Nageli (1817-1891) in his book "*Beitrage zur Wissenschaftlichen Botanik*" (Contributions to Scientific Botany – in English). The term "meristem" is derived from the Geek word, *merizein* – meaning "to divide", in recognition of its inherent function.

The meristem is a type of tissue found in plants which consists of undifferentiated parenchymatous cells that are capable of division, and the cells can develop in to all types of tissues and organs, and these continue to divide until a time when they get differentiated and finally lose their ability to divide. In older regions of the plant body, this is restricted to vascular cambium, cork cambium and the apices of the shoot and the root. The meristems are undifferentiated or incompletely differentiated and are totipotent that are capable of continuous cell division, and this provides new cells for the expansion and differentiation of tissues and the new organs needed for the basic structure of the plant body.

The meristematic cells have definite characteristic features: the cells are small and more or less isodiametric (spherical, oval or polygonal), compactly packed without intercellular spaces, with dense protoplasm that fills the cells completely, with a conspicuous large nucleus, and with no or few and small vacuoles.

The chloroplasts or the chromoplasts are not differentiated and are present as proplastids. The cells are living and have a very thin primary cell wall.

However, the meristematic cells of the cambium show certain features different from normal meristematic cells. The cells are fusiform, bear thick cell wall, may contain plastids and ergastic substances such as starch and tannin. The meristematic cells retain the property of division and continue to divide to form derivative cells and as such they are referred to as the initiating cells. The derivative cells gradually lose their meristematic property and get differentiated into mature cells (differentiation). The mature cells lose the property of division and living contents sometimes to become permanent.

There are 3 types of meristematic tissues – apical (located at the shoot and the root tips), intercalary (at the internodes and the leaf bases), and lateral (at the sides as in the vascular cambium and cork cambium). At the meristem summit, there is a small group of slowly dividing cells, which is commonly called the central zone. Cells of this zone have a stem cell function and are essential for meristem maintenance. The proliferation and growth rates at the meristem summit usually differ considerably from those at the periphery.

Classification of Meristems

I Classification on the Basis of Origin and Development

On the basis of the origin and development of initiating cells (those maintain the meristem as a continuous source of new cells and may undergo mitosis many times before they differentiate into specific cells required for a particular region of the plant body). The cells that generate from the apical meristem are arranged in lineages of partially differentiated tissues called the primary meristems, and these are of 3 categories.

1. **Promeristem or Primordial Meristem:** The term promeristem is used to refer to a group of earliest and youngest meristematic cells of a growing organ. It is the early embryonic meristem from which other meristems are derived. It occupies a small area at the tip of the stem and the root in a plant. Promeristem further divides and forms primary meristem.
2. **Primary Meristem:** It is derived from promeristem and found at the apices of root, stem and the appendages. The cells divide in all possible planes and form the permanent issues in the fundamental part or primary structure of a plant.
3. **Secondary Meristem:** It is derived from primary permanent tissues which have the capacity of division. These meristems are always lateral in their position to the axis and develop either at the time of emergency to heal the wounds or effect secondary growth or to form cork. The examples are the cork cambium, cambium of the roots and the interfascicular cambium of the stem.

II. Classification on the Basis of Position

On the basis of their position in plant body, meristems are classified into 3 types:

1. **Apical Meristems:** These are found in the apices or growth region in plants found within the root tips and the tips of the new shoots and leaves (terminal parts of the growing parts of the plant). These are completely undifferentiated or indeterminate in a plant, and differentiate into 3 kinds of primary meristems. These give rise to the primary plant body and are responsible for the root and shoot extension (increase in length). This tissue is of 2 types, shoot apical meristem (SAM) and root apical meristem (RAM). The former gives rise to leaves and flowers while the latter gives rise to future roots. This includes both the pro-meristem as well as primary meristem.
2. **Intercalary Meristems:** These are found intercalated between permanent tissues (the branches), and is considered as a part of primary meristem which has become detached due to the formation of intermediate permanent tissues. These are found at the base of the leaf in *Pinus*, at the base of the internode in grasses and *Equisetum* and below the node in *Mentha*.
3. **Lateral Meristems:** These are arranged parallel to the sides of organs and normally divide periclinally or radially and give rise to secondary permanent tissues. These grow in the girth in woody plants. These are also known as the secondary meristems because they are responsible for the secondary growth or increase in the girth of the stem and its thickness (laterally). The vascular cambium produces the secondary xylem and secondary phloem during the formation of wood in arboraceous plants. This is absent in herbaceous plants which fail to show the secondary growth. Similarly, the cork cambium gives rise to periderm which replaces the epidermis in secondary organs.

III. Classification on the Basis of Function

Haberlandt considering the nature of function of meristem, recognized 3 important categories:

1. **Protoderm:** It is the outermost layer of the young growing region which develops to form the epidermal tissue system.
2. **Procambium:** It lies inside the protoderm and is composed of narrow, elongated, prosenchymatous, meristematic cells that give rise to the vascular tissue system (like primary xylem and primary phloem). It also produces vascular cambium and cork cambium, the secondary meristems. The cork cambium differentiates into the phelloderm, to the inside, and phellem (or cork) to the outside. All the 3 layers constitute the periderm.

In roots, the procambium also gives rise to the pericycle which produces the lateral roots.

3. **Ground Meristem:** It is composed of large, thin-walled cells which develop to form ground tissue system (tissues like parenchyma, collenchyma and sclerenchyma), *i.e.*, hypodermis, cortex and pith. These meristems are responsible for the primary growth (an increase in length or height).

IV. Classification on the Basis of Plane of Division

On the basis of the pattern and plane of division, the meristems are divided into 3 types:

1 **Mass Meristem or Block Meristem**: Cell divisions in these meristems occur in all planes so that the plant body shows isodiametric or irregular increase in volume. This type of growth takes place in the cortex and pith, in endosperm and young embryo and also in the formation of spores and sperms.

2. **Rib or File Meristem**: The rib meristem divides only in one plane, by divisions perpendicular to the longitudinal axis of the plant organ resulting in parallel longitudinal rows of cells in cylindrical plant parts, *e.g.*, formation of filaments in algae.

3. **Plate Meristem**: The plate meristem divides mostly by anticlinal divisions (divisions perpendicular to the surface) resulting in the surface growth as in the leaf blades of angiosperms, and epidermis in particular.

Cytological and Molecular Aspects of the Meristems

The meristematic cells are usually thin-walled, more isodiametric in shape than the cells of the mature tissues and are relatively richer in protoplasm. But it is not possible to distinguish by morphological criterion by which the meristematic cells can be distinguished from unspecialized mature cells. The protoplasts of the meristematic cells are devoid of reserve material and crystals and the plastids are in the proplastid stage. In the apical cells of most of the plants, especially among the Angiosperms, the vacuoles are very small and are not obvious and scattered throughout the protoplast, but in Pteridophytes and in many Spermatophytes, the apical cells contain conspicuous vacuoles. In general, the larger the meristematic cell the greater the degree of vacuolization. The size of the meristematic cells varies between 10-20 μM. The wall of the meristematic cells is very thin and weak but certain cells in apical meristems have thick wall. The nucleus is always a dominant part of the cell, constituting as much as 50 per cent of the volume. In many cases it is the only organelle that is easily detected by light microscopy, because all the body organelles are so small and few. Mitochondria are present and constitute about 12 per cent of the cytoplasm. Dictyosomes and ER are usually rather sparse and only rarely produce vesicles. The organelles are rapidly growing, dividing, producing ATP, hormones and other metabolites. The apical meristems and their cells are extraordinarily dynamic parts of the plant.

In *Arabidopsis thaliana* (a Brassicaceous weed which has smallest genome and fully studied) several genes have been identified which are responsible for proper functioning of meristems. The molecular genetic analysis of meristems result in the identification of some genes such as 'CLAVATA', 'KNOTTED' *etc.*, KNOTTED I in *Arabidopsis* plays a role in leaf morphology, development and regulation of shoot apex. This gene is expressed only in certain groups of cells. The **mgo I** and **mgo II** genes are responsible for enlarged meristems and produce leaves and floral organs.

The following genes are found responsible for molecular organization in *Arabidopsis* and *Zea mays.* The site of action is also indicated in the following table:

Sl.No.	*Name of the gene*	*Site of action*
1	KNOTTED I (in *Zea mays*)	Specification of shoot meristem
2	S T M (Shoot meristemless)	Specification of shoot meristem
3	WUSCHELL	Maintenance of shoot meristem
4	CLAVATA I	Regulation of shoot meristem size
5	AINTEGUMENTA	Regulation of size of lateral organs
6	L_1 L	Specification of Tunica Layer

Apical Meristems

In higher Algae, the meristematic activity is restricted to the terminal cell of the filament. Such a cell is called apical cell. Growth by apical cell also takes place in the Bryophytes and Pteridophytes. However, in Gymnosperms and Angiosperms, the growth takes place by a group of initials and their immediate derivatives. The term apical meristem is used to refer to these initials and their immediate derivatives. The terms shoot apex and root apex are used as synonyms of apical meristem. Usually the part above the youngest leaf primordium in a shoot apex is considered as apical meristem.

Theories and Concept of Apical Organization

Wolff (1759) recognized for the first time that the shoot apex was responsible for the growth of plant body. Since then, it has attracted the attention of several workers. Several theories have been put forward from time to time to explain (the activity of apical meristem) or the mode of growth found in apical meristem.

1. Apical Cell Theory

This theory was first proposed by Hofmeister (1857) and advanced by Nageli (1878). According to this theory, a single apical cell is the structural and functional unit of apical meristem which governs the entire process of apical growth. Such organization is found only in Cryptogams (Algae, Bryophytes, Pteridophytes Figure 4.1 A). The single cell is pyramidal in shape. Typically it divides parallel to its faces (except the outermost) so that the resulting derivatives have an orderly arrangement indicating their ontogenetic relation to apical cell.

2. Histogen Theory

Hanstein (1868) proposed the Histogen theory after studying several angiosperm shoot apices and embryos. According to this theory, the body of the plant arises from a massive meristem of considerable depth comprising three precursors of tissue regions, the histogens, each beginning with one to several initials at the apex arranged in superposed stories. The histogens are: the Dermatogen (the outermost uniseriate layer, the precursor of the epidermis), the Periblem (the precursor of the cortex which surrounds the plerome), and the Plerome (precursor of the central or the inner mass, vascular cylinder).

For long time Hanstein's theory found acceptance. However, later research showed that these histogens could not be well established in the shoot apex. There are two objections against this theory: 1. In the higher plants (Spermatophytes), it is not possible to demarcate periblem and plerome, and 2. No predetermination of the future tissues can be traced to the initials of the apical meristem.

3. Tunica-Corpus Theory

Schmidt (1924) proposed the Tunica-Corpus theory based mainly on the planes of cell division in the apical meristem. According to this theory, the mass of the dividing cells of the apical meristem are of 2 types, namely: the tunica, the outer layers consisting of one or more peripheral layers of cells which divide in planes perpendicular to the surface of the meristem (anticlinal division) and (2) the corpus, the inner central mass of undifferentiated, multilayered cells which divide in various planes. The epidermis is derived from the outer layer of tunica, while the other tissues from the remaining layer of the tunica and the corpus. Hence, corpus is a core of cells that adds bulk to the apical meristem by increasing the volume, while tunica is a mantle of one or more cells which maintain their continuity over the enlarging core by surface growth.

In support of shoot apex organization, other theories have also been propounded. Dermen (1947) put forth "Histogenic Layer Concept", and according to him, there is no distinct layer of apical meristems, and he named the different layers as L_1, L_2, L_3 *etc.*, on the basis of their origin. This concept did not get any support.

Sometimes, especially among Monocotyledons, a few periclinal divisions occur in the tunica, which is a contradiction of the original definition of the tunica, and therefore, Popham and Chan (1950) introduced "Mantle Core Concept". They used the term mantle instead of tunica for all the outer layers of apex which can be distinguished histologically from the inner cell mass, and the core in place of corpus without taking into account the planes of division in these layers. They retained the term tunica for those layers in which only anticlinal divisions take place.

4. Anneau Initial and Meristeme d'attente Theory

Plantefol (1947, 1950), Buvat (1955) and Amefort (1956) propounded the Concept of French School and recognized 3 distinct regions in the apical meristem. According to these authors, the peripheral active zone was known as "Anneu initial"; the zone next to it "Mersteme d' attenete"; and the central zone was termed "Meristeme medullaire". Buvat (1955) proposed that peripheral and subterminal regions are real initiating zones whereas distal zones are inert in nature. He recognized three distinct regions in apical meristems, (i) the anneau initial (the peripheral active zone), (ii) the meristeme d' attente (the waiting meristem which becomes active only during formation of terminal flower or inflorescence) and (iii) the meristeme medullaire (the central pith region). Meristeme d'attente consists of two parts, the "promeristeme sporogene" and "promersteme receptaculair" which thus contain part of tunica (except anneau initial) and part of corpus (except meristeme medullaire).

5. Korper-Kappe Theory

Schuepp (1917) proposed that the cells at the root apex divide in two planes. The zone with inverted 'T' type of division was referred as *Korper* (cap) whereas other with straight 'T' as *Kappe* (body – outer region). This theory basically recalls the Tunica Corpus theory concept of shoot apex and similarly, it also fails to explain the differential behaviour in different species. The cells of the outer region fist divide horizontally. The lower daughter cells then divide longitudinally (at right angles in the plane of the first division). Thus, the planes form a T in a median longitudinal section of the root. The T is inverted, and this has been found among Poaceae and Fagaceae members.

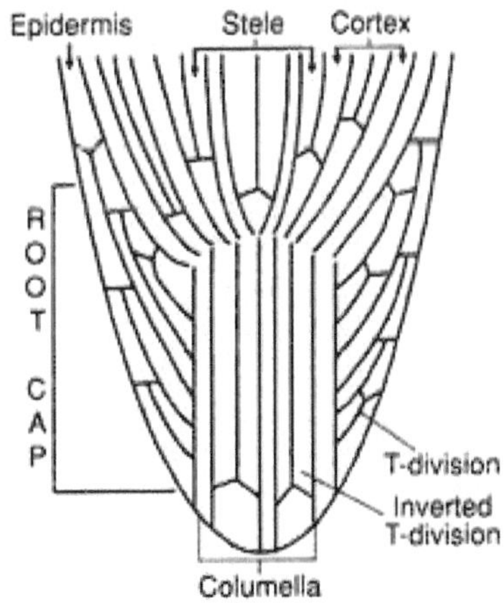

Figure 4.1: Diagrammatic Representation Showing the Root Apex Meristem According to the Korper and Kappe Theory.

Types of Root Apices in Angiosperms

A. Dicots

The root apices of Dicots are of the following types according to the number of initials present.

1. **Common type:** This is found in majority of Dicot plants and in this type, 3 groups of initials are found at the apex – outer (dermatogen – forms the epidermis and the root cap), middle (periblem – forms the cortex) and inner plerome (gives rise to the central cylinder).
2. **Ranunculus type:** This is seen in the members of Ranunculaceae, Juglandaceae, Salicaceae, Casuarinaceae, Leguminosae *etc.*, In this type, there is a single row of initials which finally, gives rise to various zones of the root including the root cap. Some of the cells of the root cap differentiate into the epidermis.
3. **Casuarina type:** This type of development of the root apex is evident in the families Proteaceae, Casuarinaceae and in some members of Fabaceae. In this type, 2 rows of initials are generally observed at the apex, and out of these, one layer of cells gives rise to the stele and the other to the cortex and the root cap.

The epidermis develops from the outermost layer of cortex, and in some members of Juglandaceae, Rosaceae, Fabaceae, Tiliaceae *etc.*, one of these two rows of the initial give rise to the stele and the inner cortex, while the other gives rise to the outer cortex and the root cap. The epidermis arises from the outermost layer of the cortex.

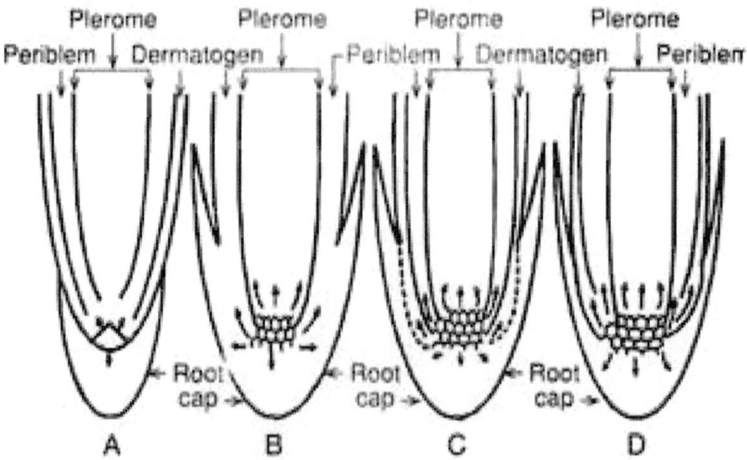

Figure 4.2: Diagrammatic Representation of the Root Apices. A. Pteridophytic type with a solitary cell; B. As found in the Gymnosperms; C. Common Dicot type and D. Monocot (Maize) type.

B. Monocots

In addition to the above types, there is a fourth type in Monocots. The first type is seen in *Zephyranthes* species; the second type which is very rare is seen in *Allium sativum, Aloe vera, Amaryllis* and *Eucharis etc.*; while the third type is found in

Haemanthus coccineus. In the additional fourth type, there are 4 rows of the initials that give rise to the root cap, the epidermis, the cortex and the stele independently. The row of initials that develop into the root cap is called the calyptrogen which produces the layers of cells outside to form a cap-like structure.

Differences between the shoot apex and the root apex:

1. The apical meristem of the root apex in contrast to that of the shoot apex, generates the cells not only toward the axis but also away from it to form the root cap.
2. The apical meristem of the root is sub-terminal in its position as the root cap remains terminal.
3. The root apex forms no branches and lateral structures such as leaves, and hence, the root apex shows no change in its morphology.
4. The root apex grows uniformly as it does not produce the nodes and internodes.
5. The branches of the root arise endogenously beyond the region of the most active growth (the pericycle).

Types of Stem Apices in Angiosperms

Wolff (1759) discovered that the new leaves and tissues of the stem arise in the very apex of the stem which he termed 'punctum vegetationis'. Today the term shoot apex is generally used. The shoot apex proper is considered as that terminal part of the shoot immediately above the upper most leaf primordium. There are great differences in the shape and size of the shoot apices. In a median longitudinal section it appears generally more or less convex. But, in *Anacharis, Elodea, Myriophyllum* and some grasses the shape of the apex is a narrow cone with a rounded tip while in a few plants like *Drimys* and *Hibiscus syriacus* it is slightly concave. Before the initiation of each leaf the apical meristem widens considerably and after the appearance of leaf primordium it again becomes narrow. This phenomenon is rhythmic, *i.e.* it recurs with the initiation of each leaf or pair of leaves. Schmidt (1924) introduced the terms minimal and maximal areas of the apex. For the period between the successive initiations of two leaves or two pairs of leaves he suggested the use of the term 'plastochron'. The shoot apices of dicots with opposite leaves like *Linociera, Ligustrum, Syringa, Vinca etc.* are suitable for the study of plastochron changes.

In the Angiosperms, the shoot apices are usually small. The diameter varies between 90 μm (in certain grasses) and 130.-200 μm (in many Dicots). In the Bananam it is 280 μm, certain Arecaceae and in *Nymphaea* 500μm and in *Trichocereus* it is 700- 800μm. They can be as large as 1500μm, in some Cactaceae. Meristems as large as 3,000 μm wide occur in *Cycas revoluta*. Generally, the difference in the diameter of the apices in Gymnosperms is much greater. The apices of the conifers are cone shaped and fairly narrow.

In vascular plants the development in stem apex follows more or less definite pattern and these appear to be a feature of the major groups of plants. These patterns manifest rather increasing complexity from lower to higher groups of plants. In the lower groups the initiating cells constitute a uniseriate superficial layer and there is no clear cut evidence of existence of diferentiated tunica and corpus layers. However, in higher plants the initiating cells are arranged in two layers. The superficial layer of cells form the tunica which is usually uniseriate with its cells dividing anticlinally and the lower layer of cells gives rise to the corpus. On the nature of the formation of the tunica and the corpus the stem apices may be classified into the following groups.

1. The Primitive Stem Apices

In species of *Selaginella, Isoetes,* Cycads *etc.*, the apices where simple surface initials on division form no distinct tunica and corpus layers, *e.g.*, *Lycopodium* stands as a very good example representing this type of stem apex. In this, the initiating layer is weakly defined and the uniseriate surface layer freely divides anticlinally and periclinally, as no morphological differentiation between the cells can be recognized. The anticlinal divisions increase the surface area and the periclinal divisions increase the inner core of cells.

2. Stem Apices with Ill-defined Tunica and Corpus

The distinction between tunica and corpus perhaps starts with some of the Conifers. In Pinaceae like *Pinus* and *Abies* the initials form a terminal uniseriate group of cells. Both anticlinal and periclinal divisions take place in the formation uniseriate superficial envelope over a central mass of cells. The former suggests, the existence of the tunica in appearance but its periclinal divisions in parts of apex makes it difficult in drawing a line between tunica and corpus. But some of the higher Gymnosperms show a special structure. In Pinaceae the initials or a small group of cells arrange in a single tier with divisions both anticlinal and periclinal. Soon a dermatogen like layer is organized covering the central mass of cells. The fundamental difference from the Pinaceae is that when the dermatogen like layer is organized the cells only have anticlinal divisions with occasional periclinal divisions. The central cells always show periclinal or irregular divisions. In this condition, the demarcation between tunica and corpus is easier than in Pinaceae. In certain Gymnosperms like *Ephedra, Araucaria etc.*, the tunica is uniseriate and the corpus forms the bulk of cells.

3. Stem Apex with Distinct Tunica and Corpus

When compared to lower plants the Angiosperms show definite and distinct apical meristem zone. The two sets of initials are placed one above the other which give rise to tunica and corpus. The tunica as we know is very rarely periclinal and ranges from one to few cells in thickness and 2 or 3 layers probably representing the common feature. A single layer of tunica as observed in *Avena* and *Triticum* probably represents a highly specialized condition. But in certain cases like *Zea,* occasional periclinal divisions are recorded in tunica. The limits of the two groups of initials is often definite although in certain exceptional cases like members of Cactaceae, *Sinocalamus, Vinca rosea*, such distinction can't be drawn out and the apices resemble those of Gymnosperms showing a weak differentiation of tunica and corpus.

4. Stem Apex with distinct Primary Zones

Haberlandt (1914) recognized three distinct primary meristimatic zones namely protoderm forming epidermis, hypodermis; ground meristem forming the medulla, pith; and procambium forming vascular elements.

5. Shoot Meristem with 3-4 Zones

Popham (1951, 1957) recognized 4 to 5 zones in the shoot meristem

1. Surface zone or mantle composed of a few to several layers corresponding to tunica.
2. A zone of subapical mother cells which are highly vacuolated and less active in their cell divisions.
3. Central region of cells which will give rise to pith.
4. Peripheral zone which will give rise to cortex and procambium. In certain plants he was able to recognize a fifth zone just below the subapical region of mother cells with the cells stretched across the axis. This is the zone resembling that of the cambial zone.

Significance of Tunica Corpus Theory

This theory has well served to understand the nature, form and pattern of stem tips of seed plants. In view of these features of meristem, the position, number and behavior of initiating cells in the stems of seed plants is now in some measure known. Although this theory does not throw much light upon morphological significance of these two layers it has got the topographical value in the detailed study of development when the distinctiveness of tunica and corpus is known. The regions are constant in structure and function showing marked variability with seasonal conditions, vigour of growth, age, position of plant and morphology of stem tip concerned. In recent years accounts also have been made in the light of tunica and corpus to aim at phylogeny of plants. In this regard Gifford's work is very much appreciated.

Newman's Classification of Shoot Apices

Newman (1965) used the term continuous meristematic residue (*cmr*) to refer to a group of permanent initials in the shoot apex, and on the basis of form of the *cmr*, shoot apices are divided into the following types.

He put forth his concept and recognized 3 kinds of shoot apices: the Monoplex type, Simplex type and Duplex type.

a. Monoplex Type

According to him, the monoplex type is found in vascular cryptogams and is the characteristic of ferns, where the shoot apex is denoted by one or more cells which divide by walls parallel to the inclined walls in the stem. In ferns, the *cmr* is in the superficial layer only. Any one cell contributes to growth in both length and breadth.

b. Simplex Type

The Simplex type is found in gymnosperms, and the *cmr*, consists of one or more initial cells arranged in a single layer, and these cells divide anticlinally and periclinally. Both the divisions are necessary for bulk growth.

c. Duplex Type

The Duplex type is found in the shoot apex of Angiosperms, and *cmr* consists of at least two successive layers of cells, and the cells of the surface layer (first layer) divide anticlinally and that of the inner layer (second layer) divides in more than in one plane (periclinally). Thus, two contrasted modes of growth is evident, hence these are known as duplex type of shoot apex.

Cyto-histological Zonation in the Promeristem of Vegetative Shoot Apices

A. Shoot Apex of Pteridophytes

In this group there are one or more initials which can be easily distinguished from the neighbouring cells. These initials give rise to all the cells of the apex. If only one initial is present it is termed the apical cell and if more than one cell they are termed 'apical initials'. The single apical cell is usually tetrahedral in shape. The single apical cell is found in Psilotales, *Equisetum* and in some other ferns. The apical cells are generally 4-sided, but in some water ferns like *Salvinia* and *Azolla* and sometimes in *Selaginella* they are 3-sided. The ferns (Filicinae) with a single apical cell are considered to be evolutionarily more advanced than those with several apical initials.

B. Shoot Apex of Gymnosperms

Since 1937 much research has been made on the structure of shoot apex in gymnosperms. The characteristic feature of all the gymnosperms is the direction of the cell division in the surface of the apex which is both anticlinal and periclinal. So, this layer represents the initiation zone of the entire apex and has been termed as the 'surface meristem'. The striking feature of the most gymnosperm apices is the occurrence of a distinct zone of central mother cells. The cells of this zone are characterized by their size, numerous large vacuoles and thick walls. Along the sides and base of the central mother cell zone, the other apical regions develop as a result of the diagonal and horizontal divisions of the central mother cells. In this way the peripheral zone or flank meristem is laterally developed and the rib meristem zone (central meristem) from the base. The term 'rib meristem' was introduced by Schuepp (1926) to describe that type of meristematic issue which shows vertical series of transversely dividing cells. According to Popham (1952), three principal types of Gymnosperms can be distinguished on the basis of the structure of the shoot apex - the *Cycas* type, the *Ginkgo* type and the *Cryptomeria – Abies* type.

1. *The Cycas Type*

In this type (Figure 4.3C), three meristematic zones can be distinguished namely a) The surface meristem, in which the cells divide anticlinally, periclinally and diagonally. The cells of this zone give rise to the epidermis and the other apical meristmatic zones. b) The rib meristem which is situated in the central region of the apex below the surface layer. These cells also divide periclinally, anticlinally and diagonally. c) The flank meristem which enlarges by cell division within the zone itself and by the addition of cells from the surface layer and from the periphery of the rib meristem. This gives rise to cortex, the procambium and the leaf primordia.

2. *The Ginkgo Type*

In this, five meristematic zones are distinguishable (Figure 4.3D). They are: a) The surface meristem which consists of the apical initials, b) The zone of central mother cells which occurs in a median position below the surface layer. The cells are large, polyhedral and are arranged irregularly. The divisions in these cells occur in various planes. c) The cambium like transitional zone' which is cup-shaped. This zone forms a transitional zone between the central mother cells and the zones below them.

3. *The Cryptomeria and Abies Type*

In this four meristematic zones can be recognizable (Figure 4.3E). This type is almost similar to the *Ginkgo* type, but the cambium like transitional zone is absent here, *e.g., Pinus, Taxus, Ephedra etc.*

The other types are: the rib meristem, which lies under the central portion of the cambium like transitional zone; the flank meristem, which forms a cylinder surrounding the rib meristem and this is a continuation of the cambium like transitional zone (Figure 4.3D) Other examples are : *Zamia, Sequoia sempervirens, Microcycas* and *Pseudotsuga.*

C. Shoot Apex of Angiosperms

Hanstein (1868) has put forward the Histogen theory. According to him, three zones can be distinguished in the shoot apex of angiosperms, *viz.,* a) an outer most zone, the 'dermatogen', a central zone 'plerome' consists of irregularly arranged cells and a hollow cylindrical zone of several layers of cells between the dermatrogen and plerome- the 'periblem'. Hanstein stated that these zones develop from independent groups of initials, which act as direct histogens. According to this theory, the meristems are destined from the beginning to produce certain tissues *i.e.,* the epidermis from the dermatogen, the cortex and the internal tissues of the

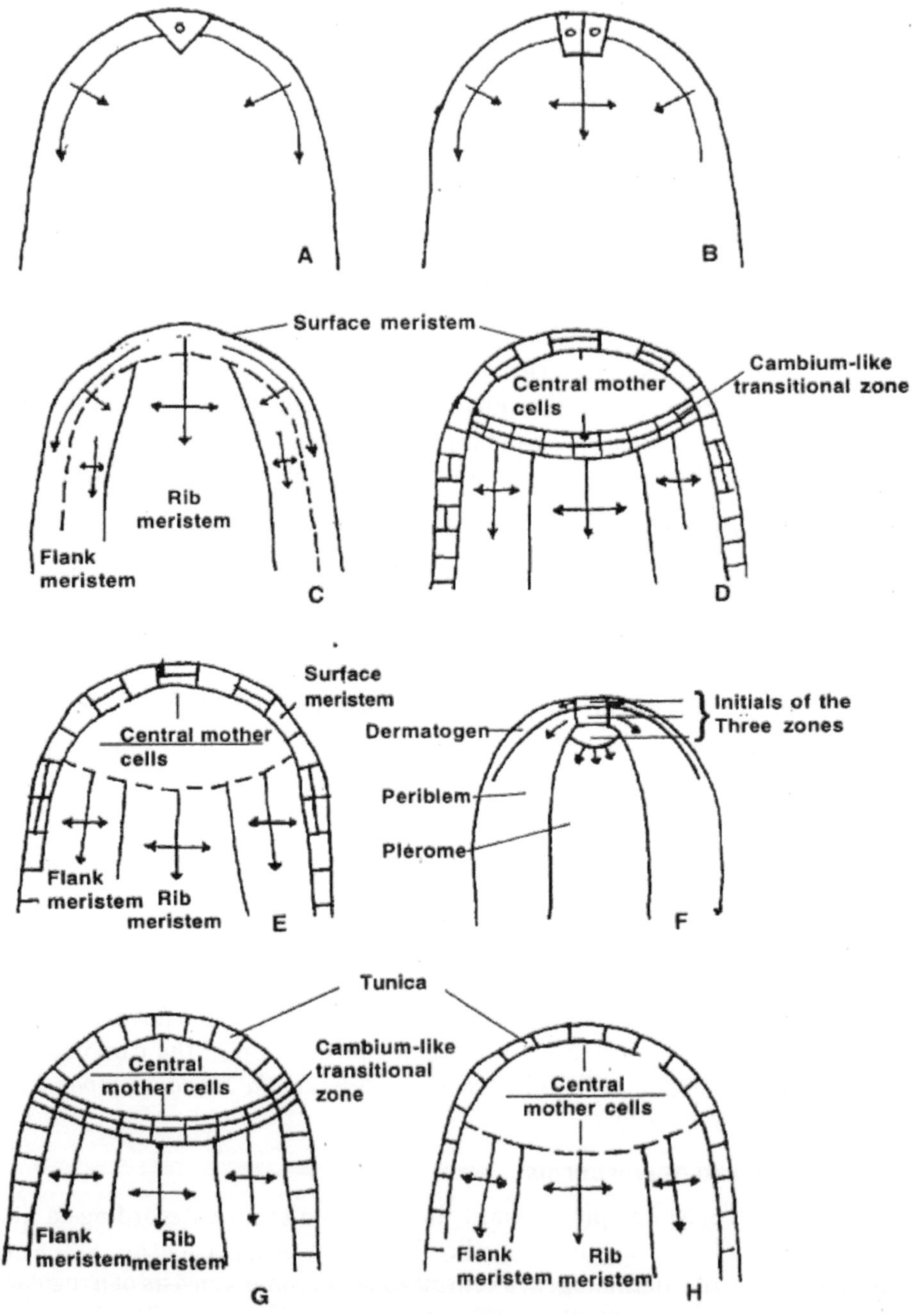

Figure 4.3A-D: Diagrams of Cyto-histological Zonation in the Promeristem of the vegetative shoot apices: A. Pteridophyte type with a single apical cell; B. *Selaginella* with 2-5 apical initials; C. *Cycas* type; D. *Ginkgo* type; E. *Cryptomeria-Abies* type; F. Schematic representation of the histogen theory of Angiosperms; G. *Opuntia* type, and H. Usual Angiosperm type (after Fahn, 1982).

leaf from the periblem, and the central cylinder from the plerome. This theory is accepted for a long time, but later researches, disproved it, because of the following. 1. In most Spermatophytes, it is not possible to distinguish clearly between the periblem and plerome and 2. No predetermination of the mature tissues can be traced in the various initials.

In 1924, the theory of Schmidt, which divided the apex into two regions – the 'tunica' and the 'corpus' was postulated. According to this theory, no constant relationship can be traced between the particular initials of the promeristem and the inner tissues of the shoot. The two regions recognized by this theory are usually distinguished by the planes of cell divisions in them. The tunica consists of the outer most layer or layers of cells which surround the inner cell mass the corpus. The plane of cell division in tunica is principally anticlinal. In the corpus, the planes of cell divisions are in all directions. The tunica enlarges in surface area and the corpus in volume. The tunica-corpus theory is very adaptable one and till today, it is generally accepted in literature.

As already mentioned, the tunica consists of a single layer or few layers of cells which surround the inner meristem. The number of layers is not always constant in a given genus or family or not even in a species. The number of layers varies in a single plant during different stages of development of the vegetative apex. The number of layers may be 1-9. Single layered tunica (monostratose) is the most common structure, followed by two to four (multistratose), but sometimes as many as 18 layers occur in *Xanthorrhoea media* (Staff, 1968). Sometimes, especially in monocots, the periclinal divisions occur in the inner layers of the tunica [if it is multistratose, but the outer most layers have only anticlinal divisions] which is a contradiction of the original definition of the tunica and therefore, Popham and Chan (1950) introduced the term 'mantle' for all the outer layers of the apex which can be distinguished histologically from the inner cell mass without taking into account the planes of cell division in these layers. They retained the term tunica for those layers in which only anticlinal cell divisions takes place. The term tunica is used in the broad sense and is equivalent to the mantle of Popham and Chan.

Although the plane of cell division is generally the same throughout the entire tunica, cytologically two zones can be distinguished in it *viz.* 1. Central apical zone consisting of one or few initials which are larger and have larger nuclei and vacuoles compared to the other tunica cells. These cells take less stain. 2. It is that region on the sides of the apex between the initials and the leaf primordia. It consists of smaller, more darkly staining cells which divide more frequently and among which periclinal divisions may occur close to the primordia. The corpus is far less homogeneous than the tunica.

Among the angiosperms two main types of corpus are recognized on the basis of internal arrangement (Popham, 1952).

The Usual Angiosperm Type

In this, three main zones can be distinguished in the corpus (Figure 4.3H), *viz.*,

a) the zone of central mother cells which probably represent the corpus initials and located below the apical portion of the tunica *i.e.*, below the tunica initials b) the rib meristem and c) the flank meristem. The latter two zones appear as continuations of the central mother cells.

The Opuntia Type

In this type, in addition to the zones found in the above type, a cambium like transitional zone can be distinguished (Figure 4.3G). This zone which is cup shaped is found between the central mother cells and the rib and flank meristems. The cambium like transitional zone differs from the other zones of the apical meristems in its height and diameter which vary considerably during the plastochron, reaching its maximum development close to the developing primordium. According to Philipson (1954), this zone is only temporary and it disappears towards the end of plastochron. Examples: *Phoenix dactylifera, Chrysanthemum, Opuntia cylindrica* and *Xanthium pennsylvanicum.*

The cells of the flank meristem are usually derived from the initials of the corpus, but in some plants, they also arise from the tunica initials. The rib meristem consists of cells arranged in rows which narrow towards the apex, they may sometimes be irregularly arranged. The majority of the divisions in this zone are horizontal, but diagonal divisions also occur. It can be concluded that the epidermis and its derivatives originate from the outer most layer of the tunica. The flank meristem contributes to the development of the leaf primordia, the cortex, all or part of the procambium and sometimes also the outer region of the pith, and the rib meristem, the reminder.

Zonation in Root Tip

Like the shoot, the apical meristem of the root also displays many patterns of growth and there exists no constant relationship between structure of initiating division and delimitation of primary tissues of the axis. In the root unlike the stem, the apical meristem produces the root cap but even away from it since it initiates the root cap and there by occupying sub-terminal position. The apical meristem of the root differs from that of the shoot in that it forms no appendages, nodes and internodes. The root branches are usually initiated endogenously far away from the meristematic region. Besides these differences, the root apex shows no change in shape and structure as seen in stems because of the absence of the leaves.

According to various investigators, it appears that there is a single central initial (Figure 4.4 A-C) or but a few initials in the root apex of spermatophyta. Other investigators such as Clowes (1950) believe that there is a larger group of initials in the median region of the root apex.

Although the Histogen theory has been abandoned for the shoot apex, the structure and growth of the root apex are generally interpreted in terms of Histogen theory. The apical meristem in the case of root tip is rather short. In number the initials range from one to many and where the initials are more than one, they are arranged either in one row or usually in four rows fairly distinct from one another as uniseriate stripes. In each group there is one to several initiating cells. Furthermore, the initiating cells of the different groups may vary in their diameter and in the rapidity of growth of the root. For example, in slender roots, the initiating layer is restricted to one, but zonation remains distinct. Where there are more layers of initiating cells the layers invariably lie one above the other along the longitudinal axis of the root.

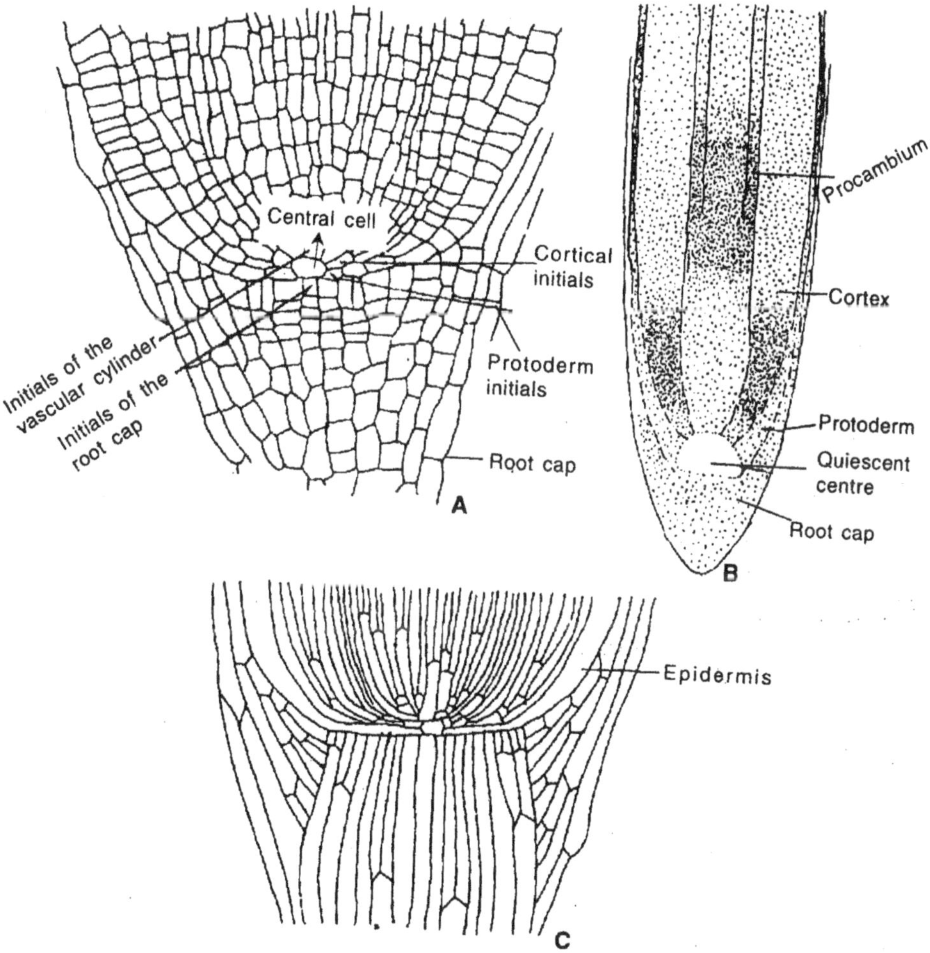

Figure 4.4A-C: (A) Longitudinal section of the root tip of *Triticum vulgare;* (B) Diagram of the root tip of *Allium cepa* showing, by means of shading, the gradation of mitotic activity in different zones; the most active zone is the most darkly shaded; (C) Diagram illustrating the Korper-Kappe pattern of the root apex of *Zea mays.*

Each of these groups quickly develops one or more growth zones which are usually marked more clearly than in stem. In many plants these groups of initiating cells represent in strict sense the histogens; although the terms dermatogen, periblem and plerome are not in general view, in the case of stem ontogeny they have been continued to indicate the general zone in the stem and root development. A fourth histogen namely, calyptrogen is added when the root cap has had an independent origin.

On the basis of the activity and distinctiveness of histogens, as in stem, the roots are also classified into different types. In the lower vascular plants all the tissues of the root are either derived from a single apical cell *e.g., Equisetum*, members of Polypodiaceae and certain species of *Selaginella* or may be formed from several initials arranged in one tier as in the members of Marattiaceae. All these plants have the same apical structural organization both in stem and root. This single apical cell or a single group of initiating cells form the entire root and the origin of the root cap from apical cell is clear. In the majority of Gymnosperms, there are two groups of apical initiating cells the one forming the central cylinder, the other giving rise to the cortex and root cap. The outermost layer of the cortex differentiating into the epidermis whose origin and growth may be traced to be slightly away from the apical region as a distal proliferation. The outstanding feature in this case is the formation of a zone from the cortex of the root and the formation of endodermis from the innermost layers of the cortex. In most of the Angiosperms the groups of initials occur at the root apex. In Dicotyledons, the terminal group forms the dermatogen and the root cap, the median one develops the cortex while the upper one gives rise to the central cylinder of the root. In this type both dermatogen and root cap have the same origin and the cap may be attributed to as specialised epidermis or dermatogen. In some dicots like Ranalian families, Amentiferae, Ranunculaceae, certain members of Leguminosae and some of monocots, all the parts of the root appear to have arisen from a common initial group. The epidermis eventually differentiating from cortex and the calyptrogen differentiating from lower part and in these forms the zonation in root tip recalls the condition encountered in Marattiaceae. In certain monocots like *Zea*, the outermost initials develop the calyptrogen or root cap, median ones dermatogen and periblem and inner ones plerome which is a condition generally seen in Angiosperms.

Variation in Monocots

In Monocots, all the three groups of root apices found in Dicots are recorded in *Allium sativum, Haemanthus* and *Zephyranthus tubis* respectively However, the first type of root apex meristem is represented by a single layer and is also recorded from *Eucharis, Amaryllis* and *Swevera* and in certain conditions all in *Allium sativum.* Commonly in most of the Monocots the outer most layer of initials produces the cap, the median one the dermatogen and periblem and the inner most the plerome. In *Agapanthus* and *Crinum*, the root cap, dermatogen and cortex arise from independent set of initials as recorded by Deshpande (1960) which feature was reported long back by Treub (1876). In all these cases where the root cap origin is independent it is termed as a definite meristem called calyptrogen. Rarely in certain members of Monocots like *Pistia* and *Hydrocharis,* the origin of Calyptrogens is from an independent meristem which forms both the initiating layers.

Deshpande (1960-61) recognised only three types of root apices in monocots.

1. The root cap organises independently and cannot be reorganized into columella and peripheral zone. This condition has been seen in certain members like *Polygonatum oppositifolium, Sanseviera thyrsifolia* and *Ruscus hypophyllum.*
2. The cap is recognised into columella and peripheral zone both representing independent histogens. In this condition the peripheral region of the root cap may be independent of dermatogen and the rest of the tissues of the root tip originate from separate initials as is found in the case of *Haemanthus coccinium*. In *Agapanthus exerinum* part of the root is organised from the dermatogen and peripheral layer may be free. The root cap is not free from other tissues of the root apex and there is no distinction of columella and peripheral layers.

Clowes (1956) while studying root tip of *Zea mays* recognises a cap like structure or region placed in between the root cap and active meristematic region and he called this region the quiescent centre (Figure 4.4B). Likewise, Deshpande has also recognised such type of quiescent centre in root apices of some members of Tiliaceae and Amaryllidaceae. Although the function of the quiescent layer is not thoroughly understood, Clowes is of the opinion that this centre becomes active at the time of formation of lateral roots or when root tip is injured. Recently it has been observed by Jensen (1966) that this quiescent centre acts as a reserve of zonal cells, which are less sensitive to damage and becomes active while the other part of the root apex is inactive. According to Deshpande, the meristematic cells below the quiescent centre form the root cap, while the initiating cells above it form the other tissues of the root.

Evolutionary Significance

From a detailed study of the structural organization of the apices in several of the plants belonging to diversified affinities, it has been now visualized that the evolutionary progress is perhaps from a single centre of growth to a group of three of four centres with progressive increase of independence of zones. From the foregoing description of different types of root apices the origin of the root cap varies. In the lower plants, the root cap has a common origin with all the other parts of the root. In Gymnosperms it is a special distally developed part of the periblem and the formation of the dermatogen in this condition is a bit delayed and is also gone up away from the root cap. In the Dicots the root cap appears to be more specialized in its development from the epidermis. In monocots although they show the same pattern of development of root cap as in Dicots but in certain exceptional cases the root cap appears to be more independent in being derived from a separate meristem as against the case seen in Dicots and majority of Monocots.

Chapter 5

Simple Tissues

Introduction

If we can keep in our mind the evolutionary history of plants, about 450 to 500 million years ago, probably in the middle of the Palaeozoic period, the plants began to colonize the land (because of the fact that the environment offered some advantages over the aquatic environment – for example, many hours of light, greater intensity of light, and more free movement of carbon-dioxide *etc.* The plants had to solve their new challenges such as – related to obtain and retain water, to keep their body erect, to disperse their seeds in the air *etc.* To solve these problems, the plants grouped specialized cells to form the tissues with specific functions to cope up with the new challenges. Such tissues form the tissue systems (Sachs, 1875) which at higher levels form the organs.

The tissues have been traditionally grouped into tissues systems – protective tissue system (which allows plants to survive in the dry and variable environments, like epidermis and periderm. This is the outermost part of the plant and organs. Depending on the species and the developmental state of the organ, this is referred to as either the epidermis or the periderm. The cells are covered by cutin and suberin, waxes which avoid or decrease the water loss from the surface. The plants develop stomata in the epidermis for the regulation of transpiration and exchange of gases between the plant and the air. This also develops trichomes with different functions) - ground tissue system (that is mainly involved in the metabolic function and mechanical support, like parenchyma, collenchyma and sclerenchyma), and vascular tissue system (like xylem and phloem for conduction of water and food, and which keep the body of most herbaceous plants erect and straight, and also give the final shape of different plant organs). This is one of the most relevant tissue system during the evolution of plants as it communicates with different organs and different parts by transporting the molecules and water. The phloem a part of the vascular system helps in the transport of organic substances in water, and xylem associated mainly with the transport of water and inorganic substances. Besides these functions, the vascular system is the main supporting tissue in bushes and trees. As such the plants increase in their size and also colonize non-aquatic territories.

The plant tissues can be classified in different ways. According to cellular diversity in different tissues, the tissues are simple tissues with only one type of cells (such as parenchyma), and complex tissues with different types of cells such as protection (collenchyma and sclerenchyma) and vascular tissues (xylem and phloem). Both the tissues and tissue systems form the plant organs which can be classified as – vegetative (like the root, responsible for water and salt uptake); the stem (responsible for substance transport, to sustain the aerial body of the plant, and to carry out photosynthesis), and the leaf (to capture the solar energy and perform photosynthesis and also responsible for the exchange of water and gas); and the reproductive organs (as the flowers and its derivatives, the seed and the fruit). The tissue systems are distributed in a characteristic pattern depending on the organ.

Plant Tissues – Definition and Types

All the living organisms in the world are classified into two types – unicellular and multicellular organisms. The former like *Amoeba, Paramoecium*, diatoms and bacteria *etc.* have only one cell, and they do not have any tissue level organization. Similarly, in some lower order animals tissue differentiation is absent (the cells form the groups, but they do not carry out a definite function). But, in the case of multicellular higher order of animals and plants (dogs, cats and higher plants), different groups of cells carry out different and specific functions. If a cell is called the basic unit of life, a group of cells make up the tissues, and such groups of cells have a common origin with a similar structure and these carry out a definite function. In simple terms, the tissues can be defined as a group of cells performing similar function.

Plant tissue is a collection of similar cells performing an organized function for the plant. It is a group of cells with common origin, structure and function. From the standpoint of number of kinds of cells making up a tissue, tissues are of two types – simple tissues and complex tissues. Simple tissues are homogeneous in nature and composed of structurally and functionally similar cells. Parenchyma, collenchyma and sclerenchyma are the common simple tissues. Complex tissues are heterogeneous in nature, being composed of different types of cells performing diverse functions. Xylem and phloem are complex tissues.

All the tissues of a plant which perform the same general function, regardless of position or continuity in the body, may be considered to form together a tissue system. The cells have physiological similarity or topographical continuity. Epidermal tissue system, ground tissue system (or fundamental tissue system) and vascular tissue system are the three tissue systems in vascular plants.

Plant anatomy divides the organism into 4 primary organs - the root, the stem, the leaf and the flower which can be divided into 3 tissue types – for example, leaves consist of 3 different tissues, dermal, vascular and ground tissues, and further these are composed of 3 cell types, parenchyma, sclerenchyma or collenchyma.

Parenchyma

The term, parenchyma is derived from Greek – *para* meaning "beside" and enchyma meaning "infusion" (visceral flesh). Anatomically, it is the functional tissue of an organ as distinguished from the connective and supporting tissue; and botanically, the cellular tissue, typically soft and succulent, found chiefly in the softer parts of leaves, pulp of the fruits, and bark and pith of stems *etc.* Originally, Erasistratus and other anatomists used it to refer to certain human tissues, but, later, it was also applied to plant tissues by Nehemiah Grew.

In the fundamental classification of plant cells, a cell with a thin primary wall is called parenchyma. In the parenchyma, there is a lot of diversity in cell size, shape, metabolism and function. In herbs, almost all the plant body is made up of one parenchyma or the other; only small portion of collenchyma and sclerenchyma are present. In shrubs and trees, large parts of the plant body consist of parenchyma.

The cells of parenchyma are living, metabolically active, and bound by a thin and flexible primary cell wall. In general, the cells account for 90 per cent of the cells found in herbaceous seed plants. These often occur in the cortex or pith of the stems or the roots, and the fleshy tissue of many fruits. Most cells retain the ability to divide and make them essential in wound healing and tissue regeneration. The cells often perform special functions like photosynthesis, storage or transport, and help the vascular tissue by forming a route to exchange the nutrients within or between the xylem and phloem.

Parenchyma cells are the oldest type of eukaryotic cells and are the first to evolve. The entire body of alga consists of only parenchyma. When the plants moved on to the land and became upright, they required special support, then collenchyma and sclerenchyma were evolved as modifications of parenchyma. In Algae, Mosses, Liverworts and lower vascular plants, all the parenchyma is involved in photosynthesis. But, in land plants division of labour became more sophisticated and various regions became specialized. Some of the generalists became specialists and new type of metabolism arose. Resin canals lined with secretory parenchyma cells became common in Gymnosperms. In Angiosperms, even more diverse types of parenchyma cells developed like mucilage cells, gum cells, laticifers, cells capable of rapid movement in plants and highly pigmented cells of flowers and fruits. Parenchyma is the most common tissue, which is morphologically and physiologically simple and unspecialized.

Parenchyma - Distribution

Almost all plant organs are occupied by parenchyma and the tissue occurs in the pith, the cortex, and the pericycle of the root and the stem. This is also found in the leaves as mesophyll tissue. It is also present in the fleshy part of the fruits and the endosperm of the seeds. It also occurs in the conducting tissue as xylem parenchyma and phloem parenchyma. The innermost layer of the periderm and the phelloderm is also composed of parenchyma.

Parenchyma - Origin

Parenchyma present in the primary body (pith and the cortex) of the plant originates from the ground meristem. Procambium and cambium give rise to xylem parenchyma and phloem parenchyma (primary and secondary tissues). Phelloderm originates from phellogen (cork cambium). In the leaf, protoderm and ground meristem give rise to epidermis and mesophyll.

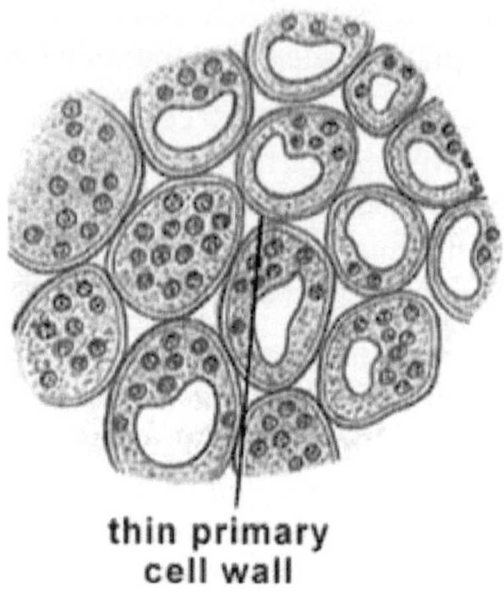

Figure 5.1: Parenchyma.

Parenchyma - Shape

Parenchyma cells vary in form, but usually isodiametric. The cells of isolated parenchyma grown in culture media are spherical. However, parenchyma cells may also be elongated and merge with prosenchyma cells. The term 'prosenchyma' was earlier used for elongated cells with tapering ends but in view of the fact that different types of cells may be elongated, the term has gone out of use. Parenchyma cells are lobed in spongy mesophyll and palisade parenchyma of *Lilium*. In the mesophyll of Xanthorrhoeaceae, the parenchyma cells have folds or projections. The cells are polygonal in cell aggregates. In relatively homogeneous parenchyma, the number of faces approach is 14 (with 8 hexagonal and 6 quadrilateral faces - *tetrakaidecahedron*), although cells with 12, 13, 15, 16 faces or fewer are also found. Parenchyma cells with inner wall protuberance occur in various anatomical structures concerned with short distance transfer of solutes such as nectarines, salt glands and vascular parenchyma. These cells are termed 'transfer cells'. The most common shapes of the cells are elongate (mesophyll of *Lilium* leaf), stellate (star-shaped as in the stem of *Scirpus* and *Juncus*) and with protuberances on their inner walls (as in phloem transfer cells of *Sherardia* leaf) *etc.*

Parenchyma - Arrangement

Mature parenchyma tissue may be tightly packed (compact) and without any intercellular spaces (endosperm of the seeds) or may have a well-developed system of intercellular spaces. For example, the parenchyma of the endosperm of most seeds is devoid, or almost so, of intercellular spaces while in the stems and leaves of hydrophytes the intercellular spaces reach maximal development. The tissue with the prominent intercellular spaces occurring in plants, growing in water-logged soils and aquatic habitats is called 'aerenchyma'. This tissue is considered to be important in aerating the plant. The development of intercellular spaces is either schizogenous or lysigenous. The resin ducts of the Coniferae, the secretory ducts of the Asteraceae and Apiaceae are formed schizogenously. Lysigenous intercellular spaces are formed by the disintegration of entire cells, *e.g.*, water plants.

Cell Wall – Structure

The cells usually have primary walls in active vegetative ground parenchyma (as in storage and chlorophyllous parenchyma) are relatively thin. They are composed of cellulose, calcium pectate and hemicelluloses. The middle lamella may or may not be distinguishable. Plasmodesmata are common in such walls. The microfibrils are variously oriented in the cell wall.

Sometimes, the primary wall of storage parenchyma of the endosperm becomes very thick due to the deposition of hemicellulose, *e.g.*, endosperm of *Phoenix dactylifera, Diospyros, Coffea arabica* and *Asparagus*. The walls of these cells gradually become thinner during germination. Parenchyma cells with relatively thick and lignified secondary walls are common, especially in the secondary xylem.

Parenchyma - Cell Contents

The parenchyma cells contain the living protoplast with a single or numerous vacuoles. The cells may also contain leucoplasts and chloroplasts (as in 'chlorenchyma' to help carry out photosynthesis). The internal structure of the parenchyma cell varies according to its function. Certain parenchyma cells contain leucoplasts (these help in storage of food). Photosynthesizing cells are usually conspicuously vacuolated. Secretory types of parenchyma cells often have dense protoplasts. Parenchyma cells may serve to store different reserve materials which may be found in solution in the vacuoles or in the form of solid particles or liquid in the cytoplasm. Sugars or other soluble carbohydrates and nitrogenous substances may be found in the cell sap. Starch, protein, oils and fats occur in cytoplasm in the form of small particles. Starch is the most common reserve material in plants and it is found in the endosperm, cotyledons, tubers, fruits, xylem and phloem parenchyma, the cortex *etc.* Protein and starch grains are found in the cotyledons of many species of the Fabaceae and protein and oil are found in the endosperm of castor and the cotyledons of soybean. In the parenchyma cells of the potato tuber amides and proteins are found in the cell sap and starch in the cytoplasm. Parenchyma cells in flowers and fruits often contain chromoplasts.

Parenchyma - Classification

Parenchyma is phylogenetically primitive and can be subdivided into 5 classes based on function: Synthetic parenchyma, Structural parenchyma, Boundary parenchyma, Transport parenchyma and Storage parenchyma.

1. Synthetic Parenchyma

This is also known as 'photosynthetic parenchyma'. In this type, the cells can often become rich in chlorophyll and hence this is known as 'chlorenchyma'. This helps in photosynthesis to convert light energy into chemical energy and store energy in carbohydrates. This is commonly found under the epidermis (where light is more intense) and is abundant in the leaves and also in the cortex of the green shoots. In the leaves, this is called as mesophyll – divided into 'Palisade parenchyma' and 'Spongy parenchyma'. The former is close to the upper epidermis and gets more intensity of light, while the latter is in the lower and the darker side of the leaf. The cells of the palisade are tightly packed with more number of chloroplasts and higher photosynthetic activity while in the spongy tissue; there are more empty intercellular spaces to facilitate the movement of gases and water.

2. Meristematic Parenchyma

All the meristematic cells are parenchymatous and these are most active type of synthetic cells. They absorb sugars, water and inorganic nutrients and use them to synthesize whole cells. The meristematic cells are parenchyma that consists of a group of cells which remain in a continuous state of cell division or they retain their power of division. They are immature cells without any intercellular spaces. The cells contain abundant cytoplasm with one or more distinct nuclei. The vacuoles are small or absent. These give rise to either primary or secondary tissues.

3. Secretory Parenchyma

These parenchyma cells secrete large quantities of substances which can move to the exterior, or to a cavity or to a duct inside. The secreted material can pass from cell to cell to the site of deposition. The secretory cells and tissues are mainly concerned with the accumulation of metabolism by-products which are not used as reserves. Most cells are specialized and are derived from elements of other tissues such as epidermis or parenchyma (secretory parenchyma or secretory epidermis). The term – 'secretion' is rather inaccurate, and includes the production and release of the molecules that are useful to the organism (hormones, enzymes and tannins, laticifers and crystals) but becomes separated from the cell

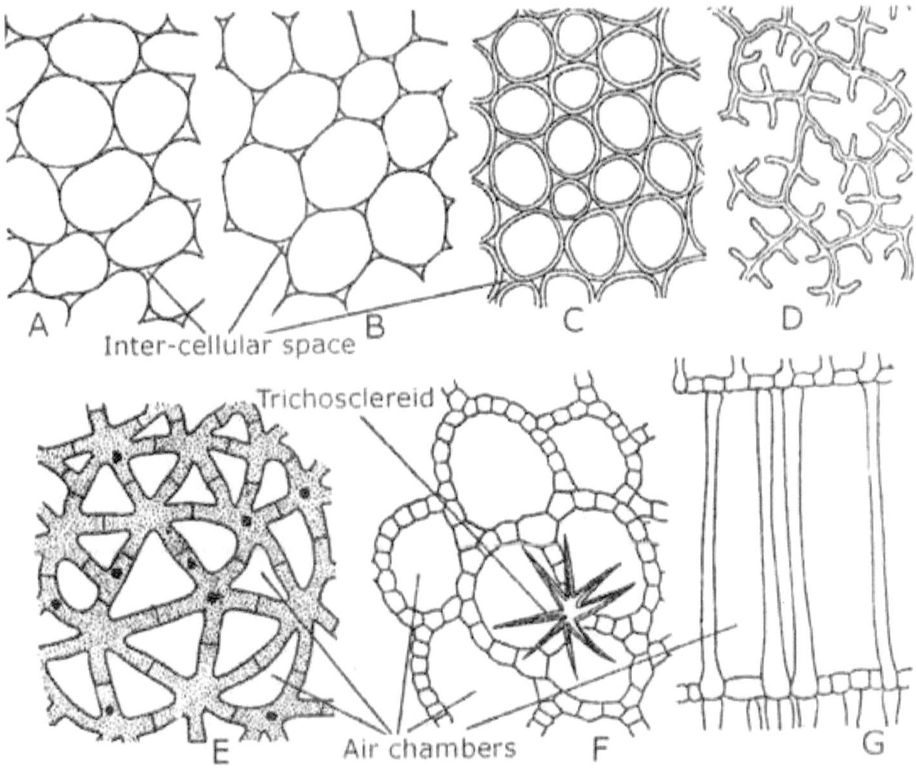

Figure 5.2: Parenchyma - Different Types. A. *Helianthus* - Thin walled parenchyma; B. Sunflower - Parenchyma from the pith; C. *Clematis* pith - Thick walled parenchyma; D. *Pinus* - Parenchyma from the Leaf; E. Petiole of *Musa* - Aerenchyma; F. Petiole of *Nymphaea* - Aerenchyma with Trichosclereid. G. *Jussiaea* stem - Aerenchyma.

4. Structural Parenchyma - Aerenchyma

This is also called as 'Aeriferous Parenchyma' or 'Aerenchyma'. This is well developed in the spongy mesophyll of leaves, submerged stems and petioles of aquatic plants like *Typha, Nymphaea, Eicchornia, Juncus etc.* Though all cells store some amount of water, the aerenchyma is specialized in this function. The cells are large, with a thin cell wall and a very large vacuole where water is stored. The cytoplasm or the vacuole contains a mucilaginous substance that increases the capacity of absorption and retention of water. Aquiferous parenchyma is present in plants that live in dry environments, the xerophytes.

In Aeriferous parenchyma, the cells are large and interconnected empty intercellular spaces into which the gases diffuse and aerate the root. This tissue is well developed in plants that live in wet or aquatic environment (hydrophytes). This is also found in non-aquatic plants under stress. Both the stem and the root can develop aerenchyma - either by schizogeny or lysigeny. The former is a process that occurs by cell differentiation during development of the organ, while the latter is a consequence of the stress and the intercellular cavities are produced by the cell death. The latter is found in wheat, rice, corn and barley. Some authors suggest that a third type, called 'expansigeny' is also seen where the intercellular cavities are produced by cell retraction but the cells do not lose the physical contacts.

This tissue is continuous from the stem to the root and the large spaces allow the gaseous movement and increase the conduction from the leaves to the roots which is vital for hydrophytes for keeping the level of oxygen normal (for respiration of the root). It is also helpful in the release of gases like ethylene from the roots through leaves. This tissue is an adaptation of plants to hypoxia of wet or flood soils. The plants with this tissue play a major part in the release of greenhouse gases to the atmosphere (methane) from the soil and funnel them through roots, shoots and leaves (rice crop).

5. Boundary Parenchyma

This is present in the interfaces between two regions of a plant or between a plant and environment. Epidermis and endodermis are two examples of this type.

a. Epidermis

It is the boundary between the plant and environment. Both plant and environment are very diverse as in the epidermis. On shoots, the function of epidermis is the retention of water inside the plant. Hence, the outer wall becomes less permeable to water due to deposition of waxes. Epidermis can be considered as 'secretory parenchyma' because it secretes certain water-proofing substances. Epidermis is the outermost layer exposed to wind and animals, hence, it must be resistant to tearing and puncture. The epidermal cells are attached to each other very firmly and are strongly interconnected and can be pulled off as a large sheet. Epidermal cells, therefore, can be classified as 'structural parenchyma'.

b. Endodermis

The conducting tissues are isolated from the cortex by a boundary of cells which constitutes the endodermis. The endodermal cells form a cylinder around the vascular tissues and like the epidermal cells abut each other very tightly with no intercellular spaces.

6. Transport Parenchyma or Transfer Parenchyma

In many parts in plants, the material must be transferred rapidly and in larger quantities over a short distance into a gland, into and out of sieve elements, and into a developing embryo. For this purpose, 'transfer cells' are present. The cell walls of these transfer cells are highly modified, having many ridges and papillae on their inner face, forming a labyrinth. Hence, these are called labyrinthine walls or transfer walls. The labyrinthine walls increase the area of absorption when compared to flat walls.

a. *Sieve Elements*

In the phloem, sugars are conducted over long distances by specialized parenchyma cells called 'sieve elements'. They have a functional structure in which the cells are elongate and have large holes to allow easy passage of phloem sap.

b. *Light Transmission*

An unusual type of transport parenchyma occurs in *Frithia, Lithops* and other members of Aizoaceae – the window leaf plants of the African deserts. The whole plant is subterranean with only the leaf tips reaching the soil surface, so that photosynthesis takes place below the ground. To accomplish this, a leaf with a transparent channel acts as an optical fibre to carry light from the surface to the photosynthetic tissue below. The plant remains in cooler, and moist place in subterranean environment, so that water storage and conservation are maximized.

7. Storage Parenchyma

Most of the plant parts used for food – seeds, fruits, tubers are composed of storage parenchyma. In these organs, the parenchyma stores starches (cereals, potatoes), proteins (beans, peas), or oils (sunflower, safflower, castor). The cells store their storage products fully that even the vacuole and other organelles are obscured. The storage parenchyma also stores amides, tannins, mineral substances, enzymes and mucilaginous substances. Succulent plants, such as cacti and Euphorbias are composed of the parenchyma cells which store water. Such cells are large and the bulk of their volume consists of greatly expanded vacuole surrounded by a thin layer of cytoplasm. The cells are much bigger than those that store starch or protein. A less common means of storage is by increasing the amount of hemicelluloses in the wall. These are deposited to such an extent that the walls become thick. Later, the hemicelluloses are digested out of the walls and the walls become quite thin as in the seeds of *Coffea, Diospyros* and *Phoenix*. Many carbohydrates, nitrogenous substances and fats are stored in the cell sap of the cells. The cells in the roots of sugar beet and scales of onion contain amides, proteins and sugars in their cell sap. The cotyledons of many legumes contain protein and starch in their parenchyma. The endosperm cells of *Ricinus communis* contain proteins and oils. Proteins and starch are present in the parenchyma of potato tuber. Starch is seen the parenchyma of endosperm, tubers, cortex, fruits, xylem and phloem. Tannins are also found in the parenchyma. Though the stored substances are solid (starch grains and crystallized proteins), they are mostly found in solution (lipids, proteins *etc.*) and these are stored in the vacuoles. In the cytoplasm, some are also stored as carbohydrates and nitrogenous substances. Some cells store only one type of substances but a mix up of different substances can also be found in the same cell.

Parenchyma - Functions

1. The living parenchyma cells are the centres/sites of various physiological activities like photosynthesis (chlorenchyma), assimilation, respiration, storage (storage parenchyma), secretion (secretory parenchyma) *etc.*
2. In succulent plants, storage parenchyma helps in water storage.
3. In aquatic plants, aerenchyma stores air in its intercellular spaces and provides buoyancy.
4. Parenchyma cell (meristematic parenchyma) of the cambium divides and form secondary tissues.
5. The tissue also is helpful in wound healing.
6. The transfer cells help in small distance transport of the solutes.
7. The thicker cell walls of *Asparagus etc.* (hemicelluloses) store reserve food to be utilized during germination.
8. Epidermal cells with their cutinized peripheral cell walls help in protection.
9. The seedlings and some hydrophytes gain mechanical rigidity from their turgid parenchyma cells.

Idioblasts

The cell in a tissue that markedly differs in its form, size, structure or contents from other cells in the same tissue is referred to as an idioblast (Gr.: *Idio*= peculiar). Specialized parenchyma cells which produce and store tannins, myrosin (*e.g.*, Brassicaceae, Capparaceae), oil substances (*e.g.*, Lauraceae), mucilaginous substances (*e.g.*, many Monocotyledons, Malvaceae, Cactaceae), resinous substances (*e.g.*, Meliaceae, Rutaceae, Rubiaceae) and crystals of calcium oxalate, *e.g.*, Moraceae are some examples. The cells also store mucilaginous substances (in Tiliaceae, Portulacaceae) and oily substances (Lauraceae) and enzymes like myrosinase (Brassicaceae). These cells significantly differ in size, structure or contents. These specialized cells are known as the idioblasts (Gr.: *Idio*= peculiar).

Collenchyma

Collenchyma in plants consists of thick-walled cells and is a supporting tissue. The supporting tissues of the plant, *i.e.*, collenchyma and sclerenchyma are designated from the functional point of view by the term "stereome" (Haberlandt, 1918). The term was derived from Greek "Collle" means "glue" and "enchyma" means "infusion". Collenchyma consists of living, more or less elongated cells which, generally, have unevenly thickened walls. Collenchyma differs from sclerenchyma in wall structure and condition of protoplast. Collenchyma has relatively soft, pliable, non-lignified primary walls, whereas sclerenchyma has hard, more or less rigid, secondary walls, which are commonly lignified. Many sclerenchyma cells lack protoplasts at maturity while collenchyma cells retain active protoplasts and can resume meristematic activity. There is a physiological and morphological relationship between collenchyma and parenchyma, and, in places where the two tissues occur side by side, transitional forms can be found between typical collenchyma and typical parenchyma.

The cells of collenchyma are also living and are elongated in structure, consisting of an irregular thick cell walls with thick deposits of cellulose and appear polygonal in cross section that provides support and structure to the plant. This may also form cylinders or occur as discrete strands and is one of the 3 ground or fundamental tissues. This is the least common cell type and the cell walls are composed of alternating layers of cellulose and pectin. The walls are basically soft and pliable, rich in water, living at maturity with a complete protoplast. Sometimes, the cells possess chloroplasts and can become meristematic. They are isodiametric in their shape to elongated which always occur as a homogeneous mass just below the epidermis in herbaceous stems, young stems of woody plants and petioles. The tissue may occur as a continuous cylinder in old stems, where due to deposition of lignin, it turns sclerenchymatous. Collenchyma is common in the epidermal tissue of young stems and leaf veins. The cells are classified on their location and pattern of cell wall thickenings – angular, annular, lamellar and lacunar.

The collenchyma is extremely plastic and the cells can extend and adjust to the increased growth of the organ. The tissue is chiefly found in the cortex of the stems and in leaves, and is the primary supporting tissue for herbaceous plants. It is temporarily functional in plants with secondary growth and as the woody tissue develops it gets crushed. Often, it constitutes the ridges and the angles of the stems and borders the veins in dicot leaves. The best examples are the "strings" in the stalks of celery.

Collenchyma - Distribution

The tissue is located either directly beneath the epidermis or one or a few layers removed from it. In stems, collenchyma frequently forms a continuous layer at the periphery of herbaceous stems, around the circumference of the axis as in *Helianthus* stem. Sometimes, it occurs in strands, often with in externally visible ridges (ribs) found in many herbaceous stems and those woody stems that have not yet undergone secondary growth, *e.g., Cucurbita*. In the petioles (*Begonia*), the distribution of collenchyma shows patterns similar to those in stems. In the leaf blade, it occurs in the ribs accompanying the larger vascular bundles (veins as in *Nerium*), sometimes, on both sides of the rib, sometimes, on one side only, usually on the lower. The tissue is also present in the floral parts, fruit and the aerial root (*Monstera*). Collenchyma is generally absent in underground roots, stems and leaves of monocots. It usually occupies the peripheral layers of cortex in dicots and may be present just beneath epidermis or below a few layers of parenchyma. It may also occur as a continuous layer in the form of hypodermis. In some cases, the cells form distinct ribs or the ridges at the outer edges of the stems (as in *Leonurus* and *Cucurbita*), as well as the petioles (as in *Apium*). It is present in the subterranean roots of *Vitis* and *Diapensia*; and the rind of the fruit in *Vitis* and *Cassia tora* is collenchymatous.

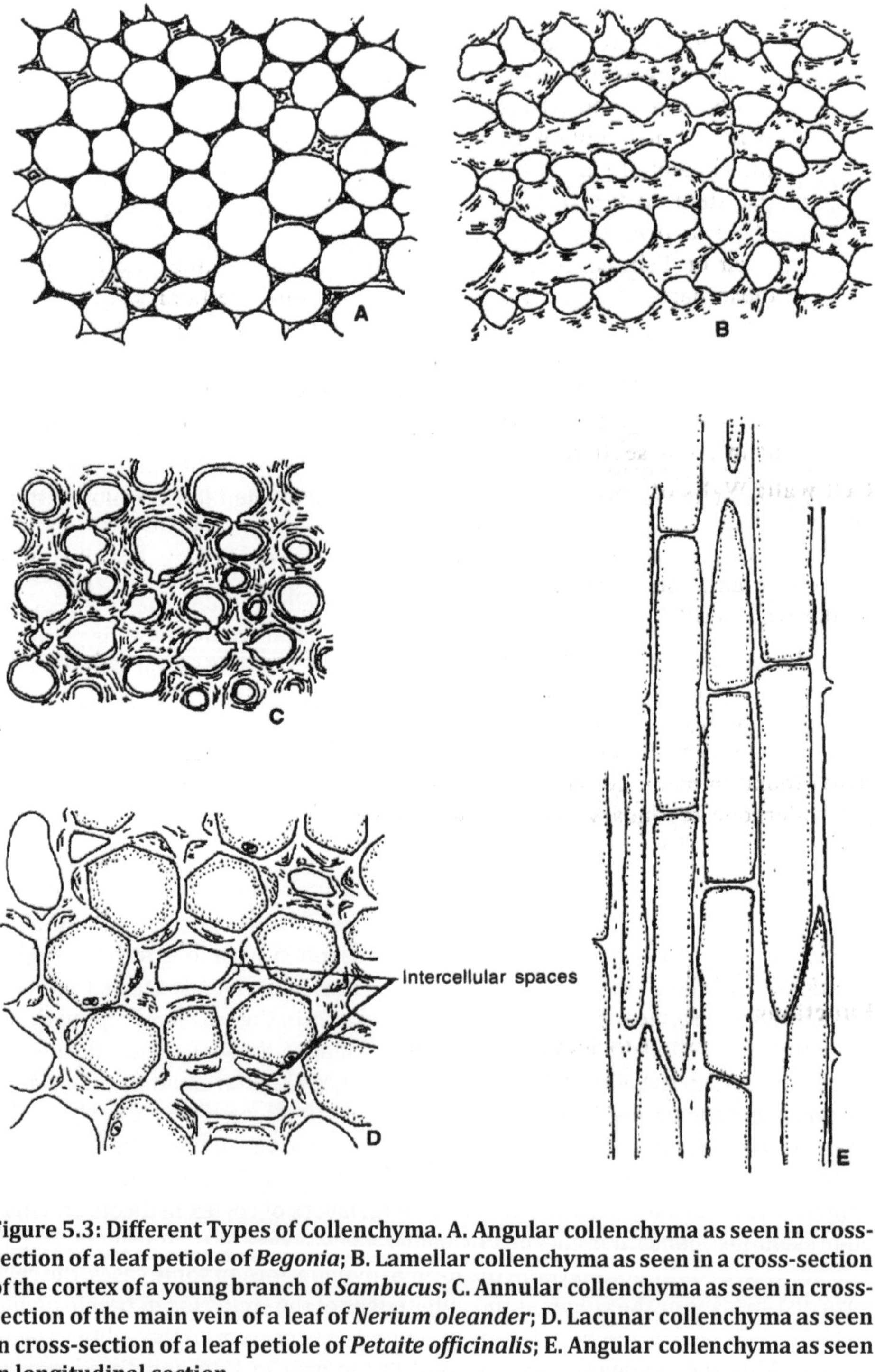

Figure 5.3: Different Types of Collenchyma. A. Angular collenchyma as seen in cross-section of a leaf petiole of *Begonia*; B. Lamellar collenchyma as seen in a cross-section of the cortex of a young branch of *Sambucus*; C. Annular collenchyma as seen in cross-section of the main vein of a leaf of *Nerium oleander*; D. Lacunar collenchyma as seen in cross-section of a leaf petiole of *Petaite officinalis*; E. Angular collenchyma as seen in longitudinal section.

Collenchyma – Origin

The tissue develops from either the ground meristem or procambium (*Apium graveolens*); and sometimes it may also originate from elongated cells which resemble procambium.

Collenchyma - Cell Shape

The size and shape of the collenchyma cells vary. Isolated collenchyma ranges from round to elliptical to elongated cells. The transitional forms are also present. The round or elliptical cells generally resemble the ground parenchyma. The cells may be short prisms and resemble neighbouring parenchyma cells or the elongated cells look like fibres with both the ends tapered which may attain a length of 2 mm rarely, and all intermediate shapes and sizes occur. The longest collenchymatous cells are found in the central portions of the strands of collenchyma, and the shorter ones on the periphery.

Collenchyma – Arrangement

The tissue appears to be more or less compact as the thickening material deposits more heavily at the corners and also on the radial walls in addition to normal uniform thickening. The nature of thickening varies, and as such 3 types are recognized (Muller, 1890):

(a) Angular Collenchyma

The thickening material gets deposited at the angles or the corners of the cells (Figure 5.3 A, E). In cross section, the thickenings appear to occur at places where several cells meet each other. This is found in the petioles of *Vitis, Begonia, Cucurbita, Beta,* and *Morus* and stems of *Solanum tuberosum* and *Atropa belladonna*.

(b) Lamellar Collenchyma or Plate Collenchyma

The thickening material gets heavily deposited mainly on the tangential walls of the cells, rather than the radial walls (Figure 5.3 B). Such lamellar collenchyma is found in the stem cortex of *Sambucus nigra* and *Rhamnus* and petiole of *Cochlearia armoracia*.

(c) Lacunar Collenchyma or Tubular Collenchyma

The thickening material appears in those places of the cell wall which are in direct contact with the intercellular spaces (that face the intercellular spaces). Such collenchyma can be seen in the petioles of the species of Asteraceae, *Salvia, Asclepias, Malva* and *Althaea* (Figure 5.3 A, E).

(d) Annular Collenchyma

Duchaigne (1955) reported another additional type, *i.e.*, Annular collenchyma in the petiole of *Nerium* (Figure 5.3 C), in which the cell lumen is more or less circular (or almost round in cross section). But, Fahn (1987) thought it to be mature collenchyma. The angular appearance of the lumen is lost due to continued and heavy deposition of the wall material.

Collenchyma – Structure and Content

The walls of collenchyma cells consist of cellulose, large amounts of pectin and hemicellulose but no lignin. The primary pit fields can be clearly distinguished in the walls of the cells. The microfibrils of cellulose are either oriented transversely or alternately transverse and longitudinal. The longitudinal microfibrils may also alternate with non-cellulose material (as in the petiole of *Apium graveolens*). The composition of the alternate layers varies; one layer is rich in cellulose and poor in pectin, while the alternate layer is pectin-rich and cellulose-poor. Since the pectic substances are hydrophilic, collenchyma walls are rich in water. Chafe (1970) has observed that the orientation of the cellulose microfibrils in the successive lamellae of the wall is alternately transverse and longitudinal. It is suggested that the visible lamellation of the cell wall is due to the composition and orientation of the microfibrils at alternate layers. Though the lignin is completely absent, in some cases, the cellulose-rich layer is impregnated with lignin (as in the stem and petioles of *Salvia officinalis, Viscum album* and *Medicago sativa*). The lamellae are also deposited on the wall and thus, the pecto-cellulosic wall may become sclerified. The cells contain living protoplast. Chloroplasts may also be present in some cells, in addition to tannins.

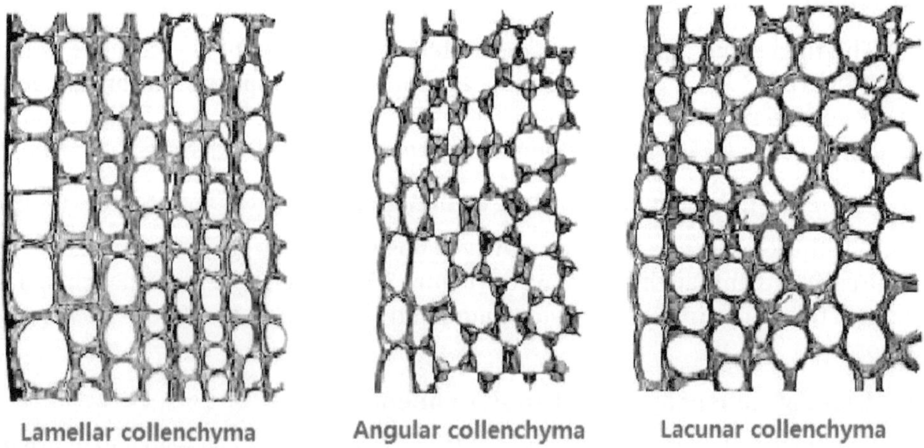

Figure 5.4: Types of Collenchyma.

Functions of Collenchyma

1. The tissue provides flexible (plastic) support by its position in organs and structure of the cell wall to the leaves and stems (as the cells are extensible with a considerable degree of plasticity and support to the organs in which they occur).

2. The tissue protects the vascular bundles of leaves by forming the cap or bundle sheath.
3. The tissue provides tensile strength for growing organs.
4. The collenchyma which contain the chloroplasts can carry out photosynthesis.
5. The peripheral thick walled collenchyma in some cases becomes thin and regains the meristematic activity (phellogen or the cork cambium) which divides to form the periderm.
6. The sclerified collenchyma serves the mechanical function in mature plants.
7. The tissue can also serve as a storage site for defensive antibacterial compounds (hydroxyproline-rich bacterial agglutinins in *Solanum tuberosum*).

Sclerenchyma

The term was derived from Greek word - "*scleros*" means `harder` and "*enchyma*" means `infusion`. Sclerenchyma can be defined as one of the types of simple permanent tissue (ground tissue) that constitutes both the primary and secondary wall. The cells are generally woody and rigid which are compactly arranged without any intercellular spaces. Its main function is to promote strength and conduction, being a dead cell due to deposition of lignin on secondary development which restricts exchange of water and gases due to which inner protoplasm degenerates. Sclerenchyma is a simple but hard tissue composed of cells with thickened secondary walls, usually with lignified walls. The cells are mostly nonliving at maturity.

Sclerenchyma - Definition

This is a simple, permanent tissue where living cells become dead during the development of a secondary wall due to the accumulation of lignin. Lignification occurs after the completion of the cell growth, and at the time of secondary thickening. The tissue predominates in the rigid (hard) areas of the plant body such as veins of a leaf, stem, branches, trunk and the bark *etc.*

Sclerenchyma – Characteristics

When mature, sclerenchyma shows the following characteristics:

This is a dead, simple, permanent ground tissue. The cells of sclerenchyma form a protective or supporting tissue in higher plants. When mature, the cells have a limited physiological activity and are dead. The cells have a cell wall with a thick secondary layer made up of cellulose, hemicellulose and lignin. The orientation of the cellulose provides a diverse combination of strength, flexibility and stiffness of plant organs subjected to different compressive and tensile forces. This occurs in 2 different forms – fibres and sclereids. The cells of sclerenchyma may differ in their shape, structure, origin and development. The fibre is a long, slender cell, many times longer than wide, while the sclereids vary in their form from isodiametric, to a considerable, elongated one and are also much branched.

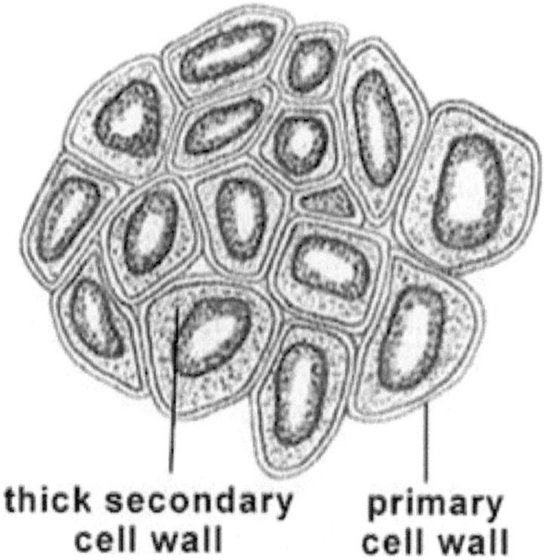

Figure 5.5: Sclerenchyma - Sclereids.

Sclereids

Sclereids arise either through secondary sclerosis of parenchyma cells or by sclereid primordia. In leaves they may develop from epidermal cells, palisade cells or spongy parenchyma. In the vascular region, the sclereids develop from the derivatives of procambial or vascular cambial cells. Macrosclereids found in seed coats arise from protoderm. Randomly distributed sclereids are formed from the parenchyma cells. The development of branched and long sclereids involves remarkable intercellular adjustments. The growth of cells is called "intrusive" or inter-positional growth and is contrasted with the coordinated growth that involves no separation of walls. (The intrusive growth is shown in detail in Figure 5.6). The sclereids are short or elongated occur in various parts of the plant, either in the layers or clusters, but often isolated and scattered cells (referred to as *idioblasts* or *crazy cells* in the stems, leaves, and fruits and serve in protection.

Sclereids – Classification

The information on the main morphological types of sclereids and their viability has reached a stage of considerable interest in recent years, in respect of other disciplines of botany. The pioneer researcher in this field was Alexander Tschirch (1885) who for the first time introduced the term "sclereid", and in 1889, recognized 4 main categories of sclereids, namely, Brachysclereids, Macrosclereids, Osteosclereids and Astrosclereids. Foster (1946) has distinguished 4 types in the genus, *Mouriria* – parenchymatous cell form with or without radiating branches,

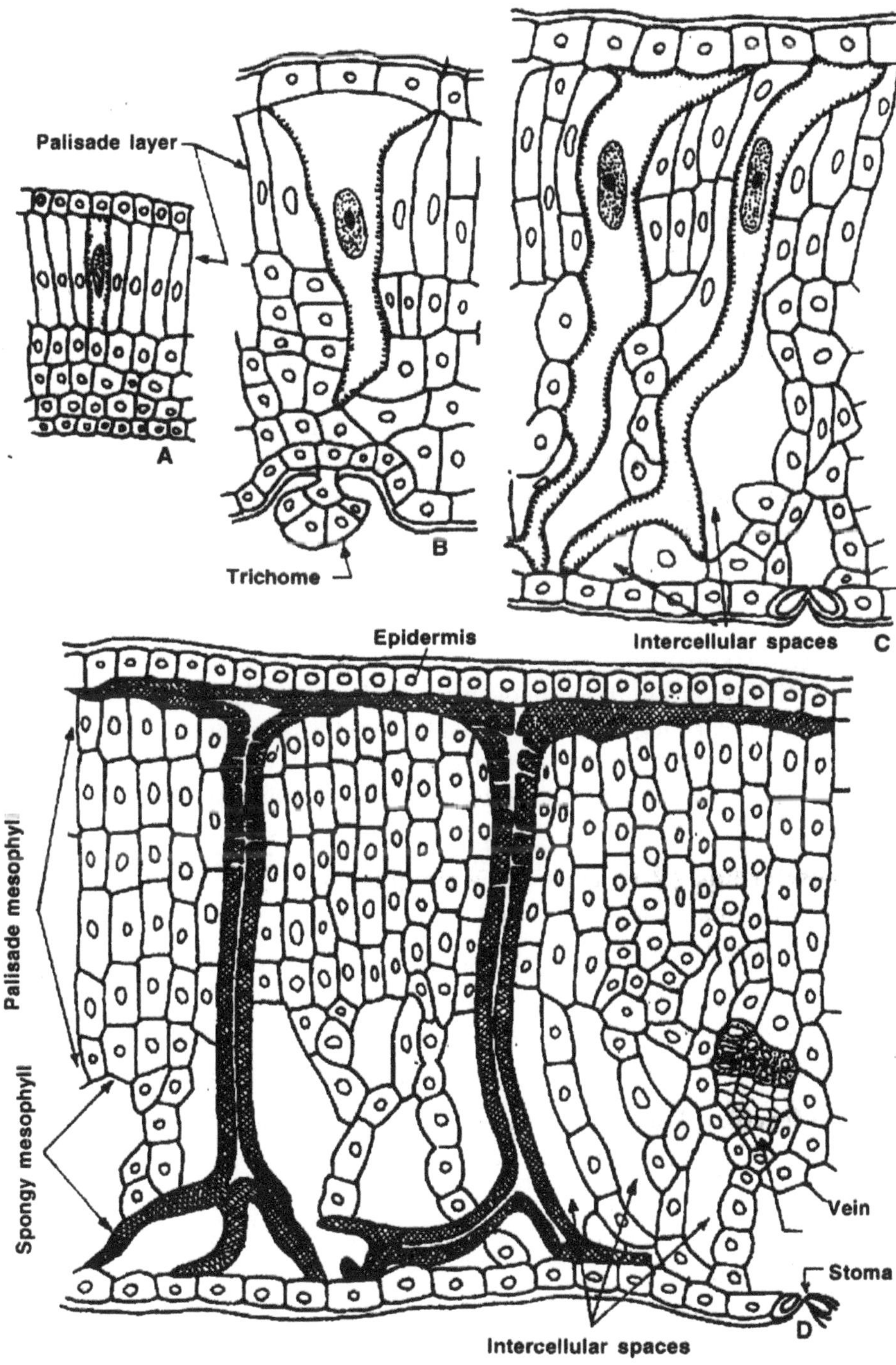

Figure 5.6: Development of Sclereids in the Leaf of *Osmanthus* (Oleaceae). A-C. Differentiating sclereids indicated by large nuclei and dots along the walls; D. Mature sclereids, indicated by cross-hatched secondary walls.

ramified sclereids of stellate and fusiform branched or unbranched types and filiform (thread-like) sclereids with or without forking. Rao (1957a) in 95 species of the genus *Memecylon*, has recorded filiform branched or unbranched sclereids and polymorpghic sclereids with their modifications and stellate sclereids. In the family, Marcgraviaceae, Richter (1920) has recognized a few categories of sclereids, namely, Ophiurosclereids, Librosclereids, Astrosclereids, Palosclereids, Rhizosclereids and Idiosclereids. Singh (1967) has categorised 4 types of sclereids in 15 species of the genus *Fagraea* – Aster, Columnar, T-shaped and Dendroid forms. Bokhari and Burtt (1970) categorised 7 types of sclereids working on the sclereids of *Cyrtandra*, namely, Vermiform sclereids, Osteosclereids, Vesicular sclereids, Astrosclereids, Polymorphic sclereids, Dendrosclereids and Tracheoids. Nicolson (1960) recognized trichosclereids in the genera, *Pothos, Scindapsus* and *Rhaphidophora*.

The sclereids may be classified into 2 main divisions – Monomorphic sclereids (Spheroidal sclereids, Vesiculose sclereids, Vermiform sclereids, Palosclereids, Osteo sclereids, Fusiform sclereids, and Filiform sclereids) and Polymorphic sclereids (Ramiform sclereids [Rhizosclereids and I-shaped sclereids], Astrosclereids [Ophiurosclereids, Librosclereids and Trichosclereids], Polyramous sclereids and Idiofibrosclerteids) depending on the constancy of the body shape.

The sclereids are a reduced form of sclerenchyma cells with a highly thickened and lignified cell walls in the form of small bundles of definite layers of tissue in majority of plants. Though these are variable in their shape, the cells are generally isodiametric, prosenchymatous, forked or elaborately branched. They group into bundles or form complete tubes at the periphery or the organs, or occur as single cells or small groups of cells within the parenchyma. Idioblast is an isolated sclereid. The stone cells are usually found in the epidermis, ground tissue and vascular tissue.

Origin of Sclereids

These are created through belated sclerosis of cells of parenchyma or they can arise from sclereid primordial that becomes individual early in its development. The process of sclerification involves the thickening of the cell wall, increasing its rigidity. In the phloem, when the tissue ceases to function in conduction, the cells may begin sclerification. But, in the vascular tissue, the sclereids will develop from the cambial and procambial cells.

General Classification of Sclereids

The sclereids may also be classified as: Stem sclereids, Leaf sclereids, Fruit sclereids and Seed scelereids, depending on the nature of the organ in which they occur

1. Stem Sclereids

The sclereids with thick cell walls and numerous pits are found in groups in the vascular region of the stem of *Hoya carnosa*, in the pith of the stems of *Hoya* and *Podocarpus* and these resemble the adjacent parenchyma cells. This typically suggests that these are originally parenchyma cells but sclerified to be considered as stone cells.

2. Leaf Sclereids

A variety of sclereids occur in the leaves. Two distinct types are found in the mesophyll. Throughout the tissue of the leaf, sclereids are dispersed in a diffuse pattern while they are concentrated in a terminal pattern about the tips of the veins of the leaf. These include: the branched sclereids of *Trochodendron*, the columnar sclereids of *Hakea*, and the hair-like sclereids (trichosclereids) that branch into the air chambers within the leaves of Water Lily and Yellow Pond Lily. The stone cells can also form a part or the entire epidermis of the foliar structures *e.g.* Clove scales of *Allium sativum*. Of special interest are the sclereids in the leaves of *Olive*. They are fibre-like (of 1 mm in their length) and permeate the mesophyll as a dense mat. These cells increase by several 100 times their original size during sclerification.

3. Fruit Sclereids

The sclereids in the fruits vary in their form as well as their use. These form concentric clusters in Pears, which grow about earlier formed sclereids. These sclereids and those within the Quince fruit often form the bordered pits when their cell wall increases in its thickness during sclerification. The layers of elongated sclereids form the endocarp that encloses the seeds in Apples.

4. Seed Scelereids

The seed coats harden during ripening of the fruit through sclerification when the secondary walls get thickened in the epidermis and beneath it, *e.g.* Seeds of legumes. The larger sclereids form columns in the epidermis of Pea, Bean, and Soybean seeds, while the bone-shaped cells occur beneath the epidermis. The sclereids in the seed coats of coconut, possess numerous bordered pits, and these larger macrosclereids found in the seed coats are of protodermal origin.

Classification on the Basis of the Size and Shape

On the basis of their size, shape, nature of cell wall and mode of deposition of secondary wall material, sclereids have been classified into five types.

1. Brachysclereids (or Stone cells)

They are shortly (small), more or less roughly isodiametric or somewhat elongated cells, with extremely thick cell walls provided with pit channels (and if branched, these are also called as "ramiform cells"). Their shape is often like that of parenchyma cells. They are found in the cortex and pith of stems, phloem and fleshy parts of fruits, *e.g.*, Pear, Apple, Cinnamon, grit of Guava, Custard apple, Sapota and Coconut.

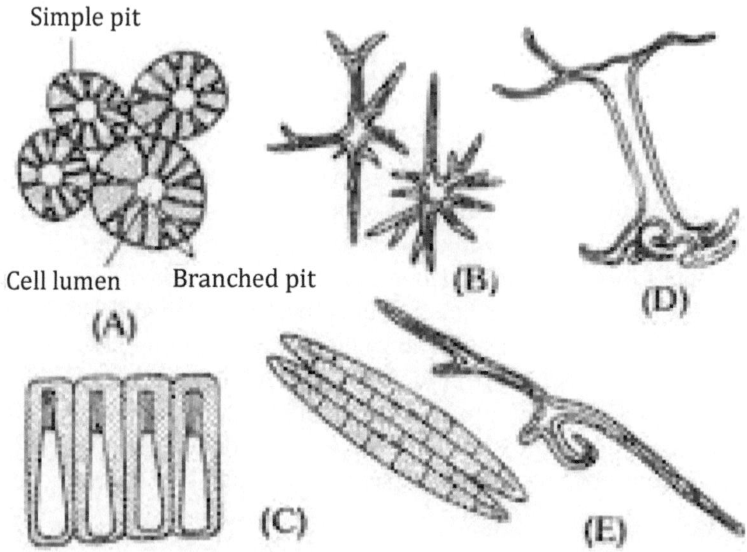

Figure 5.7: Types of Sclereids. A. Brachysclereid; B. Astrosclereid; C. Macrosclereid; D. Osteosclereid; E. Trichosclereid.

2. Macrosclereids (Malpighian cells or Rod cells)

These are elongated columnar sclereids, and are found in the seed coats (or epidermal covering) of Fabaceae, *e.g.*, Bean (*Phaseolus*), Pea (*Pisum*), Soybean (*Glycine*).

3. Osteosclereids (Prop cells or Bone cells or Columnar Cells)

These are also columnar but with swollen ends similar to a bone. They are found in seed coats (*e.g., Pisum sativum*) and sometimes, in the leaves of certain Dicotyledons (*e.g., Hakea*) and sub-epidermal covering of legume seeds.

4. Astrosclereids (Star cells)

These are branched and more or less star-shaped, and are often associated with vein endings (with no pits in the arms or branches due to intrusive growth). They occur in the petioles of leaves (*e.g., Nymphaea*, Tea) and stem cortex and leaves of *Trochodendron*.

5. Trichosclereids (Internal Hairs)

These are hair-like, branched or unbranched sclereids found in the intercellular spaces of leaves and stems of some hydrophytes, such as aerial roots of *Monstera* and leaves of *Olea*.

6. Filiform Sclereids

These are fibre-like and sparingly branched. *e.g. Olea*

Sclereids - Functions

1. The sclereids provide firmness, stiffness and mechanical strength to the part of plant where they are found.
2. Sclereids also form a protective layer in the clove scales of garlic (*Allium sativum*).

Fibres

The fibres are long and narrow, unbranched, and with tapered, thick end walls, not highly hydrated, often rather sparsely pitted, often with a slit-like, narrow lumen that is dead at maturity. Fibres are much elongated elements with acute end walls, narrow lumen and thick secondary walls. In some instances, they have septa and then designated as "septate" fibres. Sometimes they are so long that they reach up to 55 cm in length as in ramie (*Boehmeria nivea* – Urticaceae). "Ramie fibres" are the longest cells in higher plants and used in cordage. Ontogenetically fibres develop from different meristems, such as the procambium, vascular cambium, ground meristem and even from the protoderm, as in certain species of the Poaceae and Cyperaceae. Fibres may also develop from parenchyma cells, *e.g.*, in the protophloem of many Dicotyledons. The fibres formed by cambium develop from fusiform initials and elongate only little or not at all during their maturation. Fibres also show combination of coordinated and intrusive growth.

Fibres - Classification

On the basis of their position in the plant body, these are of 2 types, namely: xylary fibres and extra-xylary fibres.

a. Xylary Fibres

These are also called "wood fibres" for the reason that they are found in xylem and develop from the same meristematic tissue which gives rise to the other xylem elements. These fibres, in xylem, are believed to be derived form the tracheids (a type of conducting sclerenchyma). On the basis of wall thickness and the nature of pits, two main types are distinguished among the xylary fibres, namely: (i) Fibre tracheids and (ii) Libriform fibres. If both occur in the same wood, libriform fibres (are longer with thick cell walls than the former and simple narrow pits without borders) or fibre tracheids (with thinner walls, and pits are narrow and bordered, with a chamber and an oval inner aperture) which show intermediate characters between the tracheids and libriform fibres. Both categories of fibres may be septate. Septate fibres are distributed widely in Dicotyledons and usually retain their protoplasts in mature sapwood. Septate fibres are concerned with the storage of reserve food materials. The fibres of branch wood in the Dicotyledons are characterized by the presence of innermost layer (G-layer) of the secondary wall with much α-cellulose and little lignin. They are known as "gelatinous fibres" or mucilaginous fibres.

b. Extraxylary Fibres

These fibres are found in tissues other than the xylem (outside the xylem) such as the cortex, pericycle and phloem. Often these are associated with phloem (phloem fibres, peri-vascular fibres, pericyclic fibres and bundle cap). When encountered in the phloem they are known as "bast fibres". These fibres show no features of tracheids but they have thick walls, simple pits with circular inner apertures, heavy lignification gives the name "hard fibres" (which are typical of monocots), while no lignification gives the name "soft fibres" (which are typical of dicots for *e.g.*, linen). Some of the fibres remain living – a multinucleate fibre in Tobacco – and sepatate fibres with a thin cell wall dividing the cell into compartments – each section with its own nucleus to combine mechanical support and storage function.

On the basis of their occurrence, the extra-xylary fibres categorised into (i) phloem fibres (ii) cortical fibres and (iii) perivascular fibres. Extra-xylary fibres are long, spindle-like with acute or acuminate or blunt ends. Their cell wall is quite thick with simple or slightly bordered pits. At maturity, these fibres lose their protoplasts and become dead. Usually these fibres remain unbranched but branched fibres may be found in Monocotyledons and Dicotyledons. The fibres are long cells, with more or less thick secondary walls and usually occur in strands. These strands constitute the 'fibres' of commerce. Within a strand, the fibres overlap, a feature that imparts strength to the fibre bundles.

Fibres - Economic Importance

In commerce, the term fibre is often applied to materials that include, in botanical sense, other types of cells besides fibres and also structures that are not fibres at all. In fact, leaf fibres of the monocotyledons commonly include vascular elements. Cotton fibres are epidermal hairs of seeds of *Gossypium*. Similarly, vascular system of roots of *Muhlenbeckia* and the entire plant of *Tillandsia* are used as commercial fibres. Likewise, kapok fibres of *Ceiba pentandra* originate as hairs from the inner surface of the fruit (capsule).

Commercial fibres are divided into hard fibres (with high lignin content) and soft fibres (with little or no lignin content). The former, hard fibres are obtained from the leaves of monocotyledons, *e.g.*, Abaca or Manila hemp (*Musa textilis*), Bowstring hemp (*Sansevieria* spp.), Sisal (*Agave sisalina*), while the latter, the soft fibres (also known as bast fibres) are obtained from phloem, *e.g.*, flax (*Linum usitatissimum*), hemp (*Cannabis sativa*), jute (*Corchorus capsularis, Corchorus olitorius*), kenaf (*Hibiscus cannabinus*), sunhemp (*Crotalaria juncea*) and rozelle (*Hibiscus sabdariffa*).

References

Carlquist, S. 1988. Comparative Plant Anatomy. Holt, Reinhart and Winston, New York.

Cutter, E,G. 1978. Plant Anatomy. Part I and II. Edward Arnold, London.

Eames, A.J. and L.H.MacDaniels. 1979. An Introduction to Plant anatomy. Tata McGraw Hill, Bombay.

Esau, K. 1953. Plant anatomy. John Wiley and Sons, New York.

Esau, K. 1977. Anatomy of seed plants. Wiley Eastern, New Delhi.

Fahn, A. 1989. Plant Anatomy. Pergamon press, Oxford.

Jeffrey, E.C. Anatomy of woody plants.

Mauseth, J.D. 1988. Plant anatomy. Benjamin Cummings, California.

Metcalfe, C.R. and L. Chalk. 1950. Anatomy of Dicotyledons. 2 vols. Clarendon press, Oxford.

Pandey, B.P. 1993. Plant anatomy. S. Chand and Co, New Delhi.

Pullaiah, T., K.C.Naidu, K.Lakshminarayana and B. Hanumantha Rao. 2012. Plant development. Regency Publications.

Rudall, Paula. 1990. Anatomy of flowering plants.

Steeves, T.A. and I.M.Sussex. 1989. Patterns in Plant development (2nd ed.). Cambridge Univ. Press, Cambridge.

Chapter 6

Complex Tissues

The tissues that are composed of more than one cell type are generically referred to as the complex tissues. The complex tissues are heterogeneous in nature, being composed of different types of cells performing diverse functions. Xylem and phloem are the two most important complex tissues in a plant, which constitute the major part of vascular tissues (in most woody plants, these are produced by the vascular cambium and is often referred to as the "vascular tissues"), whose primary functions include the transport of water, ions and soluble food substances throughout the plant. The vascular bundles form a continuous and interconnected system in different organs of the plant. The term – "vascular plants" – has been used since a long time, and in recent years, a new phylum – Tracheophyta – has been introduced to include all the vascular plants, covering the pteridophyta and spermatophyta of the older classification.

Xylem

This is an important tissue of the plant as it is a part of plumbing. "Xylem" (Gr.: *Xylos* = wood) is a complex tissue. It is the principal water conducting tissue (in addition to dissolved substances) in the vascular plants. It consists of a combination of parenchyma, fibers, vessels, tracheids and ray cells. This tissue is a part of the vascular bundle and is instrumental for conduction of water and solutes, and also for the mechanical support. It is usually spatially associated with phloem, the principal food-conducting tissue. The two tissues together are called the "vascular tissue" (or tissues). The combination of xylem and phloem forms a continuous vascular cylinder throughout all parts of the plant including all branches of stem and root. The xylem is present in all the land plants, *viz.*, Pteridophytes, Gymnosperms and Angiosperms. Therefore, all these plants together constitute Tracheophyta.

The long tubes made up of individual cells, the vessels, which are open at each end. Internally the bars of wall material extend across the open space. The cells are joined end to end and result in long tubes. Both the vessels and the tracheids are dead at maturity. The secondary cell walls of the tracheids are thick and are tapered at the ends as such they do not have opening at their ends. The ends of the tracheids overlap with each other with pairs of pits which allow water to pass from cell to cell. Most of the conduction in the xylem is up and down, some side-to-side conduction (lateral conduction) occurs via the rays (the horizontal rows of long-living parenchyma cells that arise out of the vascular cambium). In trees and woody plants, ray will radiate from the centre of the stems and roots and in their cross section will appear like the spokes of a wheel.

The primary xylem originates from the procambium of the apical meristem and the secondary xylem from the vascular cambium. The xylem tissue consists of both living and non-living cells. Xylem is a complex tissue as it consists of several types of cells. The most important cells are the tracheids and tracheae (or vessels), fibres (xylem or wood fibres) and parenchyma (xylem or wood parenchyma), and xylem rays. Of these, only the xylem parenchyma and xylem rays are living and the rest are dead cells at maturity. Xylem originates from two sources, (i) Procambium of the apical meristem, which gives rise to the primary xylem of the primary body, and (ii) Vascular cambium, which gives rise to the secondary xylem. Earlier, a term - "hadrome" was used for the xylem, and included all the elements except the fibres.

Tracheary Elements

The tracheids and vessel members together represent the tracheary elements. They are concerned with the conduction of water and substances dissolved in water. They are more or less elongated cells non-living at maturity. They have lignified walls with secondary thickenings and a variety of pits. The tracheids and vessel members differ from each other in that the tracheid is an imperforate cell whereas the vessel member has perforations, one or more at each end, sometimes also on a side wall.

Cell Wall of a Tracheid

Cell wall of the tracheid is moderately thick and lignified. The secondary wall layers are deposited in different manners, which may be annular, spiral, scalariform, reticulate and pitted.

1. Annular

The lignin is deposited in the form of rings all along the walls almost at regular intervals (Figure 6.1 G).

2. Spiral/Helical

The lignin is deposited in the manner that appears spiral/helical in length. The helical thickenings may be single or more than one helix in a single element. The rings or helices may be arranged in a loose or dense manner (Figures 6.1 E, F, H, I).

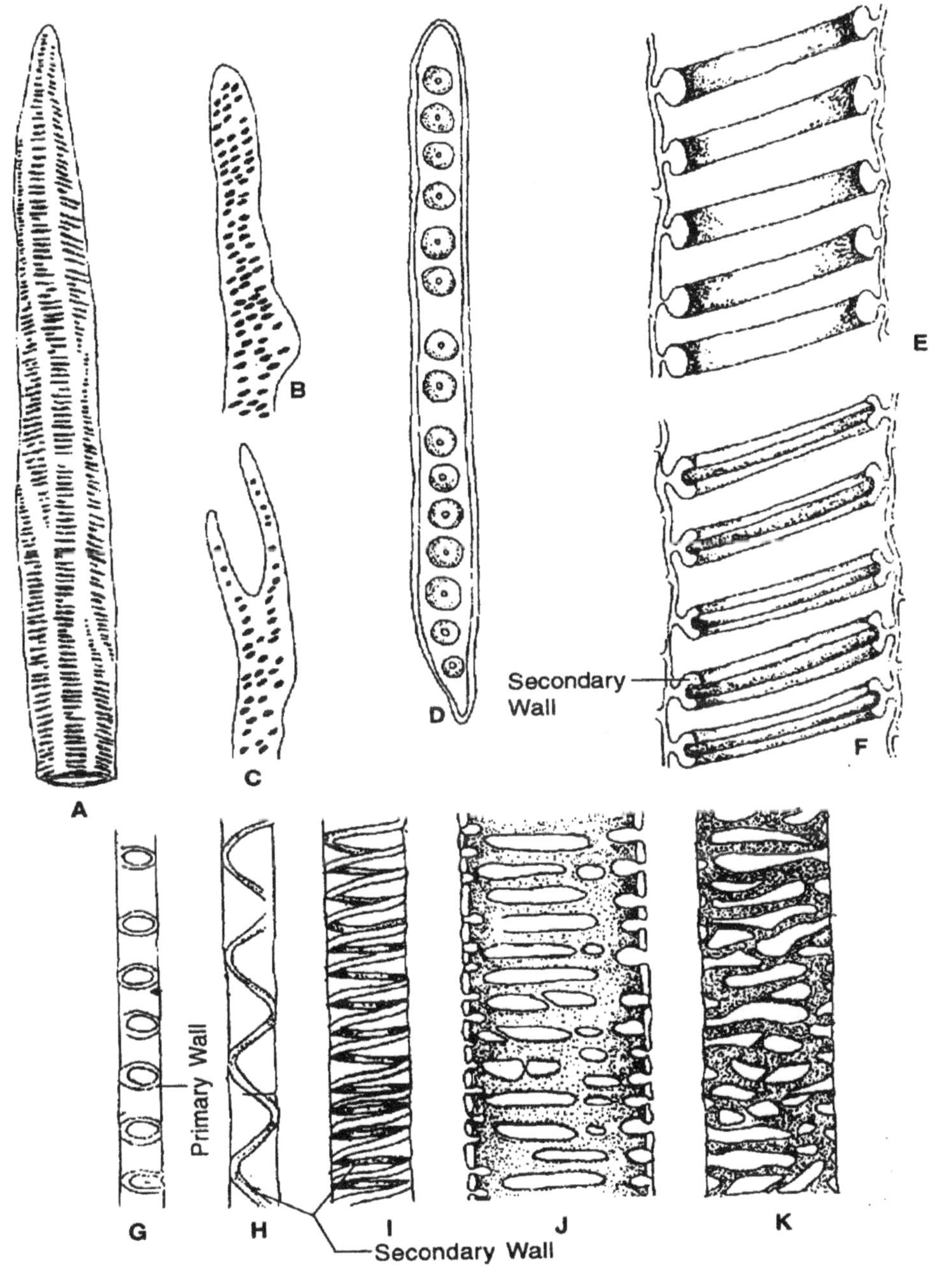

Figure 6.1A-J: A. Tip of tracheid of *Dryopteris*; with scalariform pitting; B, C. Tips of tracheids of *Kingia*; D. Tracheid of *Pinus*; E. Portion of a longitudinally sectioned tracheary element showing the helical wall thickenings and the strips by which they are joined to the primary wall; F. As in E but in which the helical thickening is deeply grooved; G-K. Different types of wall thickening in tracheary elements; G. Annular; H. Helical; I. Dense helical; I. Scalariform; J. Reticulate thickening.

3. Scalariform

The lignin deposition is in the form of closely arranged transverse bands, which give an appearance of a ladder (Figure 6.1 J).

4. Reticulate

The wall thickening is in the form of a network/reticulum (Figure 6.1 K).

5. Pitted

The entire inner surface of the cell wall is thickened except for certain un-thickened circular areas which look-like pits (Figure 6.1 A-D). The nature of pits may be variable. Of the above, the first formed ones are annular, spiral and reticulate thickenings and are found in the protoxylem. Pitted elements are characteristic of the metaxylem and secondary xylem. However, all these elements need not be present in the same plant. In addition, intermediate forms are also present.

Pitting

Transport of water and solutes takes place, in adjacent tracheids, through the pits. The pits are seen between two neighbouring tracheary elements. Between tracheary elements and fibres, very few or no pit connections exist. Tracheary elements contact with a xylem parenchyma through a half-bordered pit pair, *i.e.*, bordered pit on the tracheid side and a simple pit on the side of parenchyma.

On the basis of arrangement of pits on the tracheary elements, three patterns are discernable.

1. Scalariform Pitting

The pits are transversely elongated and arranged in longitudinal rows.

2. Opposite Pitting

The pits are circular and elliptical and arranged in horizontal rows.

3. Alternate Pitting

The pits are circular or oval and are arranged in oblique or diagonal rows. Of these, the scalariform pitting is the most primitive and found commonly in lower vascular plants (Pteridophytes).

Tracheids

In tracheids, the passage of water from cell to cell occurs mainly through pit-pairs. The term - '*tracheid*' was introduced in 1863 by Sanio. A tracheid is a very much elongated cell that occurs along the long axis of the organ. The cells are devoid of the protoplast and hence are dead. The cell has a fairly large cavity (or lumen) without any contents and tapers blunt (or chisel like ends). The end walls of the tracheids do not uniformly taper in all planes. The cells are round or polyhedral in their cross section. The tracheids are the most primitive and the fundamental cell types in the xylem of plants from their phylogenetic point of view. The wood of the ancient vascular plants was exclusively made of the tracheids alone, and this is the only type of element that is found in the fossils of the seed plants. These practically occur in the modern plants in all groups including the angiosperms though they predominate in the lower vascular plants such as the pteridophytes and gymnosperms. Vessels have evolved from the tracheids. The wall of the tracheid is hard, moderately thick and usually lignified. The secondary walls are deposited in different manners as such they may be annular, spiral, reticulate, scalariform or pitted. The bordered pits are more abundant through which these establish the communication with the adjoining tracheids and also with other cells, either living or non-living. The nature of the pits is variable. In lower vascular plants, the pits are elongated and give the scalariform appearance. In gymnosperms and the angiosperms, they have round pits with well developed borders. The tracheids occur both in the primay and secondary xylem. The tracheids are well adapted for the transport of water and the solutes and also serve as the supporting tissue due to the presence of the central lumen and hard lignified walls.

Vessels or Trachea

Longitudinal series of vessel members interconnected through their perforations are called vessels (trachea). The cells are non-living at maturity. The perforated part of a wall of a vessel member is called "perforation plate". A plate may be simple, with only one perforation, or multiperforate, with more than one perforation. The multiperforate plates are scalariform (if the perforations are elongated and arranged parallel to each other) and reticulate (if the perforations form a net-like pattern). Vessels are well suited for the conduction of water and solutes. The vessels are distinctly perforate and thus the solute translocation becomes easier in a straight line. The perforation plates are of two types – multiple plates and simple plates. In the primitive plants, the perforations remain scalariform like bars in parallel series, or in the form of a network called the reticulate or even as a group of circular holes (called foraminate perforation). In the advanced plants, the dissolution of the end walls is complete and the perforation is in the form of a single large circle (simple perforation). Similar to tracheids, these are also devoid of the protoplast and have hard, lignified cell wall with different types of thickenings. Some are intermediate between typical tracheids and vessels, called vessel tracheids (similar to fibre tracheids). The secondary walls with different patterns of deposits like – ring-like, spiral, scalariform, reticulate or pitted (bordered pits).

The vessels have originated from the tracheids phylogenetically and occur in the pteridophytes like *Pteridium* and *Selaginella*, and in higher gymnosperms (Gnetales) and in the dicots and monocots. It has been suggested that the development of the vessels was by parallel evolution (Cheadle, 1953), and these first appeared in the secondary xylem and then proceeded towards the formation of protoxylem. These are absent in dicots belonging to Winteraceae, Trochodendraceae and Tetracentrales, and do not occur in some xerophytes, parasites and aquatic plants. In monocots, the vessels first appeared in the roots and later extended to the aerial organs (Fann, 1954). The vessels are the most important elements of the xylem for easy transport of water and the solutes and also for mechanical support.

Differentiation of Tracheary Elements

The vessels arise from procambial (or vascular cambial) cells in primary and secondary xylem respectively. During the stage of cell growth and secondary wall deposition the living protoplast is concerned with the phenomenon of differentiation (Figure 6.2 A, B). The protoplast contains the full complement of organelles including a nucleus and tonoplast-bound vacuoles. The nucleus becomes polyploid and increases in size. The endoplasmic reticulum is seen in extensive profiles along and especially between the secondary wall thickenings. The microtubules are prominently displayed during the growth of cell wall. At first, they are spread evenly along the entire wall, but later, get concentrated in bands at the sites of secondary thickenings. After the secondary wall is deposited, the cell enters the stage characterized by the phenomena of lysis affecting the protoplast and certain parts of the cell wall. The cytoplasm is exposed to the hydrolases possibly through rupture of the tonoplast. The hydrolases also reach the cell walls and attack the primary wall parts that are not covered by lignified secondary wall

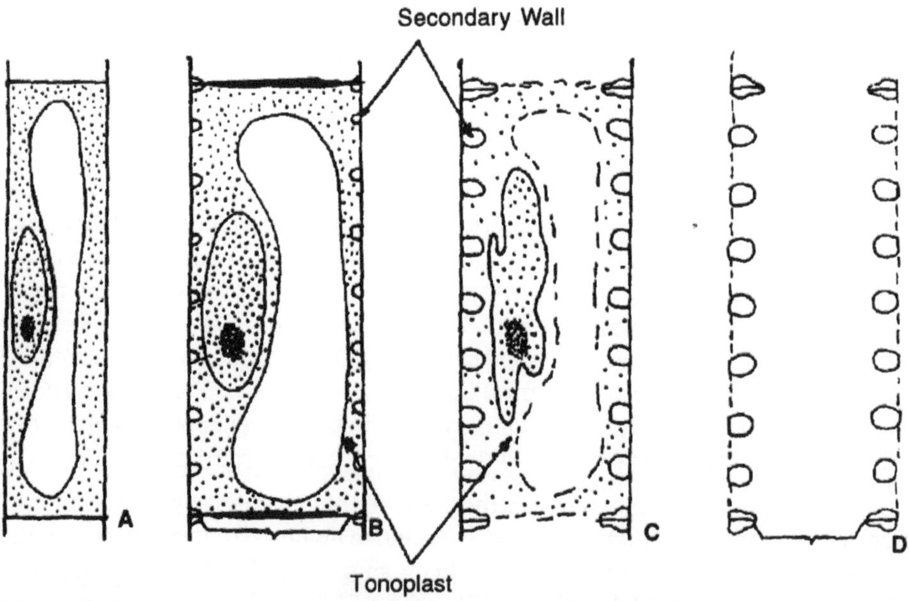

Figure 6.2: Diagrams Illustrating Development of a Vessel Member with a Helical Secondary Thickening. A. Cell without secondary wall; B. Cell has attained full width, nucleus has enlarged, secondary wall has begun to be deposited, primary wall at the pore site has increased in thickness; C. Cell at stage of lysis; D. Mature cell without protoplast.

layers. The lateral walls are partly digested (Figure 6.2 C, D). The removal of non-cellulosic components leaves a fine network of cellulose microfibrils. In the sites of perforations, the entire primary wall disappears. The wall occupying the site of the future perforation is thicker than the primary wall elsewhere. The removal of this wall appears to be gradual dissolution indicated by the appearance of small holes before the final breakdown.

Pits

The simple and bordered pits are encountered in the secondary walls of tracheids and vessel members. The number and arrangement of these pits are highly variable, even on different wall facets of the same cell, because they depend on the type of cell contiguous with the particular wall facet. Usually, numerous bordered pit-pairs occur between contiguous tracheary elements, few or no pit-pairs occur between tracheary elements and fibres; half-bordered or simple pit-pairs are found between tracheary elements and parenchyma.

The pit membrane has a torus in the centre. It is surrounded by the thin part of the membrane, the "margo". The torus is wider than the aperture in the border and consequently blocks the aperture when the pit-pair is aspirated. Aspirated bordered pit-pairs are common in the heart-wood. The bordered pits thus act as valves preventing the entry of air into conducting tracheids, which would disrupt the water column in the latter.

Xylem Fibres

A typical fibre differs from a tracheid in the more pronounced thickening of the wall and correspondingly much smaller lumen, as well as a reduction in the size of the pits. An intermediate type is called a fibre tracheid which is found in some plants. These have smaller pits with reduced (or vestigial) borders and in some cases, the protoplast persists up to the mature stage. Some may even divide as such the transverse partition walls are noticed within the original wall (septate fibre tracheids).

The xylem fibres (wood fibres) are mainly of two types – "fibre tracheids" and "libriform fibres". These are intergrade and as such it becomes difficult to demarcate them. The former are intermediate between typical fibres and tracheids – they have bordered pits but the borders are not well developed. The latter are narrower with a highly thickened secondary wall. The central lumen is almost obliterated and the pits are simple which resemble phloem fibres. These are abundant in many woody dicots.

Xylem fibres are found both in the primary and secondary xylem. The fibre-tracheids have bordered pits with cavities smaller than the pit cavities of tracheids or vessels in the same wood. These pits have a pit canal with a circular outer aperture and an elongated or slit like inner aperture. The pit in a libriform fibre has a slit-like aperture towards the cell lumen and a canal resembling a much flattened funnel, but no pit cavity. Other details of xylem fibres have been given earlier.

Phylogenetic Specialization of Tracheary Cells and Fibre Cells

The tracheary elements have developed during the evolution of land plants (Bailey, 1953). In the lower vascular plants, the function of conduction and support were combined in the tracheids. With an increasing specialisation, woods evolved with conducting elements—the vessel members being more efficient in conduction than in providing mechanical support. On the other hand, fibres evolved as principal supporting tissue. Thus, from the primitive tracheids, two lines of specialisation diverged—one toward the vessels and the other toward the fibres.

Basis for the Support of the Evolution

The following structural features may be taken as the basis in support of the evolution of the tracheary elements from primitive tracheids, usually long imperforate cells with a small diameter, angular in cross-section, having lignified scalariform pitted walls.

1. The primitive vessels are also elongate bodies like the tracheids with rather small diameter and tapering ends. Similar condition is still noticed in lower dicotyledons. With evolutionary advance, they gradually became shorter and wider, often becoming drum-shaped in appearance.
2. The wall of the primitive tracheid is rather thin, more or less of equal thickness, and angular in cross-section. The same condition prevails in primitive vessels. With progressive advance, considerable thickening appeared and the vessels became circular or nearly so in cross-section.
3. In the primitive vessels, the perforation plates are multiple, usually scalariform with numerous bars, and oblique end-walls. Progressive increase in specialisation led to gradual decrease in the number of bars and their ultimate disappearance, so that the perforation plates became simple with transverse end-walls. These are positively advanced characters.
4. The pitting of the vessel wall also changed from early scalariform arrangement, (characteristic of tracheids) to small bordered pit pairs, first in opposite (arranged in transverse rows) and ultimately in alternate (arranged spirally or irregularly) pattern. Moreover, the pit pairs between vessels and parenchyma changed from bordered to half-bordered and then to simple.

In the specialisation of the xylem fibres adapted for more efficient support, there has been a steady increase in thickness of the wall leading to decrease in cell-lumen. The pits changed from elongate to circular, the borders becoming reduced and functionless, and ultimately disappeared. Thus, the evolutionary sequence was from tracheids, through fibre-tracheids to libiriform fibres.

The vessels evolved independently in several taxa of vascular plants. In the primitive Angiospermous families such as: Winteraceae (all the seven genera – *Drimys, Pseudowintera, Bubbia, belliolum, Exospermum, Zygogynum* and *Tetrathalamus*), Trochodendraceae, Tetracentraceae, Monimiaceae and Chloranthaceae have no vessels in their xylem. The vessel members occur in the following diverse group of plants: in the most advanced Gymnosperms like the Gnetales, in Dicotyledons except for the lower most taxonomic groups (mentioned above), in Monocotyledons, in the genera *Pteridium*, in certain species of *Selaginella*, in *Equisetum* and in the root of some species of *Marsilea*. From the above, it can be assumed that vessels developed independently, by parallel evolution, in each of these groups. A few Monocotyledonous families with only aquatic species are known to include plants that lack vessels completely in all their organs, and this feature, however, may be secondary one. In some specialized Dicotyledons, such as in certain representatives of the Cactaceae, vessels in secondary xylem are replaced by so called vascular tracheids (short, imperforate cells with annular and helical thickenings). However, in such plants, the reduction of vessels cannot be cited as support for evolutionary reversal to primitive vessel-less condition.

The evolutionary sequence of vessel member types in the secondary xylem of Dicotyeldons began with long scalariform pitted tracheids similar to those still found in some lower dicotyledons. These tracheids, were succeeded by vessel members of long narrow shape with tapering ends. The cells shortened progressively, became wider, their end walls became less inclined and finally transverse. In the more primitive state, the perforation pit is scalariform, with numerous bars resembling a wall with scalariformly arranged pits devoid of pit membranes. Increase in specialization resulted in a decrease in the number of bars and finally their total elimination and appearance of a simple perforation. The tracheids were not eliminated when vessels evolved, but underwent phylogenetic changes. They became shorter than – not as short as the vessel members, however – and the pitting of their walls evolved parallel to that of the associated vessel members. They generally did not increase in width. In the specialization of xylem fibres, the emphasis on mechanical function became apparent in the increase in wall thickness and decrease in cell width. Concomitantly, the pits changed from elongated to circular, the borders became reduced and eventually disappeared. The inner aperture of the pit became elongate and then slit-like.

Ecological Aspects of Xylem Evolution

According to Carlquist (1975) who has put forward an ecological approach to xylem evolution, the structural diversity in the xylem of living plants at present, is a result of evolutionary changes, which are functionally adaptive to the various habitats. The adaptive changes were brought about through selective pressure in different environments in which the various plant taxa evolved. Carlquist expects the adapation to the anatomical features of xylem to be related to moisture availability, transpiration and to requirements for mechanical strength. Carlquist also considers that great length of vessel members with scalariform perforation plate reduces resistance to water flow. Short and narrow vessel members with thick walls would increase the strength of a vessel and make it resistant to high negative pressure which occurs in xeric habitats. Scalarifrom perforation plates are regarded as functionally disadvantageous as compared to simple perforation plates, because of resistance to water flow. Xylem/wood with scalariform perforation is rare in plants of xeric habitats and has not been found in plants growing in deserts. Vines and lianas, with a requirement for efficient water conduction, possess exceptionally wide vessels, causing minimal friction to water flow. The vessel members in these plants are relatively short, apparently enhancing resistance to negative pressure. The number of vines and lianas with scalariform perforation plates is very small.

Xylem Parenchyma

Parenchyma cells are common constituents of the xylem of most plants. The parenchyma, found in secondary xylem, is classified into two types.

(a) Axial or Wood Parenchyma

This is derived from the fusiform cambial initials together with tracheary elements and fibres. These are somewhat elongated and lie in vertical series.

(b) Ray Parenchyma

These originate from ray initials of fusiform cambium and occur in radial transverse series. The cell wall of parenchyma found in primary xylem is thin and made up of cellulose but those found in secondary xylem may often have thick, lignified and pitted cell wall.

Parenchyma is abundant in the secondary xylem of most of the plants, excepting a few conifers like *Pinus, Taxus* and *Araucaria*. These are the only living cells in xylem. The cells may be thin-walled or thick-walled. If lignified secondary wall is present, the pit-pairs between the cells and the adjacent xylem element may be bordered, half-bordered or simple. Between two parenchyma cells, the pit is obviously simple.

Xylem parenchyma cells store starch, oils and many other ergastic substances of unknown function. Sometimes, tannins and crystals may also be found in them.

In radial parenchyma, cells are divided into two categories – procumbent and upright ray cells. A "ray cell", the longest-diameter of which is oriented radially, is "procumbent"; the cell elongated axially, is upright (Figure 6.3 A, B). The two kinds of ray cells are often combined in the same ray, the upright cells typically appearing at the upper and lower margins of the ray. Rays composed of one kind of cells are called "homocellular" (Figure 6.3 C, D), those containing procumbent and upright cells, "heterocellular" (Figure 6.3 A, B). The main function of parenchyma, both axial and ray, are storage and translocation of ergastic substances.

Protoxylem and Metaxylem

Primary xylem consists of an earlier part, the protoxylem and the later part, the metaxylem. The protoxylem differentiates in the parts of the primary plant

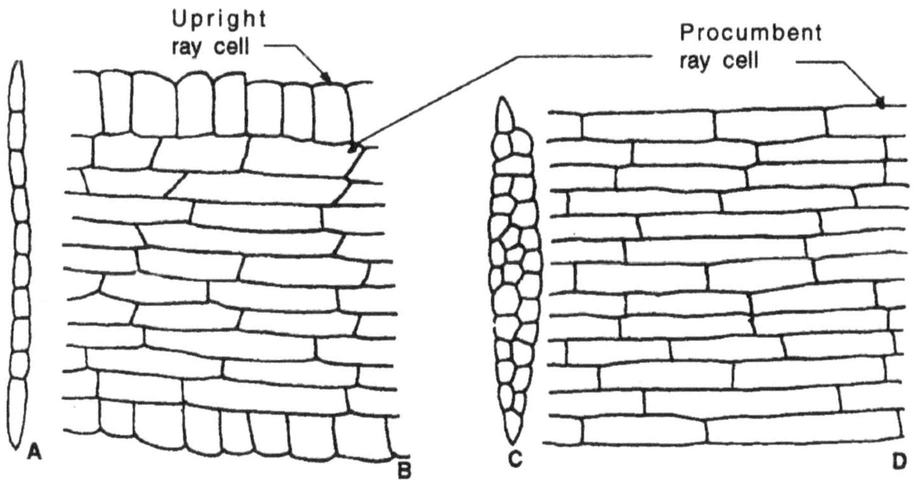

Figure 6.3: Two Types of Rays as Seen in Tangential (A, C) and Radial (B, D) Sections. A, B. *Fagus grandifolia*; C, D. *Acer saccharum*.

body that have not completed their growth and differentiation. In fact, in the shoot, the protoxylem matures among actively elongating tissues and is, therefore subjected to stresses. Its mature nonliving tracheary elements are stretched and eventually destroyed. The metaxylem is commonly initiated in the still growing primary plant body, but it matures largely after the elongation is completed. It is therefore, less affected by the primary extension of the surrounding tissues than the protoxylem. The protoxylem usually contains only tracheary elements embedded in parenchyma that is considered to be part of the protoxylem. The metaxylem is somewhat more complex than the protoxylen and may contain fibres in addition to tracheary elements and parenchyma cells

Phloem

Phloem is an equally important complex tissue as it is also a part of the plumbing of a plant, and primarily carries the dissolved food substances throughout the plant. Phloem is the principal food conducting tissue. The term "phloem", was first coined by Nageli (1858) from the Greek word for the 'bark'. The phloem was also referred to as 'bast' or 'leptome'. Similar to the xylem, the phloem originated from two sources, (i) Procambium of the apical meristem, which gives rise to primary phloem of the primary body, and (ii) Vascular cambium, which gives rise to the secondary phloem.

The phloem is composed of sieve elements (sieve cells, sieve tubes) and companion cells (without secondary walls), phloem parenchyma and phloem fibres.

Sieve Elements

Sieve cells and sieve tube members are the two types of sieve elements. These are primarily involved with the conduction of organic solutes.

Sieve Cells

These are narrow and linear with the pointed end walls. The sieve cells are joined end to end wherein their inclined end walls overlap. These are found among the members of lower vascular plants and Gymnosperms (Figure 6.4 A). The sieve cells, which may be compared to the tracheids, are narrow elongated cells without conspicuous sieve areas. They usually have greatly inclined walls, which overlap in the tissue, sieve areas being more numerous in the ends. Sieve cells are more primitive than the sieve tubes. They occur in lower vascular plants and gymnosperms. In fact, sieve tubes have evolved from the sieve cells, as vessels have evolved from the tracheids, and so sieve tubes occur in all angiosperms. In monocotyledons, unlike the xylem elements, sieve tubes first appeared in the aerial organs, the course being from the leaves to the stem and, lastly, to the roots. In spite of close ontogenetic resemblance between the tracheary elements of xylem and sieve elements of phloem, the latter unlike the former, are living.

Sieve Tubes

Sieve tube is a series of sieve tube members connected end to end in vertical rows, by means of sieve plates.

Sieve Areas and Sieve Plates

The most characteristic features of sieve elements are sieve areas in the walls and the disappearance of nucleus from the protoplast. The sieve areas are wall areas with pores through which the protoplasts of vertically or laterally adjoining sieve elements are interconnected. The pores range in width from those that are comparable to plasmodesmatal pores and are several microns in diameter. In the lower vascular plants (Pteridophytes) and Gymnosperms, the pores are commonly narrow and rather uniform in size in the sieve areas on different walls of the same cell (Figures 6.4 A, 6.5 E). In the Angiosperms, the size of pores varies considerably even on different walls of the same cell. Sieve areas with larger pores usually occur on the end walls, occasionally on the side walls. The wall parts bearing the more highly differentiated sieve areas, that is, areas with comparatively large pores are called "sieve plates". Each pore is lined with callose, a carbohydrate (β-1, 3-glucan), that stains blue with aniline blue. As the sieve element ages, more callose accumulates. The layer with in a pore thickens and some callose also appears on the surface of the sieve area. The pore is increasingly more constricted and later completely obliterated as the sieve element becomes dormant and dies. At this stage, the callose forms a pad on the sieve area.

Due to its formation, the cell to cell communication is considerably cut down or entirely prevented. The callose pad is usually formed with the approach of resting or inactive season; and it disappears when the active season (spring) sets in. This type is known as "seasonal callose" or "dormancy callose". In old functionless sieve tubes, callose becomes permanent, what is called "definitive callose". Though the term definitive callose is often used to designate the former type, it is desirable to confine it to permanent callose of old and functionless tubes.

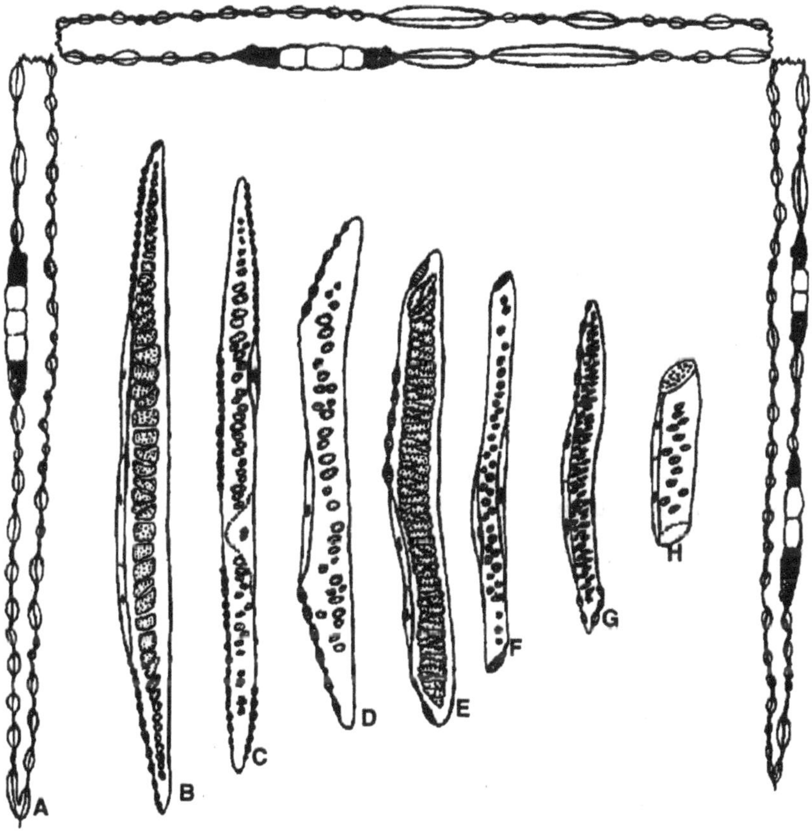

Figure 6.4: Variation in Structure of Sieve Elements. A. Sieve cell of *Pinus pinea* with associated rays, as seen in tangential section. Others are sieve-tube members with companion cells from tangential section of phloem of the following species; B. *Juglans hindsii*; C. *Pyrus malus*; D. *Liriodendron tulipifera*; E. *Acer pseudoplatanus;* F. *Cryptocarya rubra*; G. *Fraxinus americana*; H. *Wisteria* sp. in B-G, the sieve plates appear in side views and their sieve areas are thicker than the intervening wall regions because of deposition of callose.

A sieve plate is called simple, if it has only one sieve area, whereas the plate may be compound with several sieve areas arranged in scalariform, reticulate or otherwise. From evolutionary point of view, simple sieve plates on transverse end-walls are more advanced than compound plates on oblique walls in their character. The cylindrical cells which take part in the formation of the sieve tube are called sieve tube elements. Like vessel elements, the sieve tubes have also undergone decrease in length with evolutionary advance.

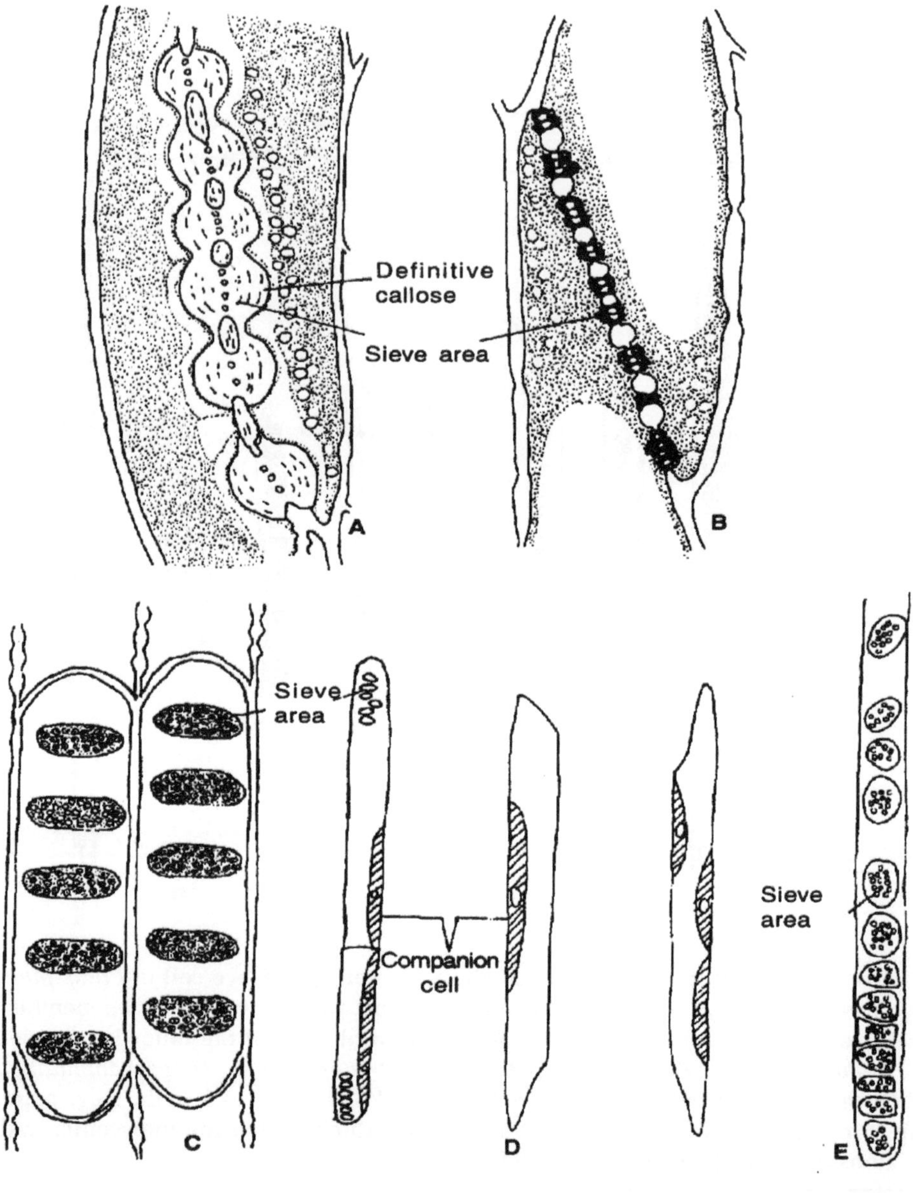

Figure 6.5: Sieve-tube Members of *Vitis*. A, B. Longitudinal sections of compound sieve plates between two elements. A. Elements of dormant state in which the plate is covered by a thick layer of callose; B. Elements reactivated after the removal of the callose. The slime which fills the sieve areas is indicated by heavy stippling; C. Surface view of the compound sieve plates; D. Sieve-tube elements with companion cells; E. Portion of a sieve cell of *Pinus*.

The callose is synthesized remarkably fast in response to injury. The deposition of callose increases during inactive season while it decreases during the active season (spring time). Sieve areas and sieve plates of Angiosperms vary widely in differentiation and arrangement (Figures 6.4 B-H, 6.5 A-D). Long sieve tube members with much inclined end walls commonly have compound sieve plates, that is, sieve plates composed of several sieve areas. The ends of such elements are wedge-shaped and the sieve areas are borne on the oblique face of the wedge. The arrangement of the multiple sieve areas may be scalariform or less orderly.

Cell Wall

The walls of the sieve element vary in thickness but are usually distinctly thicker than those of the surrounding parenchymatic cells. The cell walls have a pearl lustre appearance, hence they are called "nacreous". The wall has interconnected microfibrils in parallel orientation. The walls of sieve elements are usually only primary and consist mainly of cellulose and protein but even the greatly thickened walls are not lignified.

Protoplast

The phloem is unique tissue with many intriguing characteristics. One important feature of the mature sieve element is that its nucleus usually degenerates (Figure 6.6). The degenerated nucleus may either remain as a collapsed body or disappears completely. As the maturation of the sieve element takes place, endoplasmic reticulum (ER) diminishes in amount and the ribosomes and dictyosomes are not discernible. Though plasmalemma continues to function, the tonoplast breaks down and no distinction is maintained between the cytoplasm and the vacuole. Only plastids and mitochondria remain in the sieve elements. The plastids occur in two basic types, one with little internal organization but accumulating starch (s-type); the other containing protein as fibrils or crystals (p-type). Plastid differences in sieve elements are taxonomically useful (discussed elsewhere in this book). The sieve elements commonly form a proteinaceous inclusion, the p-protein, formerly called "slime". This protein often appears in filaments of tubular form, with subunits arranged helically. Initially the p-protein is aggregated into one or more bodies but subsequently it spreads out in the cytoplasm forming strands or networks.

Differentiation of Sieve Plate

In young sieve element, the sieve area of the incipient sieve plate is penetrated by typical plasmodesmata (Figure 6.6 B, C) with which endoplasmic reticulum cisternae are associated. These plasmodesmata mark sites of future pores. The endoplasmic cisternae located on the two sides of a pore site are interconnected by a desmotubule extending through the plasmodesmatal canal. While the protoplast is still intact, callose appears around each plasmodesma except in the middle lamella region. The paired callose masses assume the form of platelets or cones interrupted in the centre where the plasmodesmata are located. When the nucleus degenerates plasmodesmatal canals enlarge. As the pores attain their final size, the endoplasmic reticulum cisternae are removed from the pore site (Figure 6.6 D). The plasmalemma lining the original plasmodesmatal pore remains. When the plant is cut, the contents in the sieve tube readily flow out by the pressure that exists in the mature sieve element. The exudate contains some protoplast components, especially p-protein. Some of the p-protein remains lodged within the pores and forms accumulations (slime plugs) on the sieve plate.

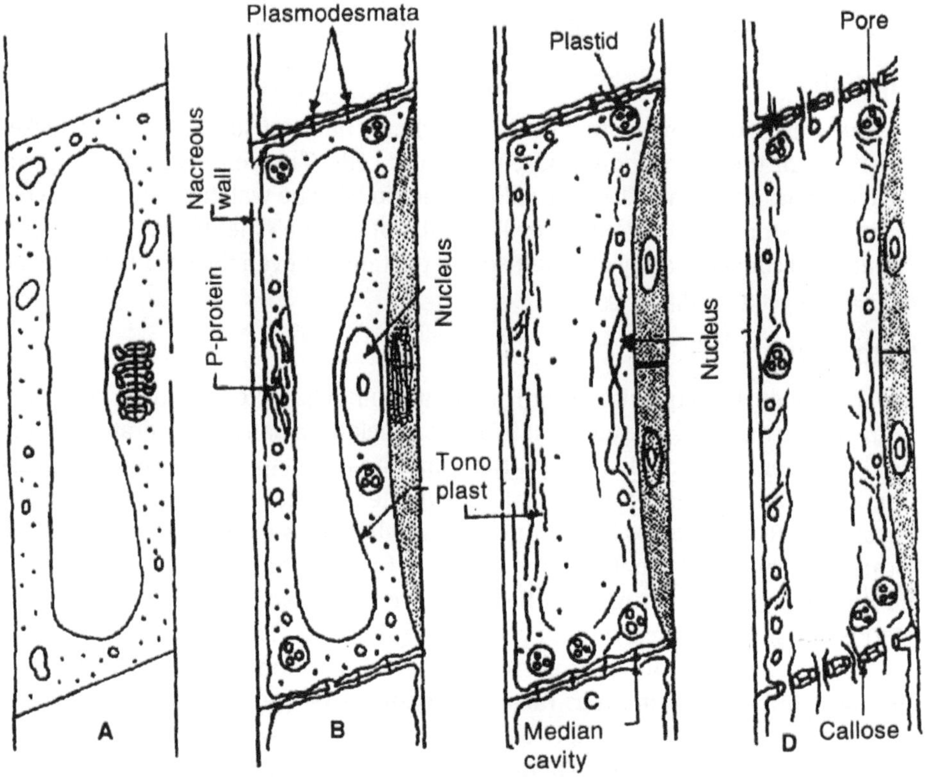

Figure 6.6: Diagrams Illustrating Differentiation of a Sieve Element. A. Precursor of sieve element in division; B. After division: Sieve element with nacreous wall and P-protein body; dividing companion cell precursor (stippled); C. Nucleus degenerating, tonoplast partly broken down, P-protein dispersed; median cavities in future sieve plates; two companion cells (stippled); D. Mature sieve element; pores in sieve plate open; they are lined with callose and contain some P-protein. No endoplasmic reticulum is shown.

Phylogeny of Sieve Elements

In the most primitive form, the sieve elements are parenchyma cells that have undergone modifications in connection with their function. This was followed by the loss of nucleus. This protoplasmic specialization apparently resulted in the development of the interdependence of the sieve elements and parenchyma cells that retain their nucleus, *i.e.*, the albuminous cells in Gymnosperms and the companion cells in Angiosperms. In the Angiosperms, the sieve elements and the companion cells develop from the same mother cell. The specialization of the sieve elements is also involved the development of thick connecting strands. In Pteridophytes and Gymnosperms, these strands are thin and resemble plasmodesmata, while in Angiosperms they are thick and conspicuous.

The specialization of sieve elements and sieve areas has taken the following course.

1. Gradual localization in highly specialized sieve areas to the end walls of the elements.
2. Gradual change in the position of the end wall from very oblique to horizontal.
3. Gradual change from compound sieve plate to simple ones, and
4. Gradual reduction of the sieve areas on the side walls of the elements.

Sclerenchyma Cells

Fibres are common components of both primary and secondary phloem. In the primary phloem, fibres occur in the outermost part of the tissue, while in the secondary phloem, in various distributional patterns among the other phloem cells of the axial system. The fibres may be non-septate or septate and may be living or non-living at maturity. The living fibres serve as storage cells as they do in the xylem. In many species, primary and secondary fibres are long cells with thick walls and are used as a commercial source of fibre (*Linum*, *Cannabis*, *Hibiscus*, and *Corchorus*).

Sclereids are frequently found in the phloem. They may occur in combination with fibres or alone, and they may be present in both axial and radial systems of the secondary phloem. Sclereids typically differentiate in older parts of the phloem as a result of sclerification of parenchyma cells. This sclerification may or may not be preceded by intrusive growth of the cells. During such growth, the sclereids often become branched or much elongated. The distinction between fibres and sclereids is not always sharp, especially if the sclereids are long and slender. The intermediate cell types are fibre-sclereids.

Companion Cells and Albuminous Cells

Companion cells are specialized parenchyma cells with regard to the functional association with sieve tube members regulating the translocation. Companion cells are formed by division of sieve tube precursors so that sieve tube members and their companion cells are ontogenetically related. One or more companion cells may be associated with one sieve tube and they may occur on one or more sides of the sieve tube member. In some taxa, the companion cells appear in vertical files (companion cell strands) as a result of divisions in their precursor. Companion cells vary in size – they may be as long as the sieve tube member to which they are related or they may be shorter. These cells are strongly attached to the sieve tube members from whom they usually cannot be separated even by maceration. The walls between sieve tube members and the companion cells are thin or possess many thin areas. The plasmodesmata on the companion cells are usually the same as that of the sieve-tube member to which they are attached. The protoplasts of the companion cells are those characteristic of metabolically active cells. Their nuclei and nucleoli are relatively large. The cells contain plastids, often differentiated as chloroplasts, numerous large mitochondria, ribosomes and rough endoplasmic reticulum. In most parts of the plant, the sieve elements are usually considerably wider than the companion cells. However, in the minor veins of the leaf, the sieve elements are conspicuously smaller than their associated parenchyma cells. These cells (companion cells or cells not ontogenetically related to the sieve elements) are called "intermediate cells" due to their supposed mediation in the transfer of photosynthates between mesophyll cells and the sieve elements. In the minor veins of many herbaceous Dicotyledons the intermediary cells were found to develop wall ingrowths. Such cells are termed as "transfer cells".

In Gymnosperms, companion cells do not occur. The functional counter-parts of companion cells in Gymnosperms are "albuminous cells" (Strasburger cells). They are rarely derived from the same precursor as the sieve cell. The albuminous cells can be distinguished from the other parenchyma cells of the phloem by the presence of connections with the sieve cells and the usual absence of starch. It was found that high respiration and acid phosphatase activities occur in the albuminous cells associated with fully differentiated sieve cells. The increased activity is restricted to those periods during which loading and unloading of sieve cells is taking place. They die in natural course when the sieve cells become functionless. Thus, the relation between sieve cells and albuminous cells is similar to that existing between sieve tubes and companion cells, excepting that they have no common origin.

Primary Phloem

The primary phloem is classified into protophloem and metaphloem. The protophloem matures in plant parts that are still undergoing extension growth and its sieve elements are stretched and soon become non-functional. Eventually, they are completely obliterated. The metaphloem differentiates later and in plant without secondary growth, constitutes the only conducting phloem in adult plant parts.

The protophloem sieve element of Angiosperms is usually narrow and inconspicuous, but they are enucleate and have sieve areas with callose. They may or may not have companion cells. They appear either in groups or singly among parenchyma cells or in many Dicotyledons, among conspicuously elongated living cells. In many species, these elongated cells are fibre-primordia. While the sieve elements cease to function and are obliterated, the fibre-primordia increase in length, develop secondary walls and mature as fibres. Such fibres are found on the periphery of phloem region in numerous Dicotyledonous stems and are often called "pericyclic fibres".

Metaphloem has more and commonly wider sieve elements than the protophloem. The companion cells are regularly present in the metaphloem of Angiosperms but fibres are usually absent. Parenchyma cells may become sclerified after the phloem ceases to conduct the solutes. The overall development and structure of phloem tissue is parallel that of the xylem. But, the phloem tissue is less sclerified and less persisting than the xylem tissue. Because of its usual position near the periphery of stem and root, the phloem becomes much modified in relation to the increase in circumference of the axis and is eventually cut off by the periderm. The old xylem, in contrast, remains relatively unchanged in its basic structure.

Chapter 7

Vascular Cambium

Introduction

It is very difficult to overemphasise the importance of the vascular cambium, a lateral meristem of great significance, which produces the secondary xylem and secondary phloem. This is composed of a layer of cells only one cell in thickness, and these cells are meristematic cambial initials from which the cells of the secondary xylem and secondary phloem are derived. The thickness of the vascular cambium varies from around 6 cells during the periods of dormancy to around 14 during the most active periods of growth. Being a meristem, the cambium consists of flattened, undifferentiated cells which possess no defence capabilities, though the cambium can be quickly reprogrammed to produce the cells that are differentiated into traumatic resin ducts.

Occurrence of Vascular Cambium (Main Cambium, Wood Cambium or Bifacial Cambium)

In the stems and roots of many plants, specifically in Dicots (Buttercup and Oak), in Gymnosperms (Pine) and in vascular plants (but not in Monocots where there is no secondary growth), this is the main growth tissue, which produces secondary xylem inwards, towards the pith and secondary phloem outwards, towards the bark. It occurs in the vascular bundles in herbaceous plants; and in woody plants, forms a cylinder of unspecialized meristem as a continuous ring from which the new tissues are grown. A few types of leaves also have a vascular cambium. In Dicots and Gymnosperms, this separates the bark and the wood (in addition to cork cambium). The vascular cambia of the root stock and scion must be aligned so that they can grow together for successful grafting.

Vascular Cambium

The vascular cambium is the meristem that produces the secondary vascular tissues. It is a lateral meristem that develops either as longitudinal strands or as a hollow cylinder. In woody Dicotyledons and Gymnosperms, the primary vascular tissues of stem and root exist only a relatively short period, and their function is taken over by the secondary vascular tissues which are produced by the cambium. In many herbaceous Angiosperms and also in most of the recent vascular plants, cambium is absent or vestigial.

The fascicular cambium (within the vascular bundle) is present in between the primary xylem and primary phloem. During the process of secondary growth, the cells of the medullary rays in a line between the neighbouring vascular bundles become meristematic to form a new inter-fascicular cambium (between the bundles), followed by the joining up of both the fascicular and interfascicular cambia to form a ring, called the "cambial ring", separating the primary xylem and primary phloem. This ring cuts off secondary xylem on inside of the ring and secondary phloem on the outside, thus pushing the primary xylem and primary phloem apart.

Vascular Cambium - Structure

In certain plants, including Monocotyledons, all the cells of the procambium undergo differentiation into primary vascular tissues. In almost all the Dicotyledons and Gymnosperms, a portion of the cambium remains meristematic even after the completion of primary growth and develops into the cambium of the secondary body. The cambium that arises within the bundles of the stem is called "vascular" or "fascicular cambium". The strip of fascicular cambium usually becomes joined by the additional strips of cambium which constitute the interfascicular cambium. The interfascicular cambium is not a continuation of the procambium, but develops from the interfascicular parenchyma. Hence, this is called as the secondary meristem from the point of view of its origin. Thus, a complete cylinder of cambium is developed. In roots of most Dicotyledons and Gymnosperms, the cambial cylinder develops between the primary xylem and phloem, a position that is retained throughout the life of the plant.

The cambium usually consists of two types of initials – 1. The fusiform initials and 2. The ray initials. The fusiform initials are responsible for the development of the vertically or longitudinally oriented elements of an organ like the tracheary elements, fibres, xylem and phloem parenchyma and the sieve elements. The horizontally or transversely (radially) oriented elements of an organ like the cells of the vascular rays, develop from the ray initials.

1. Fusiform Initials

These cells are very long (tall) with tapered ends and are axially oriented. They reach a maximum length of about 8.7 mm and spindle shaped in the old trunks of *Sequoia sempervirens*. These are approximately prismatic cells in its middle and wedge-shaped at their ends, and the tangential sides are wider than the radial sides. One of the interesting features of cambial cells is their intense vacuolation. The walls of these cells possess primary pit-fields with plasmodesmata. These cells and their derivatives constitute the axial system in the cambial zone, while the ray initials form the ray system.

Two patterns are discernible in the arrangement of fusiform initials. When they are arranged in regular superimposed tiers, the cambium is called "storied cambium". This type of cambium is found in *Tamarix, Diospyros* and *Robinia*. When the fusiform initials are irregularly aligned, the cambium is called "non-storied", *e.g.*, *Fraxinus* and *Ulmus*. This condition is apparently intermediate between the storied type and extreme non-storied type with long row of initials, such as that of *Juglans, Salix, Populus* and *Pyrus*. The Gymnosperms have cambium of non-storied type. Herbaceous plants with well-developed secondary growth, for example *Solanum tuberosum* (potato) have short initials that are not storied. The longitudinally oriented elements in an organ such as the tracheary elements, fibres, xylem and phloem parenchyma and the sieve elements, develop from the fusiform initials.

2. Ray Initials

These cells are isodiametric. They give rise to the xylem and phloem rays, which are oriented perpendicular to the long axis of the plant body. One of the interesting features of cambial cells is their intense vacuolation. The walls of these cells possess primary pit-fields with plasmodesmata. The fusiform initials are uninucleate, inspite of their large size. The fusiform and ray initials are alike in their ultrastructure and both have basic content of organelles and membranes, characteristic of the parenchyma cells. Rough endoplasmic reticulum, polyribosomes, Golgi bodies, vesicles, mitochondria and microtubules are usually associated with the active growing cambial cells. In the resting cambial cells storage materials like oil droplets and protein bodies exist.

The cambial initials together with their immediate derivatives form the cambial zone. In a cross section, the cells of the cambial zone appear to be arranged in radial rows. The cells which are positioned on either side of this zone gradually enlarge and acquire the shape and features of the mature elements of secondary xylem and secondary phloem. In the cambial zone, the fusiform initials and their derivatives constitute the axial system and the ray initials – the ray system.

Cambium - Types

Two types of cambium can be distinguished on the basis of the arrangement of the fusiform cells as seen in the tangential section.

1. Storied or Stratified Cambium

In this type, the fusiform initials are arranged in horizontal rows (in tiers or stories) so that their ends are approximately at the same level of the axis (Figure 7.1A). Tangentially, the ends of the cells in axially adjacent stories usually overlap slightly creating a zig-zag pattern. The length of these initials is found to vary between 140 μ and 520μ. This type of cambium is found in *Wisteria, Tamarix* and *Robinia*. The groups of ray initials become taller either by the loss of fusiform initials located between two groups of ray initials, allowing them to fuse; or the fusiform initials convert itself into a row of ray initials. The structural elements extend radially are produced via the ray initials.

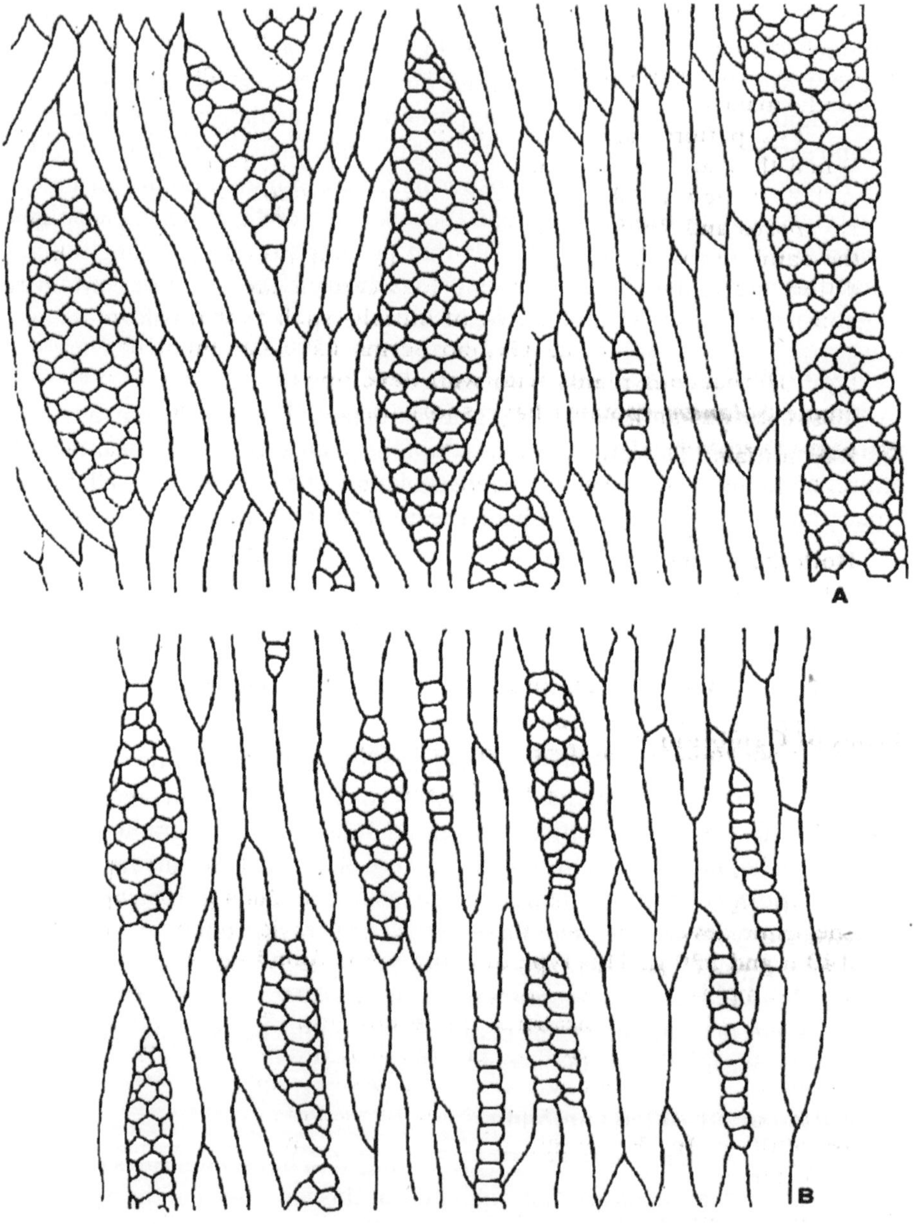

Figure 7.1: Tangential Sections of different Cambial Types. A. Stored cambium of *Robinia*; B. Non-storied cambium of *Fraxinus*.

2. Non-storied or Non-stratified Cambium

In this type of cambium, the ends of cambial fusiform initials typically overlap one another extensively in a seemingly random manner, and there is no lateral alignment (Figure 7.1 B). The initials of this type of cambium are found to vary between 320 and 2300 µm in many Dicotyledons. In vessel-less Dicotyledons, the fusiform initials may reach a maximum length of 6200 µm. In the Gymnosperms, initials of 1000-8700 µm long were reported. In many Dicots, non-storied initials are longer and are of more common occurrence.

Cambium - Divisions

The cambial initials divide periclinally and produce secondary xylem and phloem. At one time a derivative cell is produced towards the xylem, at another time towards the phloem, although not necessarily in alternation. The cambial divisions that add cells to the secondary vascular tissues are called 'additive divisions'.

As a result of the secondary thickening, the circumference of the xylem cylinder increases. Together with this the cambium also increases in circumference by division of cells. The divisions, increasing the number of initials are called multiplicative divisions. In cambia with short fusiform initials, the multiplicative divisions are mostly radial anticlinal (Figure 7.2 A). Thus, two cells appear side by side where one was present formerly and each enlarges tangentially. In herbaceous and shrubby Dicotyledonsm the anticlinal divisions are frequently lateral, that is, they intersect twice the same mother cell wall (Figure 7.2B). Long fusiform initials divide by more or less inclined anticlinal walls (pseudotransverse division) (Figure 7.2 C-E) and each new cell elongates by apical intrusive growth (Figure 7.2 F,G). As a result of this growth the new sister cells come to lie side by side in the tangential plane and they thus increase the circumference of the cambium. Increase in the number of vascular rays is also an important factor contributing to the increase of the cambium cylinder in the circumference. New ray initial are formed from fusiform initials by transverse division of an entire cambium cell or sometimes of part of the cell only.

The multiplicative and additive divisions commonly occur toward the end of the maximal growth concerned with the seasonal production of xylem and phloem. In plants with non-storied cambia, this timing in divisions means that the cambium contains, on an average, shorter fusiform initials at the end of the season than earlier. Subsequently the new cells elongate so that the average length of the initials increases until new period of divisions ensues at the end of the growth season.

The first additions from the initials layer may be made towards either the xylem or the phloem, the variation depending on the plant species. Generally, however, more cells are added to the xylem than the phloem. Cell production toward the xylem is about four times to that toward the phloem. A greater difference was observed in a conifer. In a vigorously growing *Thuja occidentalis*, 12 to 16 new cells were formed on the phloem side and 100 or more on the xylem side.

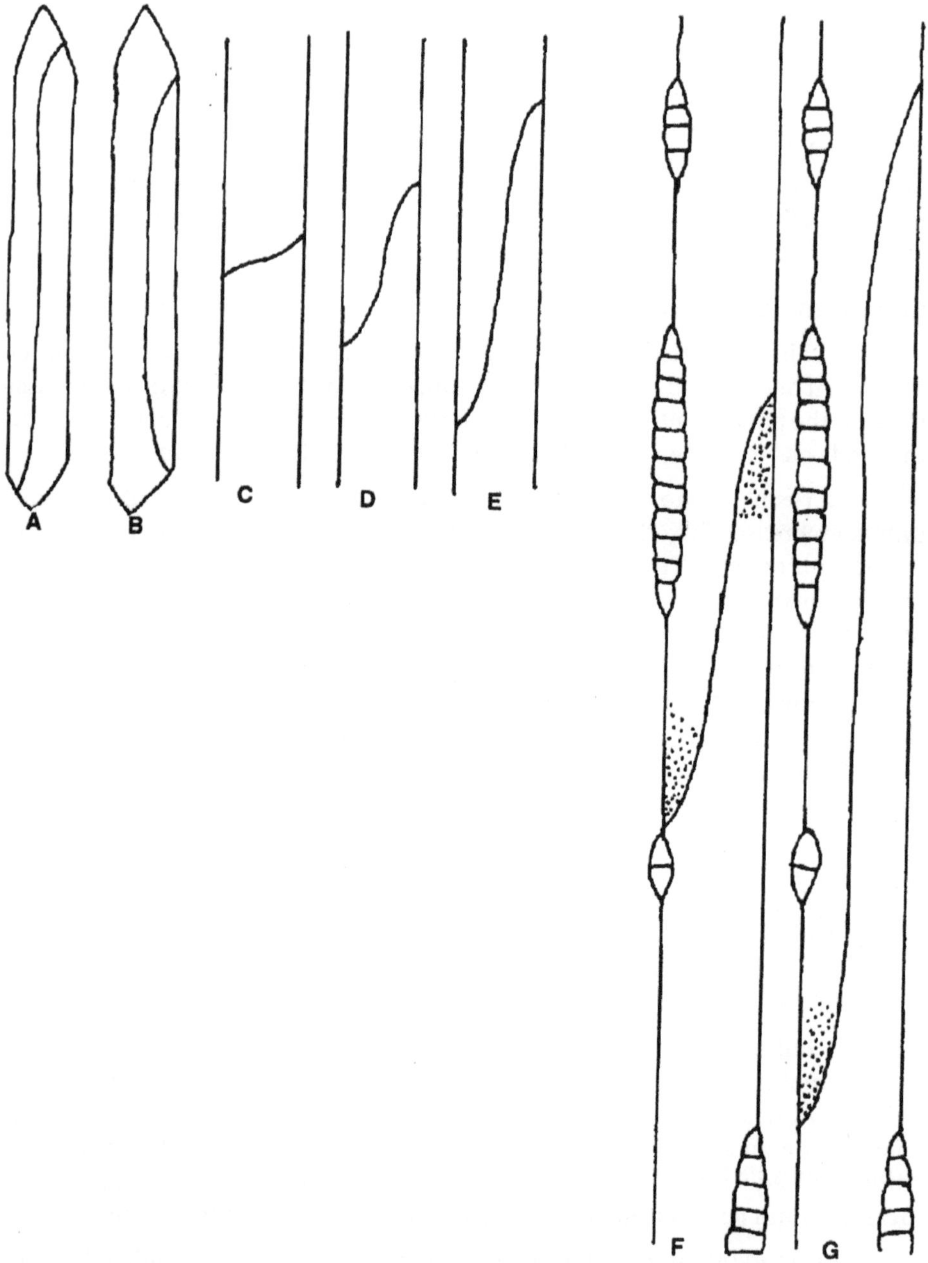

Figure 7.2: Division and Growth of Fusiform Initials. Initial divides. A. By radial anticlinal wall; B. By lateral anticlinal wall; C-E. By various anticlinal divisions; F, G. Oblique anticlinal division is followed by apical intrusive growth (Growing apices are stippled).

Cambial Activity

The pattern of radial growth is corrclated to the rate of cambial activity. The variation in the number of cells across the cambial zone indicates the balance between the rate of cell division and the rate of differentiation of the new derivatives. During the active period of cambium, the process of divisions is faster than the differentiation. As a result cambial zone becomes wide. With the beginning of differentiation, a balance is established and the width of the zone remains more or less constant. When the rate of division becomes reduced and the differentiation proceeds faster than division, the cambial zone becomes narrow.

The vascular cambium shows great variation in the period and intensity of activity. These variations are the result of many internal and external factors. There are plants whose cambium is active throughout the entire life of the plant. The cambial cells divide continuously and the resulting cells undergo gradual differentiation to form the xylem and phloem elements. This type of activity is usually found in plants growing in tropical regions. In regions with definite seasonal climates the cambium ceases its activity with the onset of unfavourable conditions, usually the autumn. It enters into a dormant state which may last from the end of summer till the following spring. In spring the cambium again becomes active. From an anatomical point of view the commencement of the cambial activity usually consists of two stages: (1) the cambial cells expand radially, and (2) the cells begin to divide.

Cambial Activity - Different Methods

Various methods are used to determine the cambial activity:

1. The case by which the bark can be peeled is often used as an indication of cambial activity, since with the enlargement of the cambial cells their radial walls usually become weakened, so that the bark of the stem and the roots can easily be peeled. The separation usually occurs in the region of young xylem cells which have already reached their maximum diameter, but which still have their primary walls.
2. The examination of a cross section cut in the region of the cambium and the tissue adjacent to it, is the second method. The number of cell layers in the cambial zone and of the not yet fully differentiated xylem elements is taken as an indication of the rate of cambial activity, in this method.
3. The rates of both cell division and differentiation can be precisely determined in this method. It involves the application of radioactive carbon to photosynthesizing plants. When the cambium is active the radioactive carbon is incorporated in the newly formed cell walls and it can easily be detected by auto-radiographic techniques.

Within the limits set by the genotype, cambial activity is dependent on factors like growth substances, temperature, photoperiod and soil moisture.

The activity of the cambium usually starts below the sprouting buds and from there it spreads downwards. In deciduous porous Dicotyledons, it was found to be slow. The divisions in the cambium may not begin at the base of the trunk until several weeks after they begin in the twigs. In ring porous Dicotyledons, the downward spread of cambial activity is very rapid. In conifers, the condition is intermediate and the spread of divisions in the cambium takes about a week.

A certain level and combination of growth regulating substances form the stimulus for the reactivation of the cambium. However, in most plants a certain period of dormancy has to be completed before the cambium can be reactivated by known treatments like the application of growth regulators and increased day length. Two main groups of regulators, gibberellins and auxins, have been shown to effect the cambial activity. Gibberellins induce rapid cell divisions which are not followed by differentiation in the cambium. Auxins cause rapid cell differentiation. The simultaneous effect of both these groups of substances seems to result in the appearance of normal cambial activity.

Cambial Meristem - Maintenance

The vascular cambium is maintained by a network of interacting signal feedback loops. At present, both the hormones and the short peptides have been identified as the information carriers. The cambial meristem receives the signals from both the xylem and phloem sides, and the signals received from the outside the meristem act to down regulate the internal factors, that promote cell proliferation and differentiation. The phytohormones involved in the cambial activity are the auxins, ethylene, gibberellins, cytokinins, abscissic acid *etc.* Each one of these is vital for the regulation of the cambial activity, and the combination of these hormones is very important in plant metabolism. The auxins stimulate mitosis, cell production and regulate inter-fascicular and fascicular cambium. The auxins also regulate ray and fusiform initials in the vascular cambium. Gibberellins stimulate the cambial cell division and also regulate the differentiation of the xylem with no effect on phloem differentiation. High concentrations of gibberellins are positively correlated to an increase of cambial cell division and an increase of auxin in the cambial stem cells. Gibberellins are also responsible for the xylem expansion through a signal travelling from the shoot to the root. Cytokinin is known to regulate the rate of cell division instead of cell differentiation.

Cambium - as Food

In most trees the cambium is edible. It was historically used in Scandinavia as flour to make the bark bread.

Wound Healing

One of the important functions of the cambium is to form callus or wound tissue over wounds. Callus may be formed by the division of parenchyma cells of the phloem, cortex or vascular rays, but it is mostly formed by the cambium. The outer cells of the parenchymatous mass become suberized or a periderm develops in them. Below this protective layer a reorganized cambium produces new vascular tissues. Callus is also developed, at the beginning of the growth season, on the circumference of wounds caused by pruning. With the continued production, by the cambium of secondary xylem in the undamaged area around the scar and because of the eventual fusion of the cambial layers, the wound becomes completely covered. The production of the secondary xylem continues so that the layer covering the wound is continually thickened.

Wound Response in Plants

The plants have developed many methods to respond to and to heal the wounds and injury. The wound response includes healing them by creating callus and depositing the molecules like suberin as well as activating the predator-disease resistance. Plants regenerate the cells and tissues after damage (wounds and injury) have occurred. There are specific cells in plants that are located at the site of wound as that retain the ability to return to a pluripotent state called "dedifferentiation" which refers to the 'ability of plants to regain and develop into specific cell types required for regeneration'. Depending on whether the wound is superficial, the site of injury will dry up and die. The deposits of suberin beneath the wound, allow for a protective layer to form over the shallow wound site. The laticifers contained within different plant parts which produce latex also perform the same properties. When the plants have deeper wounds, callus (a gel-like substance) is formed, across and underneath the damaged stem or the root, donated by the cambium. When this begins to form, the cells of the cambium produces large amounts of parenchyma.

Wound Healing – Early History

Henri-Louis Duhamel du Monceau, an 18th century French botanist, was one of the first scientists to study healing in plants, reviewed the development of the theory of plant regeneration after wounding. His interest in biology began when he studied at the Academie des Sciences to study the disease plaguing the Saffron Crocus at that time in and around the Academie. He was the first to describe the formation of callus that he had observed growing over the wound of an Elm tree.

Wound Healing – Biochemistry

Wounding allows the nutrients to pathogens and exposes their entry to infection in plants. The plants maintain ability to sense when they have an injured area and induce a defensive response. Within the wounded tissues, the endogenous molecules become released and become DAMPs (Damage Associated Molecular Patterns) including a defensive response, found at the site of the wound mediated by the hormones. When a plant senses a wound, it immediately sends a signal for innate immunity, the signals being controlled by the hormones such as jasmonic acid, ethylene and abscissic acid. Other physical factors like hydraulic pressure and electrical pulses also play a vital role in wound signaling.

Callus and Phellogen

'Callus' (the word callus can include cells with a diverse amount of differentiation) – 'wound tissue' - is group of parenchyma which can grow when needed. With in deep abrasions (tree wounds) the unharmed living cells become stimulated to divide beneath and thus callus is formed which overgrows the wound. Similarly, phellogen (the meristematic layer) appears beside the callus to divide in order to make cork wood which appears like a swelling over the surface of the covered wound, called the 'knot'. As time goes on, the tree rings will begin to cover the surface of the scarred wound and as more of these rings form, the base of the wound becomes buried deeply. Finally, the cambium layer continuously covers over the scar until the wound becomes completely covered by the secondary xylem, completely healing the wound.

Chapter

8

Laticifers

Introduction – Early History

For many centuries the conspicuous milky content in certain plants has attracted the curiosity, and the early scholars recognized the milky (or coloured) substances of a mucilaginous or resinous character restricted them. There were occasional references to the collection of these plants for their contents in the classical literature and these were utilised for many practical purposes too. Theophrastus (1916), a Greek scholar, referred these milky juices (spurges) collected for the medicinal purposes, and also referred the juices of other plants, in the family Apocynaceae (*Nerium oleander* L.), which have been well known for their poisonous character. He suggested that these juices were a fundamental and essential component of all plants, and the interpretation readily accepted by his students.

The early studies were restricted to only superficial description of the plant material (herbals). But, the studies of anatomy were pursued after the improvement of the compound microscope (Hooke, 1665), and the first topic that was investigated by the microscopists was the nature and distribution of these juices. Malphigi (1901) described these milky contents and plant juices to be composed of 2 structural units – the 'utricles' (which constitute the plant body) and the 'tubes' (distributed within the body of the utricle). In different plant materials investigated (fig, Cypress, Celery and Elderberry), Malphigi distinguished 2 types of tubes, namely – the "trachea" (which included present vascular tissues) – and the "vasa propria" (that contained the milky, resinous or mucilaginous substances). Nehemiah Grew (1682), a contemporary of Malpighi, distinguished 2 types of vessels within the parenchyma – the "air vessels" (actually the vascular conducting elements) and - the "lymphatic vessels" (which included the vasa propria of Malphigi and other components of phloem – which included the laticiferous, resiniferous and mucilaginous vessels). Grew (1682) described the occurrence and distribution of laticiferous vessels in Dandelion, Endive, Scorzonera and Sumac *etc.*, and he did not note the laticiferous substance was not present in all the plants which he studied.

Terminology

The laticiferous substance was interpreted by Grew from its analogy to milk in the animals, as it was similar to milk both in its colour and coagulability and the relationship was not accepted by subsequent workers. The term, "latex" (was derived from Latin, meaning - fluid or liquid) was superseded by Grew's term, and got established by English-speaking physicians as early as 1662 (Chandler, 1933). Schultz (1839), a German physician employed the word – "latex" – in his botanical publications in medicine, where it was used in reference to the character of the blood (as "*Her blood appeared of a good texture, otherwise than giving off a little more than its due proportion of latex*" – Spry, 1767). *The latex in the plant was thought to be contained in a vessel system like blood in animals. It also coagulated upon its removal from the plant, as did blood.* This observation was communicated by Schultz to Lindley (1848) who published them in some more detail, and he also referred the milky substance as latex. Thereafter, the term, "latex" remained as it is in the botanical literature.

During the first half of the 18th century, the many anatomical studies were performed by many investigators with some medical training, and any term with, medical implications would be introduced into botany, as such the term, "laticifer" has appeared in the literature, which was more convenient than the terms – laticiferous vessel or laticiferous structure (Esau, 1953; Jackson, 1928), and since the composition of the latex was quite variable, it became difficult to define it as a substance (Bonner and Galston, 1947). Historically, the composition of the latex was completely unknown, though it was characterized by its capacity to coagulate when it was removed from the plant. In general, it may appear nearly clear (as in *Nerium*), or white and very turbid (as in *Euphorbia*), or it may also be coloured, either yellowish-brown (as in *Cannabis*), yellowish-orange (as in *Papaver*), or red (as in *Sanguinaria*); and these observations provided no data on its chemical composition. The most precisely identified substance in latex was starch as per the early studies of latex (Hartig, 1843; Potter, 1884; and Trecul, 1865).

Laticifers – Structure and Composition

Laticifers were first described by H. A. de Bary (1877). Laticifers are highly specialized cells or series of connected cells which can produce a wide variety of proteins that includes latex. These proteins include the enzymes, proteinases and chitinases, which help defend the plant against insects and other herbivores. The laticifers may be simple or compound in origin. The simple laticifers are single cells, the compound laticifers are derived from a series of cells. Laticiferous cells are parenchymatous. The primary walls are thin, devoid of pits and are made up of cellulose and a large proportion of hemicelluloses with pectic substances. Laticifers occur in various tissues and organs of the plant body but may be restricted to phloem. Laticifers are present in about 38 Angiospermous families, *e.g.*, Euphorbiaceae, Moraceae, Asclepiadaceae, Apocynaceae, Papaveraceae *etc.*

Types of Laticifers

Laticifers - Types

Laticifer is a type of elongated secretory cell found in the leaves and stems of plants that produce the latex and rubber as secondary metabolites. These are divided into: Articulated laticifers and Non-articulated laticifers, by de Bary (1977).

1. Articulated Laticifers (Latex Vessels or Compound Laticifers)

These are composed of a series of cells joined together. They are compound in origin, *i.e.*, they develop from a series of cells whose cross walls break down completely or partially (Figure 8.2 B). The articulated laticifers arc also coenocytic. They also arise from the embryonic stage or from the meristem of developing shoot. Articulated laticifers are found in the families Asteraceae, Convolvulaceae, Papaveraceae, Euphorbiaceae, Caricaceae, Sapotaceae, Liliaceae and Musaceae.

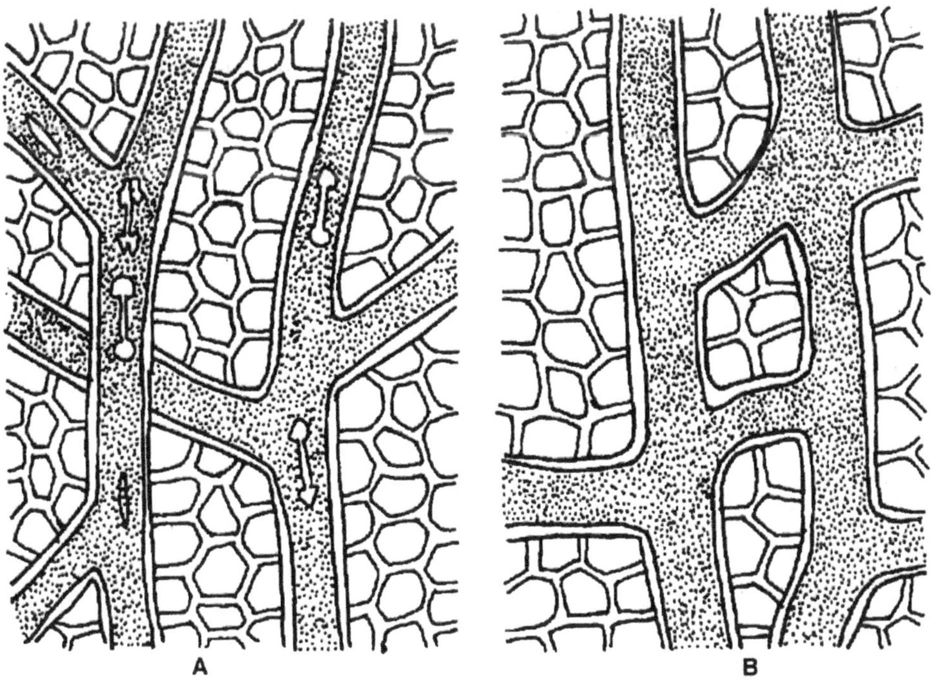

Figure 8.1: Laticifers. A. Non-articulated; B. Articulated.

2. Non-articulated Laticifers (Simple Laticifers or Latex Cells)

These are single cells and arise from some of the meristematic cells of the embryonic stage or from the meristem of the developing shoot. These consist of one long coenocytic cell. These begin their growth from the meristematic tissue of the embryo, the laticifer initial, and exhibit continuous growth throughout the life-time of the plant. As the plant grows, the latex cells also develop and enlarge. They enter other tissues. The latex cells show apical growth and branch out (Figure 8.1 A). Non-articulated latex cells show multinucleate (coenocytic) condition with scanty cytoplasm and thicker walls than the neighbouring cells. Laticiferous tubes have irregularly edged walls and a larger inner diameter than the surrounding cells of parenchyma. During the development, cell elongation occurs via karyokinesis while no cell plate develops which results in the coenocytic cells that extend throughout the plant. The unlimited elongation would result from the autonomous growth of the tip of the laticifer between the meristematic cells. The cell division without cytokinesis results in a long, multinucleate, coenocytic tube. The cells can reach a few 10s of centimetres long and they are either branched or unbranched. They have a role in the healing of the wounds, and defence against herbivory, as well as pathogen defence which are often used in taxonomy. They are found in the families - Apocynaceae, Asclepiadaceae, Euphorbiaceae, Moraceae and Urticaceae. The non-articulated laticifers may be branched (*e.g., Euphorbia, Nerium, Ficus, Asclepias*) or unbranched (*e.g., Vinca, Urtica* and *Cannabis*).

Latex – Structure and Function

Latex is a sap or suspension or in certain cases, an emulsion, of many small particles in a liquid with varying refractive index, produced by the cells called laticifers, which occur in plants of wide taxonomic diversity. The plants exude latex in response to physical damage. The chemical composition of suspended or emulsified particles of the latex differs in different species. Among the suspended material, rubber particles $(C_5H_8)_n$, waxes, resins, proteins, essential oils, mucilages and in *Euphorbia* species variously shaped starch grains can be found. Latex, like cell sap, also contains salts, organic acids, and other substances in true solution, and certain plants in their latex contain - sugars (as in Asteraceae), tannins (as in *Musa*), alkaloids such as morphine, codeine and papaverine (as in *Papaver somniferum*), and a proteolytic enzyme, papain, is present in *Carica papaya*,. The colour of the latex may be white and milky (*Euphoria, Lactuca, Asclepias, Calotropis*), yellow-brown (*Cannabis*), yellow (*Argemone*), yellow-orange (*Papaver*) or colourless (*Morus*).

The latex is the protoplast of the laticifer which contains most of the metabolites stored inside a large, central vacuole. This is highly heterogeneous and forms a suspension or emulsion of many small particles in a fluid, which is typically milky white, but it may also be red, orange, yellow, green and even colourless. Though it is a mixture of many compounds, there is predominance of terpenoids (triterpenes or tetraterpenes) while the rubber tree has up to 45 per cent polyisoprenes in its latex. In addition, the fatty acids, phytosterols, alkaloids, phenolic compounds, proteins, cardenolides, starch grains, have already been identified in the latex of many species.

The function of latex in the plant is not clear. Haberlandt (1918) believed the latex to be of nutritional value in many plants. Sperlich (1939) regarded latex as a reserve material. The most acceptable theory is that the latex contains metabolic by-products and the laticifers represent a secretory system. The compounds contained in the latex function individually or synergistically to protect the plants against herbivory and the pathogen penetration as well as the ability to seal the wounds as the latex polymerizes when it comes in contact with the air.

According to the survey, the laticifers are found in ferns (Marsileaceae), Gymnosperms (Gnetaceae) and about 38 families of Angiosperms. It is also estimated that about 14 per cent of the tropical species produce the latex compared to 6 per cent of the species in the temperate regions. Found in over 900 genera of plants, the laticifers function as internal secretory systems. Natural rubber (chemically, cis-1,4-polyisoprene) accumulates in the laticifers of many plants. Transcripts encoding enzymes involved in the biosynthesis of rubber are highly expressed in the latex of *Hevea brasiliensis* (the only plant from about 2000 rubber producing species that produce commercially viable quantities of rubber whose latex accumulates high levels of transcripts from plant defense-related or stress-related genes and hydrolytic enzymes). The natural rubber is an important raw material in commercial, defense and transportation industries, its role is not completely understood. Many alkaloids also accumulate in the laticifers. The chemical indicators have been identified the laticifers as specialized parenchyma cells as the sites of alkaloid accumulation in *Catharanthus roseus*. In Opium poppy (*Papaver somniferum*), the articulated laticifers are present in all organs of the mature plant, exclusively on the phloem region of a vascular bundle which consist of large vesicles (originate from the dilations of the rough and smooth endoplasmic reticulum during maturation) into which the alkaloids are sequestered. The latex of poppy also contains abundant low molecular mass proteins, the MLPs (major latex proteins).

Latex – Uses/Applications

Rubber is obtained from the latex of bark of *Hevea brasiliensis* (para rubber plant), of Euphorbiaceae. *Manihot glaziovii, Ficus elastica* (Indian rubber plant), *Cryptostegia grandiflora* and *Parthenium argentatum* (Guyayule rubber) also give rubber but their yields are very low when compared to *Hevea*. The latex of *Papaver somniferum* (Opium poppy) contains alkaloids morphine, codeine and papaverine which are used in Anasthesia. Guttapercha is obtained from the latex of *Palaquim gutta* which is employed in the manufacture of artificial dentures, golf balls, underground and underwater cables. Chickle (the gum) obtained from the latex of *Manilkhara achras* (=*Achras sapota*) is used in the preparation of chewing gum. Latex secreted by the plants of the Euphorbiaceae contains hydrocarbons and other substances similar to Petroleum.

Chapter 9

Epidermis

Introduction

Epidermis (Gr.: *Epi* = upon; *derma* = skin) represents the protective covering layer of the primary body of land plants. Epidermis is normally one-layered (uniseriate). However, in some plants, the protodermal cells divide periclinally in certain cases to form multiseriate (multiple) epidermis, *e.g.*, *Nerium*, *Ficus*, *Peperomia*. In *Ficus* leaves, multiple epidermis is found on the adaxial (upper) surface only but in *Nerium*, it is found on both the surfaces. In *Peperomia*, the epidermis consists of 14 to 15 layers. The velamen, the special absorbing tissue of Orchid aerial roots, is also a multiple epidermis.

Origin

The outermost layer of the apical meristematic cells, called protoderm or dermatogen divides anticlinally and by further divisions gives rise to uniseriate layer. In plants showing Tunica-Corpus organization of the shoot apex, the epidermis is derived from the tunica.

Epidermis is distinguished into three components: (1) Epidermal cell complex, (2) Stomatal complex and (3) Trichome complex.

1. Epidermal Cell Complex

It forms the bulk of the epidermis. It consists of (a) Epidermal cells, (b) Costal cells and (c) Special epidermal cells.

(a) Epidermal Cell

The ordinary cells of the epidermis vary in shape, size and arrangement, but they are always closely attached to form a compact layer devoid of intercellular spaces. The cells are tabular in cross-section and isodiametric or linear in surface view. In surface view, the walls of the epidermal cells are straight, curved, wavy or deeply sinuated (u, v or w-shaped).

Epidermis consists of living cells. Cells have a large central vacuole and thin peripheral uninucleate cytoplasm, which contains chromoplasts and leucoplasts and occasionally chloroplasts. The chloroplasts are found in the guard cells of the stomata of aquatic plants and plants growing in humid climates. The anthocyanin rich chromoplast is found in the vacuolar sap of petals and tender coloured leaves.

Epidermal cell walls vary in thickness in different plants and in different parts of the same plant. In the thin-walled epidermis, the outer wall is frequently thickest.

Cutin, a waxy substance, is usually present in the outer walls of the epidermal cells. The substance is found within the cell wall, *i.e.*, interfibrillar and intermicellar spaces of the cellulose (cutinisation), and also constitutes a special layer - the "cuticle" - on the outer surface of the cell wall (cuticularization). The cuticle is usually absent in the actively growing parts of the roots. The thickness of the cuticle varies in different plants. Plants in dry habitats show thicker cuticles. The surface of the cuticle may be smooth, rough, ridged or furrowed. The cuticle on the surface of the epidermis often penetrates, to a certain extent, between the radial walls of the epidermal cells. The cuticle is impervious to water and restricts transpiration besides providing protection. It is also resistant to microorganisms as well as decay.

Wax occurs as surface deposit in the form of 'bloom' in many glaucous leaves and fruits, *e.g.*, *Saccharum* stem, Lotus leaf, and Grapes. The wax may be deposited in large or small flakes or granules, rods, sheets *etc.* Knowledge about the wax coating of plant parts is vital in the use of herbicides and fungicides, as waxes resist wetting by sprays.

In some cases, lignified epidermal cells have also been observed as in the case of *Cycas* leaves and *Pinus* needles.

(b) Costal Cells

These cells are confined to the areas overlying the veins and differ from the other epidermal cells. They are in general, elongated in shape and oriented parallel to the long axis of the veins concerned

(c) Special Epidermal Cells

1. **Lithocysts:** The epidermal cells of plants belonging to families like Urticaceae, Cucurbitaceae, Moraceae and Acanthaceae contain deposition of calcium carbonate crystals, arranged in the form of bunches, like those of grapes. Such depositions are called "cystoliths" and cells containing them are called "lithocysts". Lithocysts are usually larger than other epidermal cells (Figure 9.1).
2. **Myrosin cells:** Certain members of the family Brassicaceae contain special sac-like cells scattered in the epidermis. These cells are filled with an enzyme, myrosin, and are called myrosin cells.

3. **Sclerotic cells:** The epidermis of seed coats of *Pisum* and *Phaseolus* of the family Fabaceae and clove scales of *Allium sativum* of the family Liliaceae is made up of layer of sclereids.

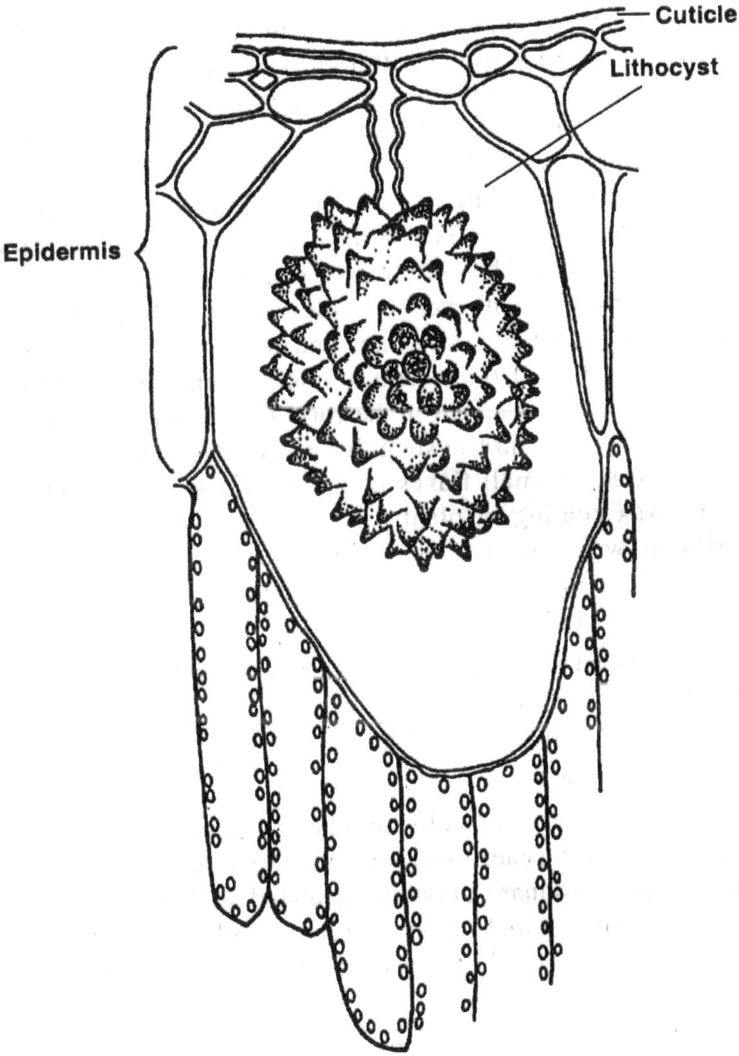

Figure 9.1: Portion of Cross Section of the Leaf Blade of *Ficus elastica* Showing Adaxial Side of the Blade in which the Multiseriate Epidermis, the Cells of which Include Lithocyst (Cystolith-containing cell) can be Distinguished.

In Poaceae (Grasses), small cells (silica cells) filled with silica solidified into bodies of various shapes and cells with suberized walls (cork cells) occur in pairs which alternate with elongated epidermal cells (Figure 9.3B).

4. **Bulliform cells:** Some epidermal cells found in groups in the leaves of certain Monocots, especially Grasses (Maize, Sorghum, Sugarcane *etc.*) are comparatively larger, thin walled and highly vacuolated. These are called "Bulliform cells" (meaning bubble like) or motor cells. These are commonly found in the upper surface of leaves (Figure 14.1 A), sometimes in the lower surface also. Bulliform cells are mainly water containing cells, poor in solid and chlorophyll contents. Rarely tannins and crystals may be found. In cross sections, these appear as a fan-like band because the median cell being large in size. There are three views regarding their function. (a) These are concerned with the unrolling of developing leaves by undergoing sudden and rapid expansion bringing about unrolling of lamina. (b) These are supposed to play an important role in hygroscopic opening and closing movements of mature leaves, due to changes in turgor. They are also called motor cells. (c) These are concerned only with water storage and have no other function.

Stomatal Complex

The epidermis shows numerous openings or pores called "stomata" or "stomates" (Gr. *Stoma* = mouth). The stomata consist of two guard cells enclosing a pore between them. The guard cells are kidney-shaped in Dicotyledons but dumb-bell-shaped in the members of Poaceae. Their walls are unevenly thickened. The wall surrounding the pore is thickened and inelastic due to the presence of secondary layer of cellulose but remaining part of the wall is thin, elastic and permeable. Each guard cell has cytoplasmic lining and a large central vacuole filled with cell sap. The cytoplasm contains a nucleus and a number of chloroplasts, often poorly developed but capable of photosynthesis. Electron microscopic studies have revealed that this chloroplast has fewer and less organized lamella than those of the mesophyll cells. The guard cells are also covered by cuticle which extends to the inner wall forming the boundary of the pore and the "substomatal chamber". Mitochondria, dictyosomes, endoplasmic reticulum and ribosomes are found in guard cells. Epidermal cells abutting guard cells are called "subsidiaries" (subsidiary or accessory cells). The stoma with its subsidiaries is termed as the "stomatal complex" or "stomatal apparatus". A cavity is present beneath the stoma, known as the substomatal chamber which communicates with the intercellular spaces of the mesophyll.

Van Cotthem (1970) divided the stomata into 8 types. They are:

1. *Anomocytic* (*Ranunculaceous type*) (Figure 9.2 A): Stoma surrounded by a limited number of cells that are indistinguishable in size, shape or form from those of the remainder of the epidermis, *e.g.*, Ranunculaceae, Malvaceae, Papaveraceae, *Tridax procumbens, Boerhavia diffusa, Cleome* spp.
2. *Anisocytic* (*Cruciferous type*) (Figure 9.2B): Stoma surrounded by three cells of which one is distinctly smaller than the other two, *e.g.*, Brassicaceae, Solanaceae.

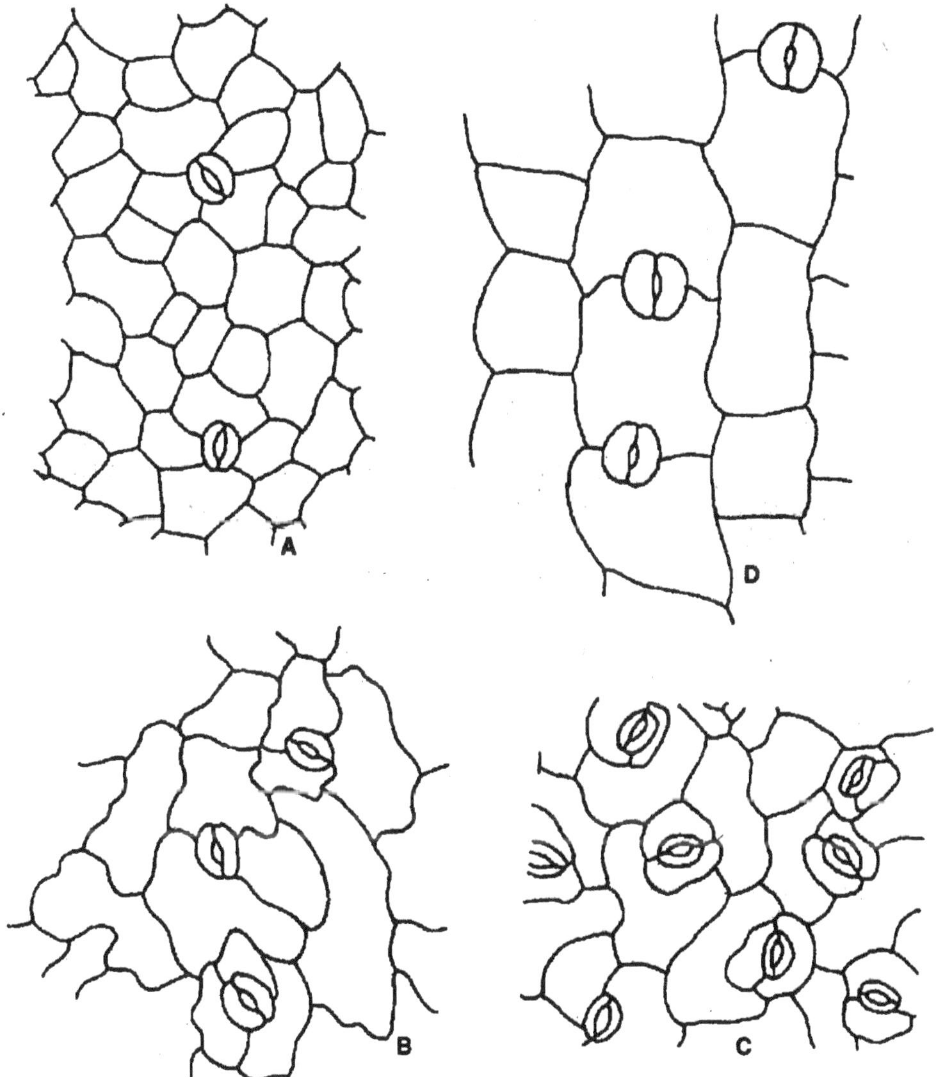

Figure 9.2: Epidermis in Surface Views Illustrating Patterns Formed by Guard Cells and Neighbouring Cells. A. *Pelargonium*-Anomocytic; B. *Brassica*-Anisocytic; C. *Dianthus*-Diacytic; D. *Vigna*-Paracytic.

3. *Paracytic* (*Rubiaceous type*) (Figure 9.2C): Stoma accompanied on either side by one or more subsidiary cells parallel to the long axis of the pore and guard cells, *e.g.*, Rubiaceae, Magnoliaceae.
4. *Diacytic* (*Caryophyllaceous type*) (Figure 9.2 D): Stoma enclosed by a pair of subsidiary cells whose common wall is at right angles to the guard cells, *e.g.*, Caryophyllaceae, Acanthaceae.

5. *Tetracytic* (Figure 9.3 A): Four subsidiary cells are present, two lateral and two terminal, *e.g.*, Most of the families of Monocotyledons.
6. *Actinocytic*: Stoma surrounded by a circle of radially elongated subsidiary cells, *e.g.*, Commelinaceae, Musaceae.
7. *Cyclocytic*: Stoma surrounded by four or more subsidiary cells which form a ring around it, *e.g.*, Tribe Laguncularieae of the family Combretaceae.
8. *Hexacytic*: Stoma accompanied by six subsidiary cells consisting of two lateral pairs parallel to the long axis of the pore and two polar (terminal) cells; the second lateral pair is as long as the stomatal complex.

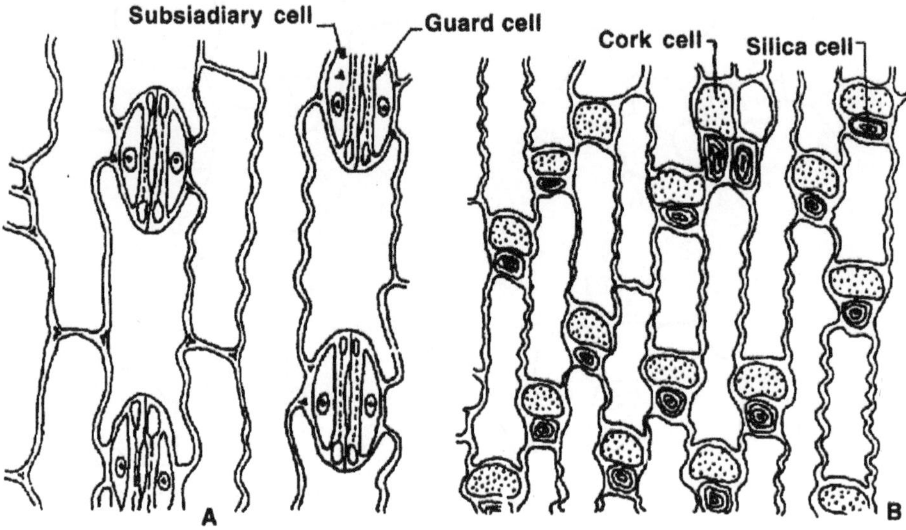

Figure 9.3: Epidermis of *Saccharum* in Surface View. A. Lower epidermis of leaf with stomata; B. Epidermis of stem with cork cells and silica cells.

Distribution of Stomata

Stomata are present on all parts of the shoot including the floral parts. They may be non-functional on some parts of the plant body. However they show their maximum concentration in the green leaves. Stomata are absent in the submerged plants. On the basis of stomatal distribution, the following three categories of leaves are recognizable (a) *Amphistomatic*: Stomata present on both the surfaces of leaf, *e.g.*, most of the herbs, (b) *Hypostomatic*: Stomata confined to the abaxial (lower) surface of leaf, *e.g.* most of the trees, (c) *Epistomatic*: Stomata confined to the adaxial (upper) surface of leaf, *e.g.*, *Nymphaea*.

Functions

Stomata play a major role in photosynthesis, respiration and transpiration. In xerophytic plants like *Nerium* stomata are confined to "stomatal crypt" and covered by hairs.

Trichomes

Epidermal outgrowth is referred to as "trichome" (Gr.: a growth of hair). Trichomes vary in size and complexity and include hairs, scales and other structures. They may be glandular or non-glandular. Trichomes are found both on vegetative and reproductive parts and may be living or dead, ephemeral or persistent. *Foliar trichomes* are important in regulation of moisture exchange with the atmosphere, *covering hairs* for plant defence against phytophagous insects, *secretory trichomes* for providing chemical defence against animals and *stinging hiars* for protecting plants from animals. A study on *vesiculated hairs* on the leaves of *Atriplex halimus* and *A. portulacoides* suggests that this type of hair may serve to remove salts from the leaf tissue and thus prevent an accumulation of toxic salts in the plant (Figure 9.4 E). Uphof (1952) classified trichomes on the basis of morphological characters. The important categories are:

Non-glandular Trichomes

These are of two types.

1. **Unicellular trichomes**: These are simple epidermal elongations which may be straight or curved, bifurcate at apex, y-shaped (*e.g., Budleja*) or T-shaped (*e.g., Lobularia, Dicranus*). Unicellular trichomes of *Cannabis sativa* are highly sclerified with warty protuberances.
2. **Multicellular trichomes**: They may be branched (Figure 9.4 A) or unbranched consisting of a single row of cells (uniseriate) or several rows of cells (multiseriate). The uniseriate trichomes may be vesiculate type (terminal cell bladder like as in *Chenopodium*) or capitate or sub-capitate type. The hairs may be star-shaped (stellate, *e.g., Althaea, Styrax,* Figure 9.4 B,C) or tree-like (dendroid, *e.g., Mimosa* and *Platanus*). Some times these hairs consist of disc-like plate of cells arranged on a short stalk such as peltate hairs or scales of *Olea* (Figure 9.5 A, B). The multicellular shaggy hairs of *Portulaca* consist of several contiguous rows of cells (Figure 9.4 D).

Glandular Trichomes

The glandular trichomes include uni or multicellular trichomes. Glandular trichomes are involved in the secretions of various substances, *e.g.,* salt solutions, sugar solutions (nectar), terpenes and gums (polysaccharides). The secreting trichomes are often called as "glands".

(a) Salt-secreting Trichomes

(i) Bladder-like hairs consisting of a large secretory cell on top of a narrow stalk consisting of one or sometimes a few cells and a basal cell is seen in *Atriplex.* (ii) Multicellular glands consisting of several secretory cells and basal collecting cells. These includes chalk glands as in *Plumbago capensis* (Figure 9.5 H) and salt glands in *Limonium* (Figure 9.6 E), *Avicennia* and *Tamarix* (Figure 9.5 C, 9.6 E). In

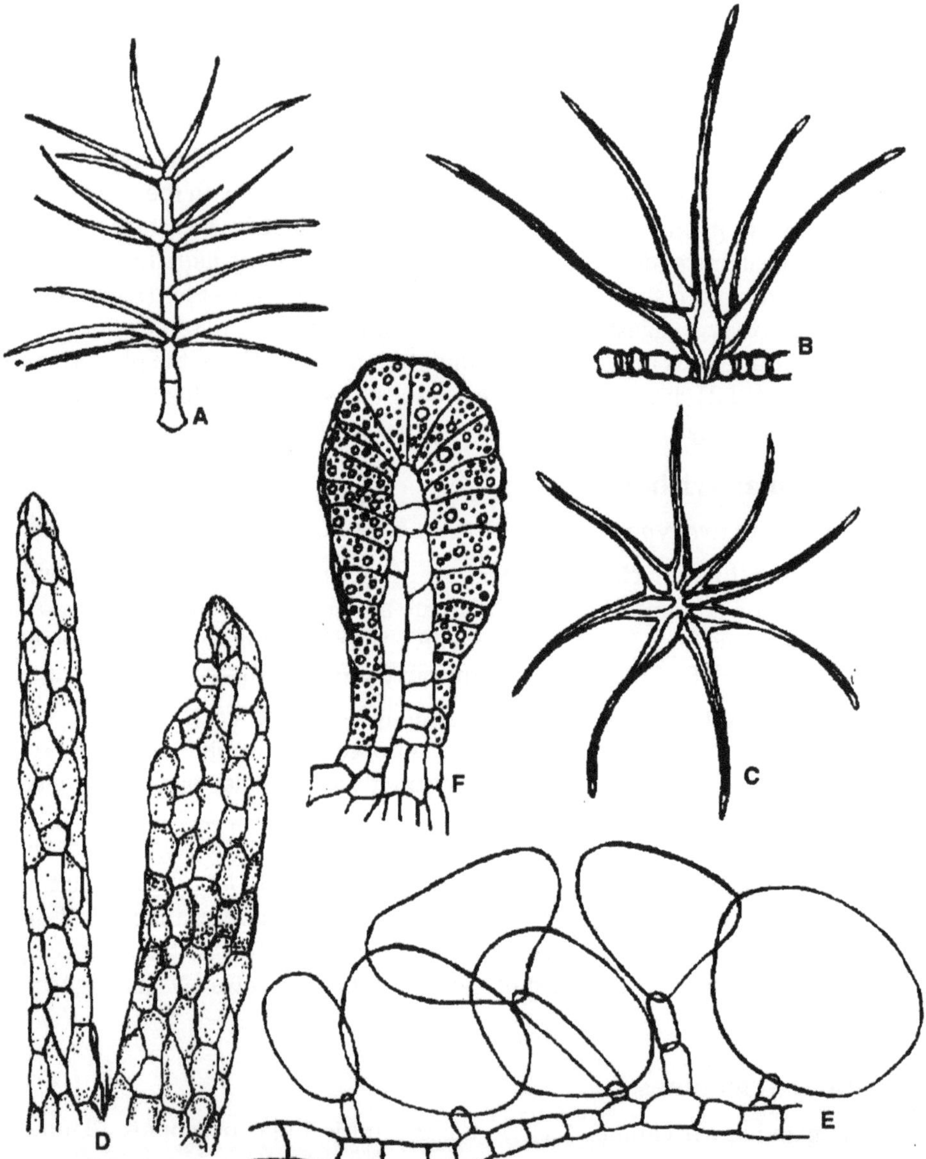

Figure 9.4: Different Types of Trichomes. A. Candelabrum-like, branched, multicellular trichome from the leaf of *Verbascum*; B. Stellate multicellular trichome from the leaf of *Styra officinalis*, lateral view; C. As in B, but surface view; D. Shaggy hairs as seen at the base of the petiole of *Portulaca oleracea*; E. Vesicular hairs of *Atriplex portulacoides*; F. Colleter on a stipule of *Viola*.

these glands, the cytoplasm is dense, rich in mitochondria, ER and Golgi bodies and has many vesicular structures. The salt solution is actively secreted onto the surface of the secretory cells. Trichomes that secrete an aqueous solution containing some inorganic and organic substances were termed as *trichome-hyduthodes, e.g., Cicer arietinum* (Figure 9.6 B).

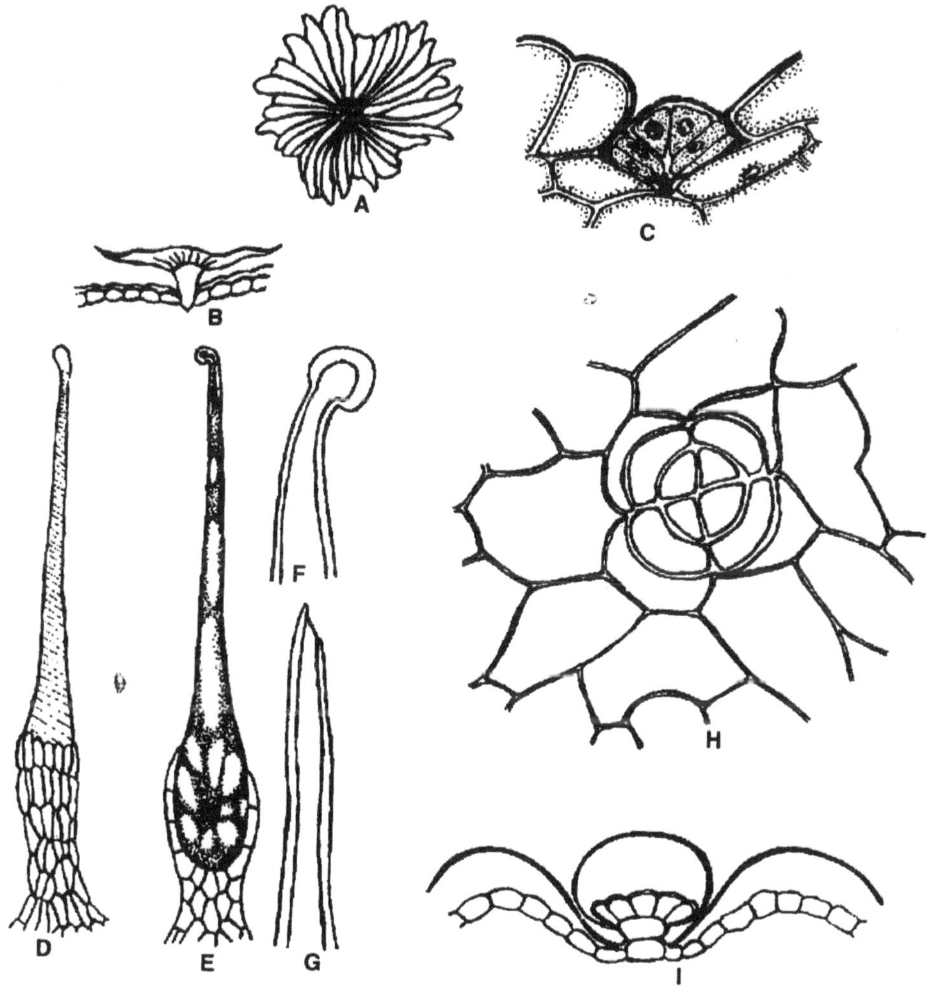

Figure 9.5: Trichomes and Epidermal Glands. A, B. Peltate hair of *Olea europaea.* A. Surface view; B. Lateral view; C. Portion of a cross-section of the leaf of *Tamarix* showing a multicellular salt gland; D-G. Stinging hair of *Urtica dioica*; H. Chalk gland of *Plumbago capensis* in surface view of the epidermis; I. Portion of a cross section of the leaf of *Thymus capitatus* showing a secretory gland.

(b) Nectar Secreting Trichomes (Figure 9.6 A)

The cytoplasm at the stage of secretion is very dense and especially rich in ER. It has been suggested that vesicles mainly of ER origin are involved in the secretion of nectar, *e.g., Abutilon*. Epidermal cells not in the form of hair may also become glandular. Such secretory epidermis occurs on special outgrowths on the teeth of leaf margins, *e.g., Prunus amygdalus*.

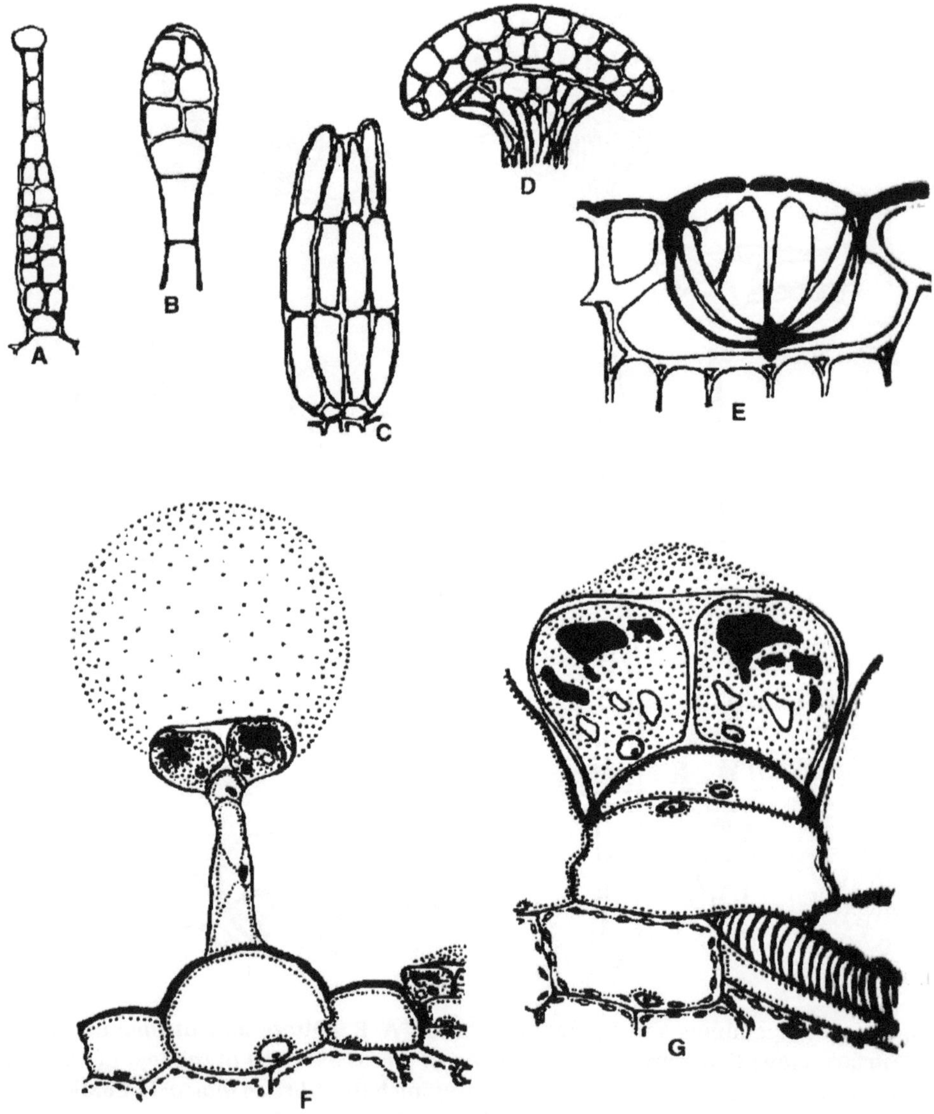

Figure 9.6: Secretory trichomes. A. Nectar-secreting trichome of the calyx; B. Hydathode trichome of *Cicer arietinum*; C. Slime-secreting trichome of *Rumax maximus;* D. A stalked gland of *Drosophyllum*; E. Salt gland of *Limonium latifolium*; E, G. Stalked and sessile gland, respectively of *Pinguicula grandiflora*.

(c) Mucilage Secreting Glands

These glands arise on the membranous sheath from the leaf base of *Rumex* (Figure 9.6 C). The secreted mucilage is mainly a polysaccharide. Golgi vesicles are involved in its secretion. The extruded mucilage accumulates in the spaces between the cell wall and the cuticle. The cuticle may rupture and the mucilage thus reaches the surface, *e.g.*, *Colleter* is a multicellular hair having a sticky secretion (Figure 9.4 F).

(d) Glands of Carnivorous Plants (Figure 9.6 D, E, F)

The capture organs of these plants are usually modified leaves. The prey - insects and often other small animals – are attracted to the capture organs by attractive colours, odour and nectar secretion. The animals are captured by the organs in various ways; by mucilages secreted by special trichomes (*Drosophyllum*, *Pinguicula* (Figure 9.6 F, G); by rapid closure of the two halves of the leaves of *Dionaea*, by attraction to "pitfalls" (*Nepenthes*) or by a complicated trap as in *Utricularia* ('mouse trap').

(e) Colleters, Trichomes Secreting Sticky Substances

Glandular trichomes consisting usually of a multicellular head and a stalk which may sometimes be lacking. All the external and often the neighbouring epidermal cells also have the ability to secrete. The secreted sticky substance, frequently a mixture of terpenes and mucilage, reaches the gland surface usually by rapid rupture of the cuticle. The secretion continues for a long period. Colleters appear mostly on bud scales (*e.g.*, *Viola*, Figure 9.4 F; *Alnus*, *Coffea* and *Rosa*).

(f) Stinging Hairs

In *Urtica dioica* specialized glandular trichomes, known as stinging hairs are found (Figure 9.5 D-G). These possess a peculiar type of cell structure for releasing the contents of glands. These are unicellular structures with broad calcified lower end and long, fine capillary like silicified upper end. Their base remains covered with a collar-like growth of epidermal cells. When the hair comes in contact with the skin of an animal, the tip of the hair breaks off. The sharp edge penetrates into the skin and injects the irritating cell contents, which are rich in histamine and acetyl choline.

(g) Terpene-secreting Trichomes

The essential oil-secreting glands are seen in Lamiaceae (Figure 9.5 I). These consist of a basal cell, a uniseriate stalk (one or several cells long) and a head of one or several secretory cells.

In young stems and leaves of *Mesembryanthemum crystallinum* (so called ice plant), some of the epidermal cells become greatly distended and serve as water reservoir. These are often called as "ice-beads".

Chapter 10

Periderm

Introduction

The process of secondary growth (the increase in the girth of plant organs) is not only a prerequisite for the efficient long-distance transport of water, solutes and photo-assimilates, but also the shape of the plant body in response to ever changing environment. Thus, the emergence of radial thickening contributed much to the success of the vascular plants during the process of evolution. According to Demura and Ye (2010), the wood represents the major source of biomass accumulation in perennial dicotyledons and gymnosperms. It is the vascular cambium (the "post-embryonic meristem") that drives this process, which consists of a ring of undifferentiated meristematic cells that differentiate into xylem (wood) and phloem (bast) upon division. Another lateral meristem, the phellogen ("cork cambium"), contributes to radial thickening and to the protection of the vascular cylinder, and divides in a bidirectional manner, and produces inwardly the phelloderm and outwardly the phellem (cork). All the three tissues are collectively called as the "periderm". The periderm restricts the gaseous exchange, water loss and pathogen attack.

Due to continued secondary growth, the stem gradually keeps on increasing in width and secondary tissues formed exert pressure on phloem, cortical and epidermal cells. Due to pressure, the phloem cells are pushed outwards, the cortical cells become flattened, the hypodermal cells get disturbed and become flattened tangentially and epidermis is put to maximum stretching. The pressure increases so much so that the epidermis ultimately gets ruptured exposing the living cells to the external atmosphere. At this stage epidermis is replaced by a group of secondary tissues which are protective in function. The term *periderm* is applied to these cells. In, majority of eudicots and gymnosperms, during the process of secondary growth, the periderm replaces the epidermis as the frontier tissue protecting the vascular tissue from both biotic and abiotic stresses. The mechanisms underlying the establishment of the periderm and its formation are largely unknown despite its importance.

Structure of Periderm

The periderm includes the phellogen (cork cambium), the meristem that produces the phellem (commonly called cork), the protective tissue formed outward by the phellogen and phelloderm, a living parenchyma tissue, formed inward by the meristem. Death of tissues lying outside the periderm results from an insertion of nonliving cork between these tissues and the living inner tissues of the axis.

The periderm is the outer layer of the bark of a tree. The phellogen, phellem and the phelloderm together make up the peirderm. The phellogen is relatively simple in structure. In contrast to the vascular cambium it has only one form of cells. In cross section, the phellogen commonly appears as a continuous tangential layer (lateral meristem) of rectangular, radially flattened cells, each with the derivatives in a radial file extending outward through the cork cells and inward through the phelloderm cells. In longitudinal section, the phellogen cells are rectangular or polygonal in outline, sometimes somewhat irregular. The protoplasts of these cells contain vacuoles of different sizes. They may also contain chloroplasts and tannins.

The cork cambium, like the vascular cambium, exhibits periods of activity and inactivity. In *Quercus,* the period of activity of the cork cambium and the vascular cambium coincide. In *Robinia,* two periods of phellogen activity occur during the single period of cambial activity. In *Acacia*, three alternating periods of phellogen activity and inactivity occur during the year. Several factors are involved in the activity and initiation of cork cambium. In *Robinia,* it is mainly active under a combination of short day and high temperature or long day and low temperature. Gibberellic acid and Naphthalene acetic acid have a distinct retarding effect on initiation of cork cambium in this plant. In *Eucalyptus*, high humidity and a continuous flush of oxygen cause early initiation of cork cambium.

Phellem (cork) cells are characterized by suberization of their walls. The suberin, a fatty substance, usually occurs as a distinct lamella that covers the original primary cellulose wall.

The cork cells are suberized and thick walled, dead, compactly packed and arranged in a radial file without intercellular spaces. They lose protoplast after differentiation and become dead and remain filled with air and coloured organic matters.

Cork used commercially as bottle cork has thin walls and lumina filled with air. It is highly impervious to water and resistant to oil. It is light in weight and has thermal insulting qualities. Mature cork of this type is also a compressible, resilient tissue. The commercially valuable properties, impervious to water and insulating qualities, also make the cork effective as a protective layer on the plant surface.

In many species the phellem consists of cork cells and non-suberized cells called "phelloid cells". Like the cork cells, the non-suberized cells may have thick or thin walls, and they may differentiate as sclereids. In *Abies* species, layers of cork cells alternate with those of cells having thick sclerified outer tangential walls. Two more types of cork cells occur; they are: (1) Hollow, thin-walled and radially widened; and (2) thick-walled and radially flattened. The cells of the second type may often be filled with dark resiniferous, tanniferous substances as in *Eucalyptus*. These two types of cork cells may occur together in the same plant as in *Betula*. Here, they occur in alternating layers and facilitate the cork to peel like sheets of paper.

The phelloderm cells resemble cortical parenchyma cells and may be distinguished from the latter by their position in the same radial files as the phellem cells. In certain plants the cells of the phelloderm contain chloroplasts and are photosynthetic. These cells may also store starch. Sclereids and other special cells are sometimes present among the cells of phelloderm.

In most woody eudicots and gymnosperms, a periderm is formed both in the root and the stem, as well as underground stems like potato tubers extensively. The number of cell layers that comprise the periderm varies among species, and usually, only one to two layers of phelloderm are present in the periderm, while several cell layers of phellem differentiate (Esau, 1977; Pereira, 2007). In trees, the phellogen can remain active for several years and/or can be replaced each year by a new phellogen (Esau, 1977). For example, the phellogen of cork oak is functional for many years and it remains viable throughout the life of the tree though its activity decreases with its age. The activity of phellogen in the trees undergoes seasonal changes and fluctuates with variations of the climate. The activity of phellogen decreases with cold spells and drought while warm weather results in higher activity. Sometimes, "wound periderm" develops close to the damaged or necrotic tissue after any mechanical injury, for example, the shedding of the branches and the leaves or even the infection of the pathogen.

Periderm - Development

The first periderm commonly appears during the first year of growth of stem and root. The subsequent deeper periderms may be initiated later in the same year or many years later (species of *Carpinus, Fagus, Quercus*) or may never appear. The first periderm of stem originates most commonly in the subepidermal alyer (*Populus, Juglans, Ulmus*) (Figure 10.1 A). Occasionally the first cork cambium is formed in the epidermis itself (*Solanum, Quercus, Nerium*). In stems of certain plants it has its origin in the second or third cortical layer (*Robinia, Aristolochia, Pinus*). In some species, however the first periderm arises rather deeply in the stem (*Berberis, Ribes,Vitis*), usually in the primary phloem. In most of the roots the first periderm originates in the pericycle, but it may appear near the surface as, for example, in some trees and perennial herbaceous plants in which the root cortex serves for food storage. The subsequent periderms arise successively in deeper layers beneath the first and thus eventually originate from parenchyma of secondary phloem, including ray cells.

The first phellogen is initiated either uniformly around the circumference of the axis or in localized areas or becomes continuously by a lateral spread and meristematic activity. The subsequent periderms appear most commonly as discontinuous but overlapping layers. The phellogen is initiated by periclinal divisions and produces the phellem and the phelloderm by periclinal divisions (10.1 A, B). The phellogen keeps pace with the increase in circumference of the axis by periodic divisions of its cells in the radial anticlinal plane.

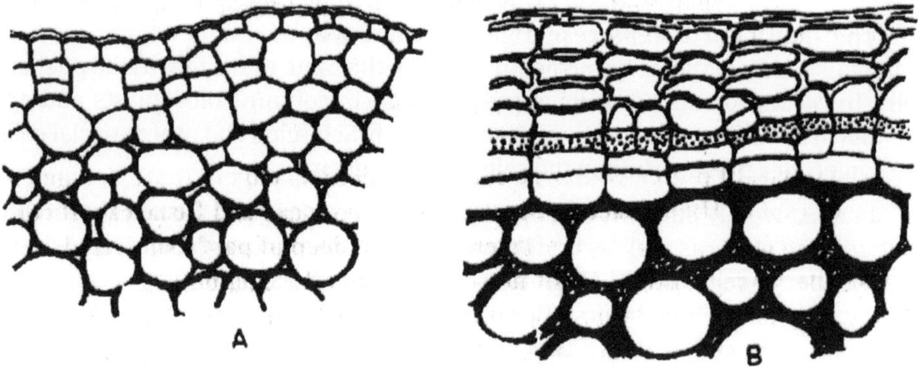

Figure 10.1: Origin of Periderm in *Pelargonium* Stem as seen in Cross Sections. A. Periclinal divisions in subepidermal layer; B. Periderm is established.

Periderm - Composition and Function

Periderm refers to a group of secondary tissues that form a protective layer which replaces the epidermis of many plant stems, roots and other parts. Even though it may develop in the leaves and fruits, its main function is to protect the stems and the roots. The periderm consists of three different layers, namely: *phelloderm* (secondary cortex), *phellogen* (cork cambium) and *phellem* (the cork). Its main function is to protect the underlying tissues from: desiccation, freezing, heat injury, mechanical destruction, disease and loss of the epidermis, the tissue that bounds the restriction of the pathogen and insects, and allows the gaseous exchange through the lenticels.

Wound Periderm

The plants do not heal their wounds in the sense that the damaged tissue is completely restored; instead wound healing involves minimizing the pathogen invasion and the loss of the fluids. The first layer of the intact cells that are adjacent to the wound is modified by the accumulation of different antimicrobial and water-impermeable substances, such as lignin and suberin, which form a sealing layer.

This is followed by the development of the periderm under the sealing layer. The formation of the periderm and suberization are one the generalized responses to wounding irrespective of the nature of the natural protective polymers of the plant tissue.

One of the important strategies of the plant's defense is the formation of the wound periderm in the boundaries of the invaded or the damaged region to isolate it from the non-wounded healthy tissue. This may prevent successive pest invasions and wall off the pathogen infection. The studies of herbivory and wound signalling have indicated the extensive overlapping of the respective sets of the induced genes, and insect feeding has been simulated using the mechanically wounded plants which proved to be too short for the periderm to form.

The wound periderm forms around any damaged tissue whether it is caused by fungal infection, attacks by the bark beetles, or mechanical injury. The formation of wound periderm is an essential final stage in a defence reaction, as it separates the damaged tissues and re-establishes a continuous surface barrier. This is produced by the activation of existing parenchyma cells which begin to divide to form a new cork cambium, which produces cork tissue outwards and phelloderm inwards. In addition, the wound periderm may also stop the spread of the invading organisms from the site of attack.

Wounding induces a series of metabolic events and related cytological responses that lead, under favourable conditions, to a complete closure of the wound. Natural wound periderms are basically alike in method of origin and growth and may have the same cellular components. The natural periderm develops beneath the surface which is effectively sealed by the cutinized epidermis. Correspondingly, the formation of wound periderm is preceded by a sealing of newly exposed surface by scar (cicatrice) tissue. This tissue includes dead (necrosed) cells on the surface and living cells beneath, which become suberized and lignified and form the so-called closing layer. The wound phellogen arises beneath the closing layer and, when cork develops, the dead scar tissue is carried outward. Successful development of wound periderm is important in horticultural practice when plant parts used for propagation must be cut (*e.g.*, potato tubers, sweet potato roots). Environmental conditions markedly influence the development of wound periderm.

Plant taxa vary with regard to the histological aspects of wound healing, as they do in details of natural development of the protective tissue. In general, Monocotyledons are less responsive to wounding than the Dicotyledons. In Dicotyledons and certain Monocotyledons (Liliales, Araceae, Pandanaceae) healing includes formation of both closing layer and wound perideerm. In other Monocotyledons no wound periderm is detectable. Among these, the Zingiberales produce a slightly suberized closing layer whereas the Arecaceae and Poaceae form a lignified closing layer.

A wound reaction occurs when the periderm is peeled off to the living cells underneath. The newly exposed cells die, and a new periderm arises below them.

This reaction is utilized in the production of commercial cork from the cork of oak. The first cork, which is of inferior quality, is removed up to the phellogen and the new phellogen developing beneath the scar tissue produces massive cork of superior quality.

Both the native periderm and wound periderm consist of 3 different tissues - phellem, phellogen and phelloderm, and it was previously determined that the phellogen walls of the immature native periderm are thin and are prone to easy fracture during the harvest which lead to periderm abrasion (also called as excoriation). Upon maturation of the periderm, the walls thicken and become less susceptible to fracture which lead to the resistance to excoriation. The histological analysis (with ruthenium red and hydoxylamine-$FeCl_2$) which stain unesterified and highly methyl-esterified pectins, indicate that the cell walls of phellogen on native and wound periderm differ significantly regardless of the stage of maturity, and the results obtained by the staining imply that the phellogen cell walls of immature native periderm contain methyl-esterified pectin but lack in unesterified (acidic) pectins. The maturation of native periderm accompanied by an apparent increase in unesterified pectin in the walls of cells of the phellogen may allow the strengthening of the cell walls via calcium pectate formation. But, the histological staining of the phellogen walls of the wound periderm implies that the walls are deficient in pectins. The maturation of the wound periderm is not accompanied by an increase in the unesterified pectins in the walls. The peroxidise staining is strong in the phellogen walls of both the immature and mature native periderm and any differences could not be detected in staining between them. The peroxidise staining was weak in the walls of the phellogen of immature wound periderm and was not detectably different in mature wound periderm. The data implies that there are distinct differences between the native and wound periderm though the data does not indicate that the changes in the activity of the peroxidase are involved in the development of resistance to periderm abrasion that occurs upon the maturation of the periderm. Most of the current knowledge on wounding periderm and suberization is derived from the wound-induced healing processes in the tuber of potato. The wound periderm formation is induced by the removal of the skin of the tuber or excising the discs of tuber flesh with the help of the cork borer and allowing the exposed tissue to heal. This enables the isolation of the significant amounts of the suberized periderm which can be made available for the anatomical and chemical characterization as well as direct measurement of water permeability. The tissue development and the anatomy are highly similar and the chemical components of suberin from other species are also found in the suberized periderm of potato. These characteristics of the induced wound periderm in the tubers of potato make them a good model to study the suberized wound periderm tissue in plants.

Polyderm

This is a special type of protective tissue called polyderm in roots and underground stems of Hypericaceae, Myrtaceae, Onagraceae and Rosaceae. It consists of alternating tissue layers, one cell deep of partly suberized cells and layers several cells deep of non-suberized cells. The polyderm may become twenty or more layers in total thickness, but only outermost layers are dead. In the living part, the non-suberized cells function as storage cells. The information about polyderm in woody roots of perennial fruit trees is barely available. When the roots start secondary growth, a protective tissue called the polyderm (composed of alternating suberized and non-suberized cell layers) develops beyond the endodermis in some families (like Myrtaceae). It is known that the layers of polyderm in the roots increased significantly under drought stress condition indicating that the formation of this tissue was affected by environmental stress.

Commercial Cork

Commercial cork is obtained from the phellem (the outer bark of an evergreen cork oak) primarily harvested from *Quercus suber* (the principle source of commercial cork) - native to the Mediterranean region, and the Chinese Oak, *Q. variabilis*, belonging to the family Fagaceae. The cork is also used commonly for the various species of the unrelated genus, *Phellodendron*, which is not a commercial cork. These sources mainly come from the Iberain Peninsula and the Western Mediterranean with most of the cork being produced from Portugal, Spain, parts of Southern France and Italy as well as North Africa. The tree is usually about 18 m (60 feet) tall with a broad, round-topped head and glossy green holly-like leaves. In the stem of this plant, the first cork cambium (rough outer bark) forms in epidermis. This is not vital to the survival of the tree and serves mainly in protection from heat and dry winds. It may remain on the plant indefinitely but in order to obtain commercial cork, this first formed periderm (rough and uneven and has little commercial value) is removed when the tree in about 20 years old and about 40 cm in diameter. After the first bark/periderm ("virgin cork" or the "male cork" for 25 year old trees) is removed, the exposed cells of the primary cortex and secondary cortex dry out and die. A new cork cambium (uniform and continuous regenerative tissue) is formed a few millimetres within the cortex, below the first formed cambium (and in 3-10 years, a uniform new cork sheathing of 1-2 inches is formed in a healthy tree and the tree lives on average for about 150 years). The stripping of this regenerated layer yields commercial cork slabs. This subsequent cambium produces cork more rapidly to outside and in about 10 years a cork layer thick enough to be of commercial value, is produced. This (second stripping) cork is of better quality and value than the first formed one, which has almost no commercial value. The stripping is still done by hand and consists of cutting slits in the outer bark which is then carefully pried loose from the inner bark and peeled away with levers and wedges. Special care is taken not to injure the deeper layer and the removed peel of cork is boiled (or steamed) to remove the soluble tannic acids from it and to increase its flexibility and its rough woody surface is scraped clean by the hand, which becomes ready for commercial distribution. By harvesting the cork only every 10 to 12 years, a tree may provide a dozen or more harvests in its life time. The tree is sometimes cultivated for ornament. The hybrids with Turkey

Oak (*Q. cerris*) are common both in the wild in S. W. Europe and in cultivation. The hybrid is known as Lucombe Oak (*Quercus x hispanica*).

The features that give commercial value to cork are its imperviousness to gases and liquids and its strength, elasticity and lightness. The cork mainly consists of suberin, a waxy substance in the higher plants. The uniqueness of this cork is its structure of air-filled cells which consist of a watertight, flexible compartment, and *en masse*, all these cells constitute an effective insulating medium which is also impervious to liquids as it is highly hydrophobic. The cork is also lighter in weight (1/5th as heavy as water) because of its internal matrix of air pockets. The commercial cork is buoyant, lightweight, adhesive, chemically inert, and elastic. The cork is traditionally important as a stopper for bottles of wine and other alcoholic beverages. It is used in bulletin boards, coasters, hot pad to protect the tables from hot pans, insulation, sealing for lids, flooring, gaskets for the engines, fishing bobbins, rollers, sound-proofing materials, refrigeration chambers, caps, insulators and shock absorbers, handles for fishing rods (as an alternative for neoprene), tennis rackets *etc.* It is also used in musical instruments (in particular, woodwind instruments) to fasten together the segments of the instruments making the seams airtight. The granules of the cork can also be mixed into the concrete, so that the composites made by this have low thermal conductivity, low density and good energy absorption.

Rhytidome

This is the most familiar part of the bark, and is the outermost layer which covers the tree trunks. This lies external to the last formed periderm. This is composed mostly of the dead cells and is produced by the formation of the multiple layers of suberized periderm, cortex and phloem. As a tree grows older, periderm often arise at successively greater depths and thus cause an accumulation of dead tissues on the surface of stem and root. This dead part of the bark composed of layers of tissues isolated by the periderm and layers of no longer growing periderms is called rhytidome. Rhytidome thus, constitutes the outer bark and is especially well developed in the older stems and roots of trees. In shrubs, early exfoliation of the older bark is common and precludes an accumulation of a thick rhytidome. This is generally the thickest and is most distinctive at the trunk (or the bole, the area from the ground to where the main branching starts) of the tree.

Bark

Bark is a non-technical term applied to all the outermost layers of tissues that lie outside the vascular cambium or the xylem in the stems and roots of woody plants. The plants with the bark include the trees, woody vines and shrubs. It overlays the wood and consists of the inner bark and the outer bark. In older trees, bark may be divided into dead outer bark (also called the rhytidome) on the surface of the stems along with parts of the outermost periderm and all the tissues on the outerside, and living inner bark, the innermost layer of the periderm, which consists of secondary phloem.

Two types of barks are found – the "ring bark" and the "scale bark". If the bark is found in the form of concentric rings surrounding the entire stem, it is called ring bark, *e.g., Vitis, Lonicera, Betula, Clematis, Mangifera*. If the bark is found in the form of separate strips or patches, it is called scale bark, *e.g., Ficus, Pinus, Pyrus, Psidium guajava*.

Bark - Chemical Composition

The tissues of the bark make up by weight between 10-20 per cent of the woody vascular plants and consist of different types of biopolymers, tannins (the condensed tannins in a fairly higher concentration inhibit the decomposition), lignin (up to 40 per cent which provides structural support by cross-linking between the different polysaccharides like cellulose), suberin (acts as a barrier to the microbial degradation and protects the internal structure of the plant), and polysaccharides. Guaiacyl lignin units present in the lignin polymers are less susceptible to the degradation, contain fewer number of aryl-aryl bonds which can form a condensed lignin structure with a lower redox potential. As such, the concentration and the type of lignin units could provide an additional resistance to fungal decay for the plants that are protected by the bark.

Bark - Products Produced and Uses

The bark produces different products such as: bark shingle siding and wall coverings, spices, flavourings, tanbark for tannins, resin, latex, medicines, poisons, different types of hallucinogens and cork. The bark has also been used in making the cloth, canoes, and ropes, and used as a surface for the paintings and map making. Different plants are also grown for their attractive colouration of the bark and surface texture for making the landscapes.

The bark was used as an emergency food in Finland during the famine during and after Civil war - 1918. The bark is thicker in some plants to provide protection, and in *Quercus suber* (Cork Oak) it is thick (more than 20 cm thick) to be harvested as a cork product without killing the tree. In some cases, the bark can be removed in long sheets, and the smooth bark of Birch trees can be used as a covering in canoes (birch canoes of N. America), as a drainage layer in the roofs, for the production of the shoes and backpacks *etc.* In some trees, the inner bark is edible, and the bark bread is made from rye to which the toasted and ground innermost layer of the bark of Scots Pine or Birch is added. In addition, the Sami people of Northern Europe use the large sheets of *Pinus sylvestris* bark (removed in the Spring) and prepare it and store it for use as a staple food, while the inner bark is eaten fresh, dried and roasted.

Lenticels

A lenticel is a porous tissue of cells with large intercellular spaces in the periderm of the secondary organs and the bark of woody stems and roots of dicot flowering plants. Normally it functions as a pore and provides a pathway for the direct exchange of gases between the internal tissues and atmosphere through the bark. The name is derived from its shape similar to a lens (lens-like).

Lenticel - Structure

The lenticels are found as raised, circular, oval or elongated areas on the stems, petioles and fruits. In woody plants, these appear as rough, cork-like structures on young branches. The lenticels also occur on the fruits (like apples and pears). The lenticels on the pome fruits can be derived from non-functional stomata, breaks in the epidermis from the removal of the trichomes *etc.* the number varies from 450 to 800 or from 1500 to 2500 in fruits. These also occur on the tubers of potato. Underneath, the porous tissue creates a number of large intercellular spaces between the cells. This tissue arises from the cell division in the phellogen (sub-stomatal ground tissue). In mangoes and other fruits, discolouration may occur due to the amount of the lignin in the cell walls. These help to promote the gaseous exchange of oxygen, carbon dioxide and water vapour. These also account for up to 21 per cent transpiration in fruits of apple. The formation is directly related to the growth and strength of the shoot and the hydration of the tissue (internal moisture).

A lenticel may be defined as a limited part of the periderm in which the phellogen is more active than elsewhere and produces a tissue that, in contrast to the phellem, has numerous intercellular spaces. The lenticels of the phellogen itself also have intercellular spaces. Because of this relatively open arrangement of cells, the lenticels are regarded as structures permitting the entry of air through the periderm.

Lenticels are usual components of periderm of stems and roots. Outwardly, a lenticel often appears as a vertically or horizontally elongated mass of loose cells that protrude above the surface through a fissure in the periderm. Lenticels vary in their size from structures barely visible without magnification to that of 1 cm and more in length. They occur singly or in rows. Vertical rows of lenticels frequently occur opposite to wide vascular rays.

The phellogen of a lenticel is continuous with that of the corky periderm but usually bends inwards so that it appears more deeply situated. The loose tissue formed by the lenticels of phellogen toward the outside is complementary of filling tissue, and the tissue formed toward the inside of the phelloderm.

Two types of complementary tissues are distinguished: that in which connection between the cells is relatively strong (as for example in *Sambucus nigra, Salix* and *Ginkgo*) and that in which the cells have almost no attachment between them, giving the tissue a powder-like appearance (as for example, in the stems of *Pyrus, Prunus* and *Robinia* and in the roots of *Morus*). In the latter type, the cells of the complementary tissue are held in place by closing layers. The first type of lenticels frequently appear under the stomata. The parenchyma cells beneath a stoma undergo some preparatory divisions; then, a phellogen pushes the overlying cells outward and ruptures the epidermis (Figure 10.2 A).

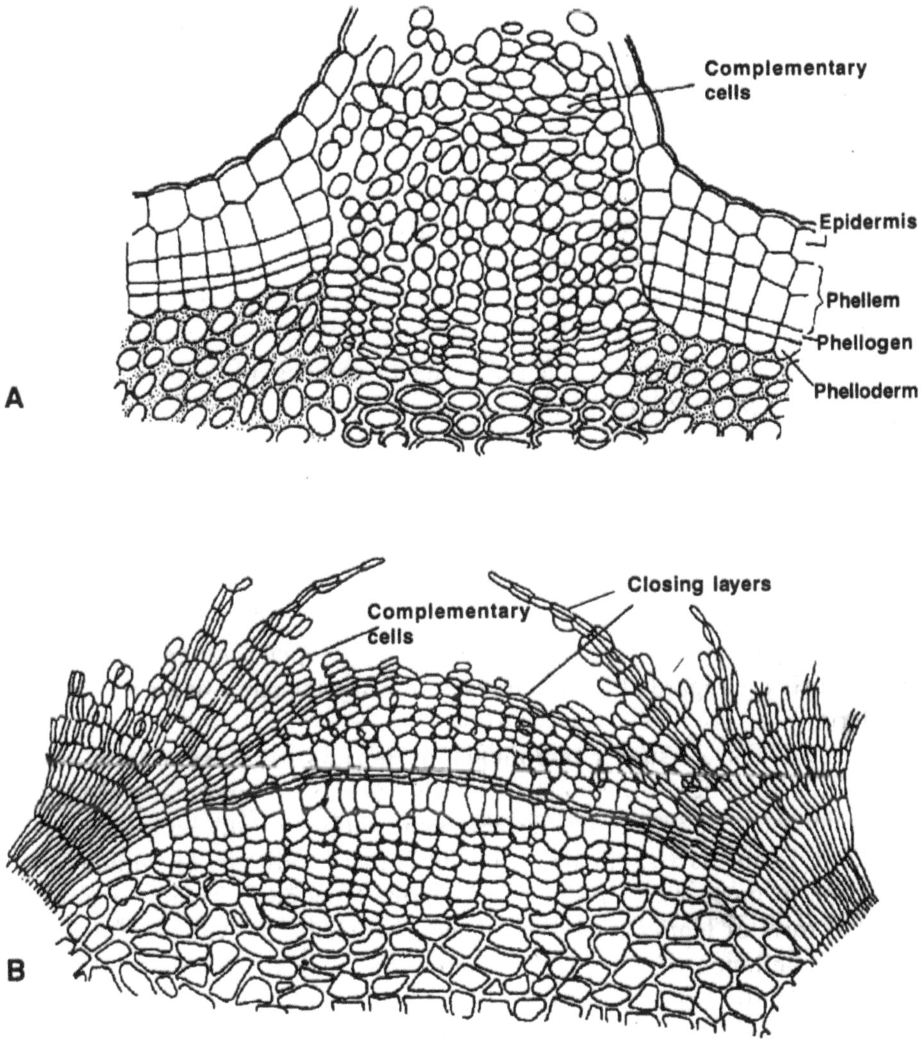

Figure 10.2: A. Young lenticel of *Sambucus nigra*; B. Mature lenticel of *Prunus avium*.

Lenticels are maintained as long as the periderm continues to grow. The deeper periderm also has lenticels. In dicotyledons, three structural types of lenticels are recognized. In the first type, the lenticel has a filling tissue composed of suberized cells, *e.g., Magnolia, Salix, Malus etc.* In the second type, the lenticels consists of a mass of more or less loose non-suberized filling tissue, succeeded at the end of the season, by a more compact layer of suberized cells, *e.g., Quercus, Sambucus, Tilia etc.* In the third type, the filling tissue is layered because loose non-suberized tissue regularly alternates with compact suberized tissue, *e.g., Betula, Fagus, Prunus etc.*

Protective Tissue of Monocotyledons

In herbaceous monocots, the epidermis which has a thick cuticle is the only external protection on the plant axis. When the epidermis is ruptured, the cortical cells below become suberized as in typical cork cells (Poaceae, Juncaceae, Typhaceae). In perennial monocotyledons, different types of protective tissues are formed on the stems. In *Roystonia* (palm), the bark is white and smooth and possesses a hard periderm, which is similar to that of Dicotyledons. It remains on the stem throughout the life of the plant. In *Curcuma, Cordyline,* and many palms, a special kind of protective tissue is formed by the secondary activity of a storied meristem which develops in the outer cortex. The initials of this meristem undergo three to eight periclinal divisions and gives rise to radially arranged layers of cells, which become suberized. The cork of this type is called storied cork. The initials here do not form a regular uninterrupted cylinder. In species of *Dracaena, Aloe arborescens* and *Yucca*, the secondary protective tissue developed is almost similar to those of Dicotyledons, which is produced by a typical cork cambium.

The protective tissues are developed in the perennial monocots by periderm and storied cork formation. The periderm is formed with the origin of the cork cambium (or phellogen) and the subsequent formation of the phellem and phelloderm from it like dicots. The periderm formation of *Dracaena, Aloe* and *Yucca* has no difference than those of the dicots.

Storied cork is a special type of protective tissue composed of cells which are radially arranged in files and have suberized walls. This is observed in *Curcuma, Cordyline* and in many palms. This is formed by the activity of the storied meristem which originates at the outer cortex. The storied meristem does not form a regular uninterrupted cylinder in contrast to phellogen. The cortical parenchyma cells destined to be the initials of this meristem, become 3-8 layered by pericilinal division, and the derivatives of this meristem become suberized and these are termed as the storied cork. The cells are arranged in irregular tangential bands and the radial files. In some cases, these bands may coalesce tangentially and radially when they enclose some un-dividing cells of the cortex which also get suberized. Hence, the storied cork along with the enclosed subertised cells forms a hard protective tissue.

The cork is the outermost layer of a woody stem that has been derived from cork cambium. This serves in protection against the damage from the parasites, herbivorous animals and diseases, and also dehydration and fire. The cork also contains antiseptics like tannins which protect the plant against fungal and bacterial attacks which result in decay.

Chapter 11

Root

The root constitutes the lower portion of the plant axis and it usually develops below the soil surface, although there are roots that grow in air. The differences in the development and arrangement of the primary tissues in these two organs are many. The histogenesis of the epidermis of the root differs from that of stem. In seed plants, the primary xylem of the root is exarch and that in the stem is endarch. The xylem and phloem strands in the root do not form common bundles but are arranged alternately, on different radii while in the stem, the vascular bundles are collateral, or bicollateral (or amphivasal). Roots bear no appendages that are comparable to the leaves of the stem; roots are devoid of stomata and their branches originate in the relatively mature tissue of the pericycle in contrast to the stem where the branches originate from the apical meristem. Roots also possess a root cap which has no parallel in stems.

The radicle present in the seed consists of root meristem and it gives rise to the first root on germination of the seed. In Gymnosperms and Dicotyledons, this root develops into the tap root, known as "seminal root", with its branches. In Monocotyledons, first root usually dies out early in the growth of the plant and the root system of the mature plant consists of adventitious roots. The roots of mature Monocotyledons are usually adventitious and develop from the stem. They may branch several times or they may be unbranched. Generally, such roots do not show secondary thickening.

A tap root system generally penetrates the soil more deeply than the fibrous roots, but the latter bind more firmly the superficial layers of the soil. The tap root and its larger branches undergo secondary growth but the small absorbing roots remain in the primary state and are often ephemeral.

The root system is concerned with anchorage, absorption, storage and conduction.

Primary Structure

The primary structure of the root is as follows: It consists of a root cap externally, and the epidermis, cortex and vascular cylinder anatomically in its cross section.

Root Cap (Calyptra)

The root cap is situated at the tip of roots; it protects the root promeristem and aids the penetration of the growing root into the soil. The root cap consists of living parenchyma cells derived from the apical meristem by divisions contributing cells away from the apex. This part of the apical meristem is called calyptrogen. As the new cells are produced, the cells on the periphery of the root cap are sloughed off. In *Avena,* the time interval between the origin of the root cap cell and its shedding from the cap was calculated to be 5 to 6 days. The environmental conditions affect the development of the root cap. When plants normally growing in soil are raised in a water culture, their roots may fail to develop root caps. Aquatic plants, however, often have particularly large root caps.

The root cap has long been regarded as an organ controlling geo-reaction of the root. When roots normally growing downward are placed horizontally, the subsequent growth in the elongation zone occurs along a curve until the normal downward orientation is attained (Figure 11.1). When the tip is excised or the root cap is removed, growth of the root is not stopped, but the geotropic reaction is lost until the root cap gets regenerated.

The site of perception of gravity in the root cap is the central column of cells in which starch-containing amyloplasts act as statoliths, the gravity sensors. In vertically growing roots, the amyloplasts are sedimented on the distal horizontal walls of the cells in the central column (Figure 11.2 A). A few minutes after the root is placed horizontally, the amyloplasts slide towards the longitudinal walls directed downward (Figure 11.2 B). With in 24 hours the root assumes its normal vertical position by growing along a curve and the amyloplasts return to their previous location next to horizontal walls.

1. Epidermis

The epidermal cells of roots are thin–walled and are usually devoid of cuticle. The epidermis of roots is usually uniseriate. However, in the aerial roots of tropical Orchidaceae and epiphytic Araceae, the epidermis is multiseriate and it is specialized to form a special tissue called "velamen". The velamen consists of non-living cells, compactly arranged and often bearing secondary wall thickenings in the form of numerous narrow strips. The velamen is commonly interpreted as an absorptive tissue, but some physiological studies on orchid velamen indicate that the principal roles of this tissue are mechanical protection and reduction in the loss of water from the cortex.

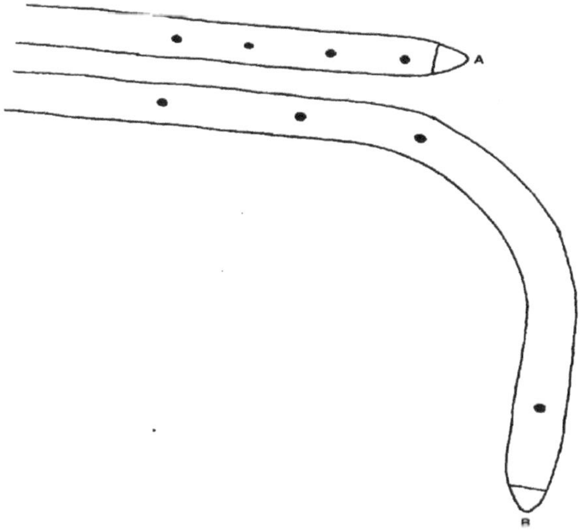

Figure 11.1: Diagram Illustrating the Response of Root Tip to Gravity. A. Root tip of root normally growing downward is placed horizontally; B. 23 hours later the root tip grows along a curve and returns to a vertical orientation with regard to the soil.

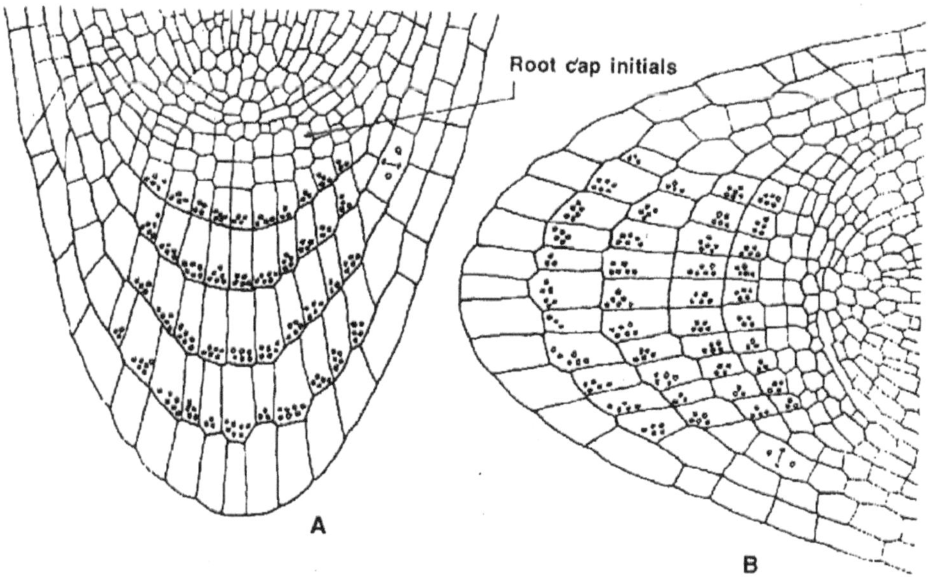

Figure 11.2. Diagram illustrating the response of amyloplasts with starch (small circles), the statoliths, to gravity. (A) Amyloplasts are sedimented on distal horizontal walls in root cap of root growing vertically downward; (B) In the same root placed horizontally, the amyloplasts slide toward normally vertical walls that are now parallel with the soil. This change in position is involved in the geo-reaction shown in Figure 11.1.

In young roots, the epidermis is specialized as an absorbing tissue and usually bears root hairs, which are tubular extensions of epidermal cells. The region of root hairs is usually restricted to one or a few centimeters backwards from the root apex, where elongation has ceased. Root hairs are absent close to the apical meristem and they usually die and dry out on the more mature portions of the root. Certain herbaceous plants, especially water plants, lack root hairs. Plants that usually grow in soil and which produce root hairs fail to do so when they are grown in water.

Epidermal cells of the young roots and root hairs of some genera secrete fibrillar mucilage forming a mucilaginous layer. In plants grown on soil, this layer was found to be colonized by bacteria.

2. Cortex

The Cortex of the root consists of mainly parenchyma cells. In many Monocotyledons, cortex consists of much sclerenchyma in addition to parenchyma. In aquatic plants, the cortex is largely aerenchymatous enclosing air cavities. The air spaces may have either a schizogenous or a lysigenous origin or both.

The cortical cells are usually devoid of plastids but accumulate various amounts of starch. Some aerial climbers such as *Tinospora* have got photosynthetic or assimilatory roots, and the chloroplasts are found in their cortical cells. The root cortex is usually wider than the stem cortex and therefore it plays a larger role in storage. The arrangement of cells may also vary, for example, in *Avena sativa,* cortical cells are arranged in distinct radial rows. In *Commelina,* the cells of inner cortex are arranged in the form of concentric rings. In most Monocotyledons, some Dicotyledons and Gymnosperms, the sub-epidermal or outer few cortical layers differentiate into a protective tissue called "exodermis". The walls of these cells become suberized, and lignin also is deposited in the walls of these cells and, in certain cases, casparian strips have also been distinguished.

The innermost layer of the cortex is differentiated as an endodermis. In the absorbing region of the root the endodermal cell wall contains suberin in a band-like region extending completely around the cell within the radial and transverse walls (Figure 11.3 A, B). This band, called "Casparian strip", is not merely a wall thickening but an integral part of the primary wall, for the deposition of suberin is continuous across the middle lamella. In the Electron microscope, the casparian strip appears as a slightly thickened, homogeneous electron opaque wall region. If plasmolysis is brought about in endodermal cells, the protoplast withdraws from the tangential walls but remains attached to the casparian strips. This type of plasmolysis is called band plasmolysis. Plasmodesmata occur in all portions of the endodermal cell walls except for the casparian strips in which the plasmalemma adheres. In transpiring plants, the solute moves through the cortex in the cell walls (free space of apoplastic pathway). It is blocked in its apoplastic progress by casparian strip and is forced to pass through the protoplast of the endodermal cells (symplastic pathway), and resume the apoplastic course in the vascular cylinder until the solute enters tracheary cells. Thus, the movement of solutes is subjected to the control of the differentially permeable protoplast membranes within the endodermis.

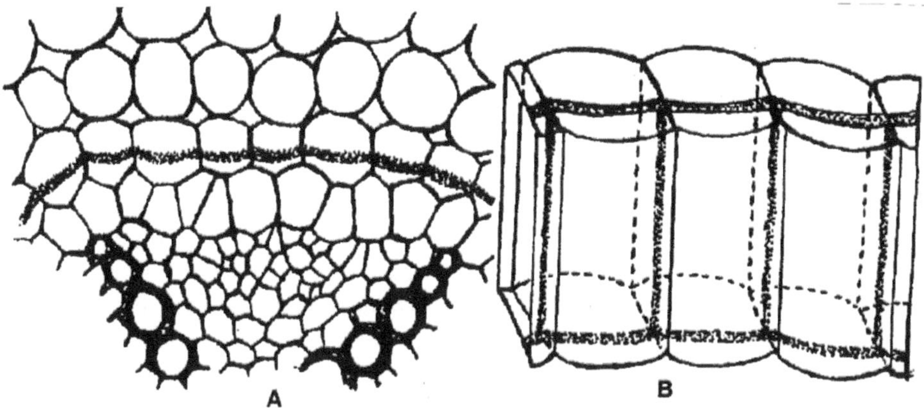

Figure 11.3. Structure of Endodermis. A. Cross section of part of *Convolvulus arvensis* showing position of endodermis with regard to xylem and phloem; B. Diagram of three connected endodermal cells oriented as they are in A; Casparian strip occurs in transverse and radial walls (that is, in all anticlinal walls) but is absent in tangential walls.

In roots with considerable amount of secondary growth, the endodermis is usually cast off together with the cortex, but in those remaining in primary state (commonly root of Monocotyledons), it often develops thick secondary walls. These walls typically consist of a suberin lamella covered by layers of lignified cellulose. The formation of the secondary wall may be delayed in the endodermal cells opposite the xylem. Such thin walled cells (without casparian strips) in an otherwise thick-walled endodermis are called *passage cells.*

The endodermis is usually uniseriate, but in many Asteraceae, certain young endodermal cells divide tangentially, and schizogenous secretory canals develop in the two-layered zones.

The exact and basic function of endodermis is still not clear. Some functions, on the basis of its relation to water and vascular tissues have been attributed to it. (i) It is considered as a water-tight jacket between xylem and cortex, (ii) It acts like an air-dam due to presence of compactly set thickened cells, and prevents diffusion of air into xylem which would otherwise get clogged, (iii) It is a protective layer, sometimes referred to as "accessory epidermis", (iv) It regulates and maintains root pressure, (v) It provides regulated and systematized flow of water through its passage cells, (vi) The thin walled cells retain power of division, as such, the lateral branch and adventitious buds often originate through this layer.

In many plants, the walls of the cells of the outer sub-epidermal layers of the cortex become suberized. In this way, the exodermis, a protective tissue, is formed. This exodermis is similar in structure and cytochemical characteristics to the endodermis. Its cells may have casparian strips, but more commonly they appear to have a suberin lamella covered by a cellulose wall. The exodermis contains either one kind of cell or short and long cells. The exodermis may be several layers in thickness. The cortex is sclerified beneath the exodermis in Poaceae, Arecaceae and Cyperaceae. In *Phoenix,* the exodermis is fibrous.

3. Vascular Cylinder

The primary vascular tissue is surrounded by a region of cells which is termed the pericycle. The pericycle is commonly a single layer of cells in thickness but may be multiseriate. The pericycle may be composed entirely of parenchyma or it may contain sclerenchyma. It may be interrupted by the elements of protoxylem. The primordia of the lateral roots in all Spermatophytes and the phellogen and portions of the vascular cambium in the Dicotyledons develop from the pericycle. Schizogenous secretory canals occur in the pericycle of the Apiaceae. In roots of certain water plants and parasites, the pericycle is absent.

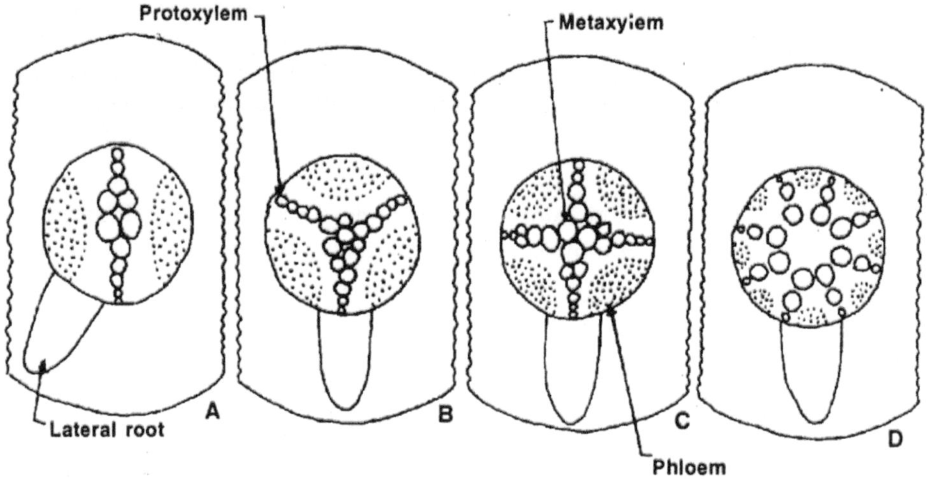

Figure 11.4: Different Patterns Formed by Primary Xylem in Cross Sections of Roots and the Position of Lateral Root with Regard to Xylem and Phloem of Main Root. A. Diarch; B. Triarch; C. Tetrarch; D. Polyarch.

The xylem usually forms a solid core with ridge like projections, seen as radial files of cells in transections, extending towards the pericycle. The strands of phloem alternate with the xylem ridges. The number of xylem groups varies in different species and in relation to this variation the roots are called monarch, diarch, triarch, tetrarch *etc.* or polyarch. *Trapa natans* has a thin root which is monarch. Diarch roots (Figure 11.4 A) are found in *Lycopersicon, Nicotiana, Beta, Raphanus, Daucus* and *Linum*. In *Pisum* the root is triarch (Figure 11.4 B), while in *Cicer, Vicia, Ranunculus* and *Gossypium,* it is tetrarch (Figure 11.4 C). Polyarch arrangement (Figure 11.4D) is characteristic of the adventitious roots of Monocotyledons. In

Monocotyledons, the number of xylem strands in the seminal root is small, as in Dicotyledonous roots, but the adventitious roots are polyarch and the number of strands in the Arecaceae and the Pandanaceae may be 100 or more.

The primary vascular structure of Monocotyledon roots is highly variable and often complex. In some, the center is occupied by a single metaxylem vessel, (such as *Triticum*), in others a circle of such vessels surround pith (*Zea* and *Iris*). The number of protoxylem points may exceed a hundred in large palm roots, and in some Monocotyledons, phloem strands are scattered throughout the xylem core.

The tracheary elements in the root mature centripetally and therefore the xylem is exarch, *i.e.* the protoxylem is situated on the outer side of the metaxylem. The differentiation of the phloem is also centripetal so that the protophloem is closest to the pericycle, while the metaphloem is closest to the axis of the root.

The companion cells are characteristic of the metaphloem but may be absent in protophloem. In roots having secondary growth, the cells located between the xylem and the phloem eventually functions as a vascular cambium. In roots without secondary growth, the cells in the same location mature as parenchyma or sclerenchyma cells.

The primary phloem of the roots of most plants does not contain fibers but in certain plants such as those of the Papilionaceae, Malvaceae and Annonaceae, fibres are found in the primary phloem.

Mycorrhizae

The epidermis and cortex of roots of many plants are often associated with soil fungi. This association between fungal hyphae and the young roots of higher plants is known as *mycorrhiza* (Greek: *Mykes* = Fungus, *rhiza* = Root). Usually this association is symbiotic, both the higher plant and the fungus derive benefit from the association.

Mycorrhizae are classified according to the relationship of the fungus to the cortical cells of the host. The two principal forms of mycorrhizae are: (i) "Ectomycorrhizae", in which the fungus envelops the entire root tip with a dense sheath called hyphal mantle and enters intercellular spaces, (ii) "Endomycorrhizae", in which the fungus forms an inconspicuous mantle but invades the interior of the cells.

Ectomycorrhizae

In ectomycorrhizae, fungus remains predominantly on the outside of the root, with a small amount of penetration between the outer cortical cells. Ectomycorrhizal roots are short, branched and appear swollen. On the surface of the root, the fungal mycelium forms a thick, dense sheath or "mantle". From the sheath, hyphae penetrate into the intercellular spaces of the epidermis and cortex. No intracellular penetration into epidermal or cortical cells occurs, but an extensive network called the "harting net" is formed between these cells. In ectomycorrhizal roots, the development of root hairs is depressed and the volumes of the apical meristem and root cap may be reduced. The ectomycorrhizae occur in many plants, *e.g., Pinus, Picea, Abies, Cedrus, Larix, Quercus, Castanea, Fagus, Betula, Populus, Carya, Salix* and *Eucalyptus.*

The ectomycorrhizas are the most advanced symbiotic association, between the higher plants and the fungi, that involves about 3 per cent of the seed plants including a majority of the forest trees. In this type of association, the root system of the plant is completely surrounded by a sheath of fungal tissue which can be more than 100 μm thick (usually up to 50 μm thick). The hyphae penetrate between the outermost layers, called *Hartig net,* from which a network of hyphal elements (hyphae, strands and rhizomorphs) extend out to explore the soil domain and interface with the fungal tissue of the root. The fungi are mainly Basidiomycota which include the common wood land mushrooms (like *Amanita* species, *Boletus* species, *Tricholoma* species (which can be highly specific like *Boletus elegans* with Larch, and non-specific like *Amanita muscaria* with 20 or more of tree species). Similarly, about 40 fungal species are capable of forming the mycorrhizae with Pine.

The fungi can also link together many trees (the mycelium acting as "Wood wide web" (according to Wohlleben, 2017) or more formally "Common Mycorrhizal Networks" (according to Gilbert and Johnson, 2017). These fungi depend on the host plant for their carbon sources, most being uncompetitive as saprotrophs. The fungi are unable to utilize cellulose and lignin with few exceptions, such as *Tricholoma fumosum,* but the fungus provides enhanced mineral uptake for the plant and the fungus is able to capture the nutrients, in particular phosphate and ammonium ions, which the root cannot access. Without ectomycorrhizas, the host plants grow poorly. This group is reasonably homogenous, but a subgroup, "ectendomycorrhizas" has been appended.

Endomycorrhizas

In the endomycorrhizae, there is little surface mycelium but an extensive network within the root. The endomycorrhizal roots are similar to uninfected roots in form but are darker in colour. The endomycorrhizae are divided into two groups – those involving septate fungi and those involving non-septate fungi. The Septate fungi are rare and form associations only with orchids and some members of the order Ericales. The orchids are extremely dependent on their mycorrhizal associations. If proper fungi were not invaded, the seedlings would die. So they are essential for successful plant growth. The endomycorrhizae occur in the Orchidaceae, Ericaceae, *Acer, Liriodendron, Ornithogalum umbellatum etc.*

(i) VAM (Vesicular-Arbuscular Mycorrhizae)

The endomycorrhizae involving non-septate fungi are the common type present in most of the terrestrial plants. After penetrating, the host produces characteristic vesicles and arbuscules. So these are called VAM (Vesicular-Arbuscular Mycorrhizae). In the soil, the fungus extends as a thin mycelium and penetrates either into the cell or between the cells, when a hypha contacts a suitable root. They do not break either the plasmalemma or the tonoplast of the host while they extend into the cell. The hypha usually coils extensively, creating large surface area that produces contact with the cell. Other hyphae ramify throughout the cortex, pushing the cells slightly apart. Some of these hyphae swell, forming the vesicles for which they are named as VAM. In cortex, one of the hyphae invade the cells, ramifies and dichotomizes, forming an arbuscule that looks as a miniature tree. The hyphae of the arbuscules get filled up with phosphate granules which are absorbed by the plant. Once the arbuscule has completed its transfer of phosphate to the plant, it collapses and the plant cell returns to a state similar to that of an uninvaded cell. While this happens, the plant cell responds by increasing the amount of cytoplasm it has. The protoplastids become chromoplasts. They do not accumulate starch, as the sugars are being converted to the fungi. Plant also responds by increasing the amount of vascular tissue in the area and increasing the lignin content. The fungus never penetrates the endodermis and stele.

The endomycorrhizas comprise 3 major and 2 minor groups. The commonest and the first to evolve are "AM endomycorrhizas", the members of Glomeromycota. These are obligate biotrophs and are associated with the roots of about 80 per cent of plant species including many crop plants. The association is endotrophic and this has been earlier referred to as VAM.

(ii) Ericoid Endomycorrhizas

The Ericoid endomycorrhizas are those of *Erica*, *Calluna*, and *Vaccinium*. The fungi are the members of Ascomycota. The rootlets are covered with a sparse network of hyphae, and the fungus digests the polypeptides saprophytically and passes the absorbed nitrogen to the host. The fungus may even provide the host with carbon sources in extremely harsh conditions. Two subgroups may be separated out of this group: Arbutoid endomycorrhizas and Monotropoid endomycorrhizas.

Arbutoid Endomycorrhizas

These are a type of endomycorrhizal fungi that appear similar to ectomycorrhizal fungi.

They form a fungal sheath that encompasses the roots of the plant. However, the hyphae penetrate the cortical cells of the plant root differentiating it from ectomycorrhizal fungi.

Monotropoid Endomycorrhizas

These occur in the species of *Monotropa*, a non-chlorophyllous plant that grows under the forest trees like *Fagus, Pinus, Quercus* and *Salix* as an epiparasite depending on the fungal partner. Both *Monotropa* and the associated trees are connected by the mycelium of a common mycorrhizal fungus *Boletus*. This association is very similar to orchidoid mycorrhiza, and the host cells are penetrated by the hyphal pegs that are restricted to epidermal layers. In *Monotropa*, the root system is highly adapted to gain all its nutrients through mycorrhizas. The roots are surrounded by a dense fungal sheath from which the hyphae spread into the soil. The sheath encloses the root apex in *Monotropa* and *Pterospora* while the apex remains free in *Sarcodes sanguine*. In all the three above genera, a Hartig net surrounds the outer epidermal layer, which fails to penetrate underlying cortex. The individual hyphae from the Hartig net grow out into the outer cortical cells, the walls of which invaginate to accommodate the growing hyphae. Such intrusions that begin to develop from the Hartig net and extend from the invaginated plant cell wall into the cortical cell (intracellularly) walls are called as the "fungal pegs".

Orchidaceous endomycorrhizas

The Orchidaceous endomycorrhizas are similar to EM but the carbon nutrition is even more dedicated to the supporting host as the young orchid seedling is non-photosynthetic and depends on the fungus utilisinhg complex carbon sources in the soil and making the carbohydrates available to the young orchid. The best example is the basidiomycete genus *Rhizoctonia*.

Ectendomycorrhizas

In addition to the ectomycorrhizas and the endomycorrhizas, there is still another important group referred to as the Ectendomycorrhizas, which is purely a descriptive name for the roots that show the characters of both ecto- and endomycorrhizas. These are restricted to the genera, *Pinus, Picea* and to a lesser extent, *Larix*. These have the same characteristics as EM but also show extensive intracellular penetration of the fungal hyphae in the living cells of host root.

The chief role of mycorrhizal fungi appears to be conversion of minerals of the soil and of the decaying organic material into forms accessible to the host. The host is presumably secreting sugars, amino acids and other organic materials, making them available to the fungus. Mycorrhizae also make the host plant less susceptible to drought. Some mycorrhizae increase the resistance of the host to pathogenic infections.

Through this association, the host plant as well as the fungus are benefited.

Benefits to the Plant from the Fungus

Mycorrhizal associations are particularly beneficial in areas where the soil does not contain sufficient nitrogen and phosphorus, as well as in areas where water is not easily accessible. Because the mycorrhizal mycelia are much finer and smaller in diameter than roots and root hairs, they vastly increase the surface area for absorption of water, phosphorus, amino acids, and nitrogen—almost like a second set of roots! As these nutrients are essential for plant growth, plants with mycorrhizal associations have a leg-up on their non-mycorrhizal associated counterparts that rely solely on roots for the uptake of materials. Without mycorrhiza, plants can be out-competed, possibly leading to a change in the plant composition of the area. The additional studies have found that the plants with fungal association are more resistant to certain soil-borne diseases, because the fungi can be an effective method of disease control as they create a physical barrier between the pathogens and the plant roots.

The mycorrhizas also thicken the cell walls of the root through lignifications and the production of other carbohydrates – compete with the pathogens for the uptake of the essential nutrients – stimulate the production of metabolites by the plant that increase the resistance to disease - stimulate the flavonolic wall infusions which prevent the formation of the lesions and prevent the pathogen invasion – increase the plant root concentrations of ortho-dihydroxy-phenol and other allo-chemicals to deter pathogenic activity. In addition, the fungi can also impart resistance to toxicity and the resistance to the insects to the host improving the fitness and vigour of the plant.

If the relationships are more complex, the fungi can connect the individual plants within a mycorrhizal network, and such a network functions to transport the material (like water, carbon and other nutrients) from plant to plant, and also provides defence communication via chemicals signifying an attack within the network. Such signals are used to start producing the natural insect repellents, and also use them to producing an attractant to bring in natural predators of the pests. In some cases, the fungi also allow the host plants to bypass the necessity for the soil uptake (like trees in dystrophic forests where the phosphates and other nutrients are taken directly from the leaf litter through the hyphae).

These fungi are also interact with and change the environment in favour of the host, by improving soil structure (important in highly compacted or sandy soils for the survival of the plants) and soil quality. The mycorrhgizal filaments create humic compounds, polysaccharides, and glycoproteins which bind the soil, increase the soil porosity, and promote the aeration and water movement into the soil. Some ectomycorrhizal associations create structures that host nitrogen fixing bacteria which contribute to the amount of nitrogen taken up by the plants in nitrogen poor environments, and also play an important part in the nitrogen cycle (the fungi do not fix nitrogen by themselves).

Benefits to the Fungi from the Host Plant

As the host is provided with enough water and the nutrients, it produces glucose and sucrose by photosynthesis, and some of this is made directly accessible to the fungus. The fixed carbon thus provided to the fungus functions as a trigger for the uptake of nitrogen and transport, which is necessary for the growth and reproduction of the fungus.

Root Nodules

The root nodules are found on the roots of plants, in particular, the legumes, that form a symbiotic association with the nitrogen fixing bacteria. Under nitrogen-limiting conditions, the plants form a symbiotic relationship with a host-specific strain of bacteria known as "rhizobia", and this process has evolved multiple times within the legumes (beans, peas, and soybeans) and other species within the Rosid clade.

Within the "root nodules", nitrogen in gaseous state from the atmosphere is converted into amino acids (the building blocks of the proteins) and nucleotides (the building blocks of DNA and RNA as well as ATP) and other cellular constituents like vitamins, flavones and hormones. The ability to fix the gaseous nitrogen makes the legumes ideal organisms as their requirement for nitrogen fertilizers is reduced. The high nitrogen content blocks the development of the nodule as there is no benefit for the plant to form a symbiotic association. The energy that is required to split the nitrogen gas in the root nodule comes from the sugar that is translocated from the leaf (a photosynthetic product). A breakdown product of sucrose, malate is the direct carbon source for the bacteroid. The process of nitrogen fixation in the root nodule is very oxygen sensitive. The root nodules also harbour an iron-containing protein, "leghaemoglobin" (related to animal protein, myoglobin) to facilitate the oxygen diffusion for respiration.

Roots may have an association with bacteria that is beneficial to the plant. The association leads to a development of root nodules, a phenomenon particularly characteristic of Fabaceae. The bacteria capable of infecting the roots of legumes and stimulating nodule development belong to several strains (sometimes given the status of species) of the genus *Rhizobium*. Each strain of the bacterium is able to infect the roots of only a particular group of legumes, or in the case of the root nodule forming bacteria of soyabean, only one species. The bacteria are free living in the soil, but fix nitrogen only in symbiotic association with the host legume. The seedling roots may be infected by *Rhizobium* at any time after root hairs are present on them. The bacteria in the soil are attracted by secretions from the root, especially of tryptophan, which they convert to Indole Acetic Acid (IAA). The presence of this auxin may cause a root hair to grow around the bacteria nearby. The bacteria invade the root, chiefly through the root hairs, and in multiplying, form an infection thread by becoming enclosed in a sheath of gum like material. The thread penetrates deeply into the root and induces a proliferation of the inner cortical cells. This proliferation, superficially resembling a branch root primordium becomes root nodule. The mature nodule has a characteristic shape, depending upon the species of legume; it consists of an outer layer of small cells containing a vascular trace, and an inner region of much larger cells which become vacuolate and are filled with bacteria. The bacteria fix atmospheric nitrogen, using carbohydrates from the host legume as an energy source for the reduction of N_2 to NH_3 by the enzyme Nitrogenase. After 8-10 weeks the nodule begins to break down, and eventually falls from the root and disintegrates. The bacteria are then released into the soils as free-living coccoid cells.

Symbiosis

The most commonly limiting nutrient in plants is nitrogen. The legumes use nitrogen fixing bacteria, the symbiotic rhizobia bacteria, with in their root nodules counter the limitation. The bacteria convert nitrogen gas to ammonia during nitrogen fixation, and ammonia is then assimilated into nucleotides, amino acids, vitamins and flavones which are essential to the plant growth. The root cells convert sugar into organic acids which then supply to the rhizobia in exchange. Hence, a symbiotic relationship exists between the rhizobia and the legumes.

Symbiosis in Leguminaceae

The plants that continue nitrogen fixation include the family Leguminaceae (or Fabaceae) with the taxa such as clovers, beans, soybeans, alfalfa, lupines, peanuts and others contain symbiotic bacteria called the rhizobia present within the nodules help in producing the nitrogen compounds that help the plants to grow and compete with other plants. When the plants die, the fixed nitrogen is released and is made available to other plants and also helps to fertilize the soil. A few genera such as *Styphnolobium* do not possess this association. In many traditional farming practices, the crops are rotated through various types which include one main legume crop such as clover.

Symbiosis in Non-legumes

There are some exceptions to the families of plants that posses the nitrogen-fixing root nodules other than legumes, and these are represented by *Parasponia*, a tropical genus of Cannabinaceae which interacts with rhizobia to form nitrogen-fixing nodules. In addition, actinorhizal plants like alder and bayberry can also form nodules in symbiotic association with *Frankia* bacteria, and these belong to about 25 genera distributed among 8 families. The ability to fix nitrogen is not universal. For example, of 122 genera in Rosaceae, only 4 genera are capable of nitrogen fixation. These belong to the orders Cucurbitales, Fagales and Rosales which together with Fabales form a clade of Eurosids. The Fabales were the first lineage to branch off. As such the ability to fix nitrogen may be plesiomorphic and subsequently lost to the descendants of the original nitrogen fixers. It may be that the basic genetic and physiological requirements in an incipient state in the last common ancestors and have been evolved to full function in some of them. Other families are represented by Betulaceae (*Alnus*), Coriariaceae (*Coriaria*), Myricaceae (*Myrica*), Rhamnaceae (*Ceanothus*), Rosaceae (*Cercocarpus*), Cannabinaceae (*Trema*), Casuarinaceae (*Casuarina*), Datoscaceae (*Datosca*), and Elaeagnaceae (*Elaeagnus*) *etc.*

Types of Root Nodules

There are two main types of nodules: Determinate and Indeterminate.

Determinate Nodules

These are found on certain tribes of tropical legumes (genera *Glycine, Phaseolus,* and *Vigna*) and some temperate legumes (like *Lotus*). These lose their meristematic activity shortly after the initiation, as such the growth is due to cellular expansion which results in mature nodules that are spherical. Another type is found in a wider range of herbs, shrubs and trees (like *Arachis*), which are associated with the axils of the lateral or adventitious roots. These are formed following the infection through the cracks where the roots emerge. These do not use the root hairs with a different internal structure from those of soybean.

Indeterminate Nodules

These types are found in a majority of the legumes from all the 3 sub-families in temperate and tropical regions (these can be seen in Faboideae like *Pisum, Medicago, Trifolium*, and *Vicia*; and mimosoid legumes like *Acacia*; and a few caesalpinoid legumes like Partridge pea). Because these maintain an active apical meristem which produces new cells for the growth over the life of the nodule, earn their name as "indeterminate". These nodules manifest zones which demarcate different stages of development/symbiosis during their active growth.

The root nodules on the non-legume genera such as *Parasponia* in association with *Rhizobium*, and those with Actinobacteia *Frankia* in genera like *Alnus*, vary significantly from those formed in the legume-rhizobia symbiosis. In these symbioses, the bacteria are never released from the infection thread. *Frankia* nodulates approximately 200 species in families (such as: Coriariaceae, Datiscaceae, Betulaceae, Casuarinaceae, Myricaceae, Rhamnaceae, Ealeagnaceae and Rosaceae). Some fungi produce nodule like structures, called "tubercular ectomycorrhizae" on the roots of their hosts (*Suillus tomentosus* with its host, *Pinus contorta* var. *latifolia*). These host nitrogen-fixing bacteria help in nitrogen fixation and allow the pines to colonize nitrogen poor sites.

The indeterminate nodules are characterized by persistent nodule meristem and these get elongated due to cell division in the nodule. These show a development gradient, which shows different zones due to different stages of development. These are as follows;

Zone I: This refers to the active meristem where the new tissues are formed.

Zone II: This is the infectious zone which shows infectious threads with the bacteria.

Zone III: This is the nitrogen-fixing zone, that shows the bacteroids. The cells contain a large central vacuole. The pigment Leg-haemoglobin (LegHb) in the cells gives the nodules the characteristic pink colour.

Zone IV: The senescent zone is the region where the plant and the bacterial cells are degraded together. The LegHb is broken down and gives the geenish colour to the base of the nodules.

Such types of nodules are more common in pea, clover, alfalfa *etc.*

Root Nodule Formation - Different Steps

The root nodules which are common in the roots of legumes are formed due to the association with a nitrogen-fixing bacterium, *Rhizobium* (the general term used for different genera like *Rhizobium, Bradyrhizobium, Azorhizobium etc.*). As the plants cannot take atmospheric nitrogen directly, the bacteria fix the atmospheric nitrogen to ammonia which can be taken up by the plants and utilized in the synthesis of the essential macromolecules like amino acids and nucleotides *etc.* This is an effective way to supplement the soil and increase its nitrogen content. It is also used as a bio-fertilizer which reduces the use of the chemical fertilizers. The crop rotation also helps in the release of nitrogen into the soil which is made available for other plants.

The different steps associated with the formation of the root nodules are as follows:

When the soil has a low level of nitrogen, the nodule formation is initiated. Both the partners of symbiosis use cell signalling for the association and developing nodules. The different steps are:

1. The legume roots secrete flavonoids to attract the rhizobia towards the root and congregate around the root hairs.
2. The rhizobia secrete *nod* factors which cause the curling of the root hairs around them.
3. The *nod* factors stimulate many developmental changes (for example, the membrane depolarization, the curling of the root hairs, the cell division in the root cortex and the intra-cellular calcium movement).
4. The *nod* factor attaches to the receptors that are present on the plasma membrane of the root hairs, which lead to the formation of the infection thread. The rhizobia can also enter through the cracks in the cells of the root.
5. The infection thread provides the passage to the bacteria to enter the cells of the epidermis.
6. The rhizobia then enter the cells of the cortex, and each bacterium gets surrounded by a plant-derived membrane, "symbiosome".
7. The formation of the nodule is initiated by the chemicals produced by the bacteria, and this is a result of calcium dependent signal transduction pathway that triggers the biochemical changes which lead to cell division and the formation of the nodule.
8. The cytokinin also plays an important role in the formation of the nodule.
9. The bacteria within the nodules get differentiated into bacteroids which fix nitrogen and vascular tissues are developed in the nodules for the exchange of the nutrients

The Process of Root Nodulation

The legumes release flavonoids (the organic compounds as secondary metabolites) from their roots which attract the rhizobia which activate *nod* genes (initiate root hair curling) in the bacteria to produce *nod* factors that initiate the formation of the nodules. The process of curling begins with the very tip of the root hair curling around the *Rhizobium*. This is followed by the formation of a small tube, the infection thread within the root tip. This provides a pathway for the bacterium to travel into the epidermal cells as the root hair continues to curl. The partial curling can be achieved even by nod factor alone, which was demonstrated by the isolation and the application of *nod* factors to the parts of the root hair, and even the application on the lateral roots also caused curling. This proved that it is the *nod* factor which is responsible for the stimulation of curling. A number of biochemical and morphological changes happen when the *nod* factor is sensed by the root. The very first change is that the cell division gets triggered in the root to create the nodule, followed by the redirection of the growth of the root hair to curl around the bacterium multiple times until it completely encapsulates one or more bacteria. Such encapsulated bacteria divide multiple times to form a micro-colony, from which the bacteria enter the developing nodule through the infection thread, that grows through the root hair into the basal part of the epidermal cell, and into the cortex of the root onwards, which are then surrounded by symbiosome-membrane that is plant-derived to differentiate into bacteroid which fix nitrogen. Approximately 4 weeks after plating the crop, effective nodulation takes place, with the size and the shape of the nodules dependent on the crop. Soybeans or the peanuts have larger nodules than the forage legumes (like red clover or alfalfa) since their nitrogen needs are higher. The status of nitrogen fixation in the plant is indicated by the number of the nodules and their internal colour.

The Factors that affect Root Nodulation

The factors that control nodulation in the roots are both external (such as heat, acidity, drought and nitrate content) and internal (such as auto-regulation and ethylene). If the soil is rich in its nitrogen content, it interferes with the formation of the nodules and symbiotic association as the plants have already enough nitrogen and they need not need more. The fixation of nitrogen is an oxygen-sensitive process, and the diffusion of oxygen is facilitated by LegHb, a pink-coloured, haeme pigment of the root nodule. The nodulation is also regulated by auto-regulation by the tissues of the leaf. The plants have evolved defense mechanisms to even check the infection. Ethylene also regulates the nodule formation internally and the exogenous application has shown to inhibit the formation of the nodules.

The auto-regulation of nodulation controls the number of the nodules per plant through a systemic process involving the leaf, because tissues of the leaf sense the early nodulation events in the root through an unknown chemical signal, and restrict further development of the nodules in the new root tissue. The LRR (Leucine Rich Repeat) Receptor Kinases (like NARK in soybean [*Glycine max*]; HAR1 in *Lotus japonicus*; SUNN in *Medicago truncatula etc.*, are essential for AON (auto-regulation of nodulation); and the mutation leading to the loss of function in these AON receptor kinases lead to super-nodulation (or hyper-nodulation). Very often, the growth abnormalities accompany the loss of AON receptor kinase activity, and this suggests the nodule growth and the development of the root are linked together finally. The investigations into the mechanisms of formation of the nodule also showed that the ENOD40 gene, that codes for a 12-13 amino acid protein is up-regulated during the formation of the nodule.

Connection to Root Structure

The nodules have apparently evolved 3 times within Fabaceae but are rare outside the family, and the propensity of these plants to develop the nodules seems to relate to the structure of the root. The scientists are of the opinion that the tendency to develop the lateral roots in response to abscissic acid may enable the later evolution of the nodules.

Legumes and Rhizobia

Many legumes provide a home for the symbiotic nitrogen fixing bacteria, rhizobia in their root nodules, and such a relationship is common in nitrogen limited conditions. The bacteria convert atmospheric nitrogen into ammonia, which is used in the formation of amino acids and nucleotides. The bacteria normally live in the soil and can exist without a host, and when the legumes encounter low nitrogen conditions in the soil and want to form a symbiotic association with the bacteria, release flavonoids into the soil. The bacteria respond and release the nodulation factor (*nod* factor) and this stimulates the formation of the nodules in host roots. Such an exposure to nod factor triggers the formation of deformed root hairs which permit the bacteria to enter into the plant, which form an infection thread (an intercellular tube that penetrates the cells of the host, as such the bacteria enter the cells of the host through the deformed root hair. Sometimes, the bacteria can also enter the root by inserting themselves between the cracks of the root cells (such an infection is called "crack entry"). The bacteria enter the cells of the root from the intercellular space during which they use the infection thread to penetrate the cell walls. The infection triggers rapid cell division in the root cells and results in a nodule of tissue.

Root Development

Root Apical Meristem (RAM) is a complex structure. To understand its development it is better to analyze simple systems. In this view point, *Arabidopsis* root is one of the ideal systems for the study of plant organogenesis. Using molecular genetic approaches, rapid progress has been made in identifying some of the pathways regulating root development. The primary root grows mainly by cell division within its meristematic zone. Newly produced cells elongate the root that was built during embryogenesis. When cells leave the meristem they are incorporated into the zone of elongation. Here they elongate along the longitudinal axis of the root. Cell elongation also contributes to the root growth. Finally, the cells enter the zone of differentiation, where they begin to produce the features that will later characterize them.

Theories Relating to Root Apex Organization

Several theories have been proposed to describe the root apex organization.

(a) **Apical cell theory:** Nageli (1868) proposed that a single tetrahedral apical cell in the root apices brings about growth. It has three cutting faces for giving rise to the tissues of the body of the root. The root cap is derived from its base. But, such a type of development is not found in Angiosperms.

(b) **Histogen theory**: Hanstein (1968) put forth Histogen theory proposing that in root apex regions, three distinct histogens are found. The dermatogen giving rise to the epidermis, the periblem to the root cortex and the plerome to the vascular cylinder. Haberlandt (1914) proposed the terms protoderm, ground meristem and procambium, respectively for these histogens.

(c) **Korper-Kappe theory:** Schuepp (1917) proposed that the cells at the root apex divide in two planes. The zone with inverted 'T' type of division was referred as Korper (cap) whereas the other with straight 'T' as kappe (body). The theory basically recalls the Tunica corpus theory concept of shoot apex and similarly, it also fails to explain the differential behaviour in different species.

Organization of the RAM

Developmental zones: Primary root growth occurs at the root apical meristem (RAM). The growing root tip of *Arabidopsis* is organized into a series of concentric cell layers, comprising of an outer epidermis, and inner cortical, endodermal and pericycle layers surrounding the vascular tissue in the middle of the root. In these layers the number and arrangement of cells are invariant. The root tip is also divided into several morphological zones along the longitudinal axis.

(a) The most distal part of the root tip is the root cap. This root cap protects the root as it pushes through the soil.

(b) Next to it is the zone of root apical meristem (RAM). This is also known as "zone of cell division".

(c) Behind the RAM is the zone of elongation (where cell division ceases and there is rapid cell growth by elongation).

(d) Lastly, the zone of differentiation, where cells assume their final fate. Cells in this zone are provided by the root hairs.

Structure of RAM

Root apical meristems (RAM) in Dicotyledonous plants have two organizational schemes; "closed" (with highly organized) and "open" (tiers lacking or disorganized). These schemes are commonly believed to remain unchanged during the growth of the root. In many root meristems, cell files can be traced as a small group of initials. Often these initials are organized in a tiered arrangement, with each tier giving rise to particular tissues. In radish, tier 1, produces root cap and epidermis, tier 2, cortex and tier 3, vasculature. In maize, tier 1, produces only root cap, tier 2, epidermis and cortex, tier 3, vasculature. Other plants show no tiered arrangement or only a single initial layer. *Arabidopsis* roots are extremely small and simple. The radial pattern consists of concentric arrangements of 6 different tissues, each of a single cell layer thick.

In *Arabidopsis*, the majority of the RAM is made up of proliferating cells and it is divided into two sub-functional zones. The base of the RAM is called promeristem. This promeristem comprises of three tiers of initial cells that give rise to root. The lower tier comprises of a central group of 12 cells that give rise to the tip of the root cap. This tip of the root cap is called "columella". The columella initials are surrounded by a ring of 16 cells, which give rise to the lateral root cap and epidermis. The middle tier of cells consists of central group of four cells known as the "quiescent centre" (QC). QC cells have a characteristic structure. QC acts to maintain the initials in an undifferentiated state. This strategy ensures that after each initial cell division, one daughter cell is disconnected from the QC and allowed to differentiate. The quiescent centre contains RAM's stem cells and hence it is equivalent to the central zone of SAM. This is surrounded by a ring of 8 endodermal and cortical initials, which give rise to cortex and endodermis. The cells in the upper tier give rise to pericycle and vascular bundles.

The pericycle is made up of cells that can initiate the formation of lateral roots. All the cells in the pericycle appear to be capable of lateral root initiation, but under normal circumstances only the pericycle cells nearest to the internal xylem poles perform this function. In the center of the root, the vascular tissue has bilateral symmetry with xylem on the axis of symmetry and phloem on both sides of it.

Specification of Cell Fates and Tissue Differentiation in RAM

The cell fate and tissue differentiation in RAM of *Arabidopsis* depends upon the cell lineage, cell-to-cell interaction, position of the initial cell and regulation of the quiescent centre.

Cell lineage analysis has shown that the organization of the root is generated by a stereotypical series of cell divisions in the promeristem. Here, all these patterns of cell division are invariant.

Although pattern of cell division in the RAM of *Arabidopsis* may be invariant cell lineages, this does not mean that the fate of each cell is determined from the outset of the development. In *Arabidopsis* roots, cell-to-cell interaction is also important in determining cell fate.

Laser ablation studies in the RAM of *Arabidopsis* indicate that position rather than lineage is the important factor for determining the cell fate. For example, 1. If a cortical/endodermal initial cell in the second tier of the promeristem normally gives rise to daughter cells that divide to produce cells of the cortex and endodermis. 2. If the cortical/endodermal initial is ablated with a laser beam, the adjacent pericycle initial undergoes an unexpected division and replaces it, which assumes the pattern of cell division characteristic of cortical/endodermal initial and gives rise to cortical and endodermal derivatives. Thus, the cell changes fate from pericycle to cortical initial when its position is altered. 3. If one daughter cell is ablated, the underlying cortical/endodermal initial undergoes an additional division to generate a new one. 4. If three cortical cells are ablated, the cortical endodermal initial cell underlying the middle daughter cell is no longer able to produce any more daughter cells that divide. Therefore, this experiment reveals that signals from the cortical daughter cell control the developmental potency of the underlying initial cells.

The quiescent centre also regulates the differentiation of neighbouring cells. If one or two QC cells were ablated, the columella initials that were no longer in direct contact with a QC cell ceased to divide and underwent differentiation as a columella cell. Similarly, cortical initial cells that were no longer in direct contact with the QC cell behaved as cortical daughter cells and underwent an asymmetric division to generate cortical and endodermal cell files. Therefore, direct contact with the QC inhibits the differentiation of initial cells in the RAM (van den Berg, 1997).

Molecular Basis of RAM Development

The developmental potency of endodermal/cortical initial cells is found to be interfered by several *Arabidopsis* mutants like *short root mutant, scare-crow mutant* and *pinocchio mutant.*

1. In short root mutant, the endodermal and cortical initial cells fail to divide and produce a single file of cells that differentiates as cortex. Here, the endodermis is absent.
2. In scare-crow mutant, the endodermal and cortical initial cells fail to divide and produce a single tier of cells that is an intermediate in character between the cortex and endodermis.
3. In Pinocchio mutant, the initial cells fail to divide but here endodermis is present while the cortex is missing.

Most laser ablation experiments result in normal development of the root. Because, when the ablated cells are crushed, they are replaced by the divisions of the adjacent cell. This shows that cell-cell interactions (or) cell-cell signaling play a very important role in the normal development of the root. The *fass mutant* confirms this.

1. The *fass mutant* consists of fat roots with more number of cells than extra cell layers. Therefore, this *fass mutant* can partially overcome the deficiency of *scare-crow mutant* phenotype, as the *scare-crow mutant* is deficient in the generation of extra cell layers. When these two mutants are combined, the double *fass + scare-crow mutant* has both cortex and endodermis.
2. When the double *fass* mutant and short root mutant are combined, even though extra layers are formed, they still lack endodermis. By this it is clear that in short root mutant, the defect must be in molecular pathway

that determines the identity of the endodermal cells but not the lack of extra cell layers.

3. In another root mutant, where the phenotype is analogous to the *shoot meristemless* and effect the proliferation in RAM, the corresponding genes are named as - "Root meristemless-1" and "Root meristemless-2".

These genes do not show any defect in SAM proliferation, whereas the cell proliferation is abolished in short root mutant where endodermis is absent. This indicates that the cell proliferation of RAM is linked with the ability of the endodermal/cortical initials, which divide, and produce endodermis (or) in short mutant the endodermal initials fail to divide and do not give rise to endodermis. As the cell proliferation of RAM is linked with the ability of the endodermal initials, no cell proliferation of RAM could be seen in short root mutant.

Biology and Development of the Root Hairs

Root hairs are tubular-shaped, tip-growing cells that arise from root epidermal cells known as "trichoblasts". Root hairs are thought to increase the absorption capacity of the root. Root hairs are normally 0.1-1.0 cm long. They can be extremely numerous with more than 20,000 hairs/cm^2 of root and billions per plant increasing a vast surface area for absorption. Root hairs are short lived and no hairs are seen on older part of the root. The root hair zone is therefore extremely important for water and mineral nutrient uptake. They are also sites of attachment of soil mycorrhizae. Like all tip growing cells (pollen tube tip, fungal mycelia, moss protonema), root hairs grow by the addition of cell wall precursors to the tip delivered by vesicles from the endomembrane system. Because these non-dividing cells elongate outside the body of the root, their internal contents are easy to observe and experimentally manipulate. In recent years, root hairs are being used as experimental models to study biological processes.

Root Hair Initiation

Dolan *et al.* (1994) reported the details of root hair initiation in *Arabidopsis*. The root epidermal cells can be differentiated into two cell types, root hair cells (trichoblasts = *hc*) and hairless cells (atrichoblasts = *nhc*). In *Arabidopsis*, epidermal cells are arranged in 16 to 22 cell files. Cells in 8 symmetrically positioned files differentiate into the hair cells, and the cells of the other files, into hairless cells. The position of root hair cell file is determined by the underlying cortical cells, which are eight in number. Root hairs always are formed from the epidermal cells positioned over the radial cell wall between cortical cells; and atrichoblast is formed from the epidermal cells positioned on top of cortical cells. Root hair cells can also be cytologically distinguished by their dense cytoplasm than neighbouring epidermal cells at early stages. Thus, cell fate has been specified prior to cell elongation.

After specification, the epidermal cells that are committed to the root hair fate become wider, longer and deeper by diffuse growth. Before the hair begins to grow small GTP-binding proteins from the ROP family appear at the growth site. Within minutes of ROP protein localization the root hair cell wall begins to bulge out. ROP protein remains at the tip of the developing hair until growth ends. As the bulge enlarges, the endoplasmic reticulum within it condenses and actin (a component of microfilaments) accumulates. Under optimal conditions, it takes about 30 minutes for formation of root hair swelling on the surface of the cells.

Root Hair Tip Growth

Once a swelling is formed, the rest of the hair is developed by tip growth. During tip growth new cell wall material is synthesized and transported to the cell tip. The tip is packed with vesicles containing cell wall polysaccharides and cell wall proteins. These vesicles are produced by smooth and rough endoplasmic reticulum and dictyosomes. The nucleus enters the tube and moves along the tip until the tip is growing. Actin is concentrated at the tip throughout the tip growth. Root hair tip growth requires calcium. The calcium gradient at the tip of the hair controls the direction of the growth. If the calcium ion concentration is artificially increased towards one side of the tip, the hair will re-orient to grow in that direction. Root proteins that are involved in root hair initiation also have a strong effect on the direction of the tip growth. Microtubules also control the direction of the growth.

Cessation of Root Tip Growth

Root hair tip ceases its growth when hairs reach a predictable length. The end of the growth is precisely controlled and produces a symmetrical, dome-shaped tip with the same diameter as the hair shaft. When hairs stop growing, the cytoplasm at the tip disperses and the vacuole enlarges into the dome. ROP protein, the calcium gradient and calcium channel activity are lost from the tip.

Regulators of Root Hair Growth

Phytohormones like kinetin, abscisic acid (at 10^{-9} to 10^{-5} M), brassinolide (10^{-10} to 10^{-6} M) and the synthetic hormone 2,4-D (10^{-14} to 10^{-5} M), stimulate root hair elongation(Izumo *et al.,* 1994). Marked promotion of growth was observed with 2,4-D (10^{-7} M) and kinetin (10^{-6} M). Further it was found that kinetin and abscisic acid cause branching and other morphological changes such as swelling of the tip or the base, twisting, curling and bending.

Dolan *et al.* (1994) have suggested that ethylene may induce trichoblast differentiation in the epidermis. Smith *et al.* (1994) have shown that the protein-phosphate inhibitors kadaic acid and calyculin-A (at concentrations as low as 3 nM) inhibit root hair growth and alter the cortical cell shape of *Arabidopsis* roots. The results indicate that these protein phosphatases have a role in controlling growth and development of root hairs.

Determination of Root Hair Development

The determination of which epidermal cell of the root will become root hairs is dependent upon whether GL2 gene is up or down-regulated. If GL2 is up-regulated the cell retains the atrichoblast fate. If GL2 is silenced, the cell becomes trichoblastic and a root hair develops. Here CAPRICE (CPC) protein promotes differentiation of hair-forming cells by controlling a negative regulator. GLABRA 2 (GL2) is preferentially expressed in hairless cells. It is observed that CPC is also predominantly expressed in the hairless cells, but not in neighbouring hair-forming cells. Then, the CPC protein moves to the hair-forming cells through plasmodesmata and represses the GL2 expression. If the bHLH protein interacts with CPC protein, GL2 expression takes place. GL2 transcription stimulatory substance is thought to pass from the underlying cortical cells into the overlying epidermal cells. At the edges of the cortical cells the apoplast between the adjoining cells results in a region incapable of generating this message and transporting it into the overlying epidermis. Hence, epidermal cells positioned over a cell-cell junction in the cortical cells receive less of this stimulatory substance, resulting in the formation of trichoblast.

Galway *et al.* (1994) have shown that the TTG gene (transparent testa glabrous) of *Arabidopsis* is required to specify epidermal cell fate and cell patterning in the root, because the normal positioning of root hairs over anticlinal walls of cortical cells is absent. They suggest that the TTG gene normally provides positional signals, so that differentiating epidermal cells to misinterpret their position and differentiate into inappropriate cell types. The TTG mutations were also associated with abnormalities in the morphology and organization of cells within and surrounding the root apical meristem and the gene may therefore have a fundamental role in development. The functions of root hair are: (a) Root hair plays a role in water and mineral uptake. (b) Root hair is an entry point for oxygen diffusion, and (c) Root hair is a site for mycorrhizal interaction.

Quiescent Centre

The middle tier of cells in the RAM comprises a central group of four large non-dividing cells known as the Quiescent Centre (QC). QC cells have a characteristic structure. QC maintains the initials in an undifferentiated state. This strategy ensures that after each division of the initial, one daughter cell is disconnected from the QC and allowed to differentiate. Clowes (1958) in his autoradiographic studies of DNA synthesis in the root tip of *Zea mays*, noted central cup-like region, the cells of which had fewer mitochondria and ER, very small nuclei and low rate of DNA and protein synthesis. He referred first this inactive region as 'quiescent centre'. Since then such regions have been reported in many plants. The cells of the quiescent centre do not divide but may become active when active initials are damaged. Thus, this region acts as a reservoir of initial cell producing cells. A ring of eight cortical/endodermal initials, that give rise to the cortex and endodermis, surrounds Quiescent centre.

Extensive research on root apex has shown that a group of cells at the pole of the stele and cortex has a very low mitotic activity. This group of cells, which appears in the shape of a hemisphere or a disc, was termed "quiescent centre". The concept states that the most distal cells in the body of the root (the former initials of plerome and periblem) divide infrequently, show little change in size, and have low rates of synthesis of nucleic acids and protein. In apices with 125,000 to 250,000 meristematic cells, the quiescent centre contains 500 to 1000 cells.

Various views have been expressed on possible causes of the appearance of a quiescent centre in roots. One view suggests control by hormones which may be synthesized in the quiescent centre, a substance may stimulate division at low concentration and inhibit at high concentration. Another view is that there is a competition between cells for supplies of nutrients or hormones. It was also suggested that the root cap controls quiescence in the particular portion of the meristem. One of the regulator of growth in plants is pressure. Pressure exerted by the rapidly dividing neighbouring cells was thus suggested to cause the inactivity in the Quiescent Centre. The way the files of cells in the root tip run towards the Quiescent centre also suggests that the pressure extended by the expansion of the temporary initials (around the quiescent centre) and their derivatives prevents the quiescent centre from expanding and thus, its cells from dividing.

Primary Differentiation

At various distances from the apical meristem, cells enlarge and develop their specific characteristics in relation to their position in the root; the cells become differentiated. The meristems of the three regions of the root, the epidermis, the cortex and the vascular cylinder, are delimited close to the apical meristem.

The protoderm of Monocotyledon root is set apart as a derivative of the initials that also give rise to the cortex. In the Dicotyledons, the protoderm is separated by a stepwise periclinal divisions (T divisions) from the derivatives of the layer of initials that forms the root cap. In roots with no distinct stories of initials, the protoderm has a common origin with the root cap and cortex and is not differentiated in early stages of development.

In the root region where vascular tissues are differentiating, the young epidermis proceeds with the formation of root hairs. Root hair-forming cells may not differ from those remaining devoid of hairs. In some plants, however, the protodermal cells that give rise to root hairs are smaller than their sister cells and show other features indicating specialization as hair-forming cells, or trichoblasts. Root hairs arise as small papillae in the region where cell division is subsiding and reach their full development at a root level where at least the first xylem is mature.

Cortical ground meristem consists of longitudinal files of cells attaining their maximum number by periclinal divisions (T divisions) close to the apex. The increase in circumference of the cortical layers occurs by anticlinal divisions. In many roots, the periclinal divisions progress in a centripetal order, that is they are repeated several times in the innermost layer of the nascent cortex. After the periclinal divisions in the inner cortex are completed the inner most layer differentiates as the endodermis.

The vascular cylinder is derived from procambium, or provascular tissue, appearing as a column in the centre of the root. The pericycle is usually the first cell layer in the vascular cylinder that can be identified as such. It delimits the vascular core close to the apical meristem. Within the core, enlargement and increasing vacuolation of certain cells block out the xylem pattern (diarch, triarch *etc.*) (Figure 11.4). Because the metaxylem cells exceed the protoxylem cells in width, they are the first to become recognizable at the onset of enlargement and vacuolation. However, the deposition of secondary walls and functional maturation occur first in the protoxylem and later in the metaxylem. Thus, the primary xylem differentiates centripetally and is exarch. Phloem differentiation occurs in the same direction, that is, centripetally, for the protophloem appears next to the pericycle and the metaphloem deeper in the vascular core.

The longitudinal differentiation of the primary vascular tissues in the root is acropetal, the first phloem maturing closer to the apical meristem than the first xylem (Figure 11.5).

Lateral Roots

The lateral root primordia are formed from the pericycle cells just behind the root hair zone. Although all the cells in pericycle are capable to initiate lateral root growth, only those that lie close to the xylem strands do this. The lateral root primordia usually arise in partially differentiated pericycle cells that have large vacuoles. But, by that time the adjacent endodermal cells have developed casparian strips. In Corn, the cells of the pericycle and endodermis are thick-walled and lignified. The first visible change associated with the initiation of the primordium is a thinning of the wall and loss of lignin. The lateral root primordia do not arise from single cells in the pericycle but always form a group of cells. These are about 30 in *Raphanus* and 24 in *Vicia faba*. In *Arabidopsis*, approximately 14 pericycle cells become a lateral root primordium before the secondary root meristem appears. When they begin to divide, the cells become more densely cytoplasmic.

The pericyclic origin of the lateral root places it in close juxtaposition with the vascular tissues of the parent root with which the vascular tissues of the lateral root eventually become connected. The position of the lateral root with regard to xylem ridges of the parent root varies in relation to the vascular pattern of the parent root but is stable in a root with a given pattern. In a diarch root, the lateral root arises between phloem and xylem, in a triarch, tetrach and so forth, opposite the xylem, in a polyarch Monocotyledon root, opposite the phloem (Figure 11.4).

When the phloem and xylem begin to differentiate in the lateral root these tissues become connected with the equivalent tissues in the parent root by a differentiation of the intervening parenchyma cells into vascular elements.

Lateral Root Initiation

One of the first events in root initiation is a periclinal division in a group of pericycle cells resulting in formation of two layers. They are designated as inner layer (IL) and outer layer (OL). Cells of the primordia begin to differentiate immediately after initiation.

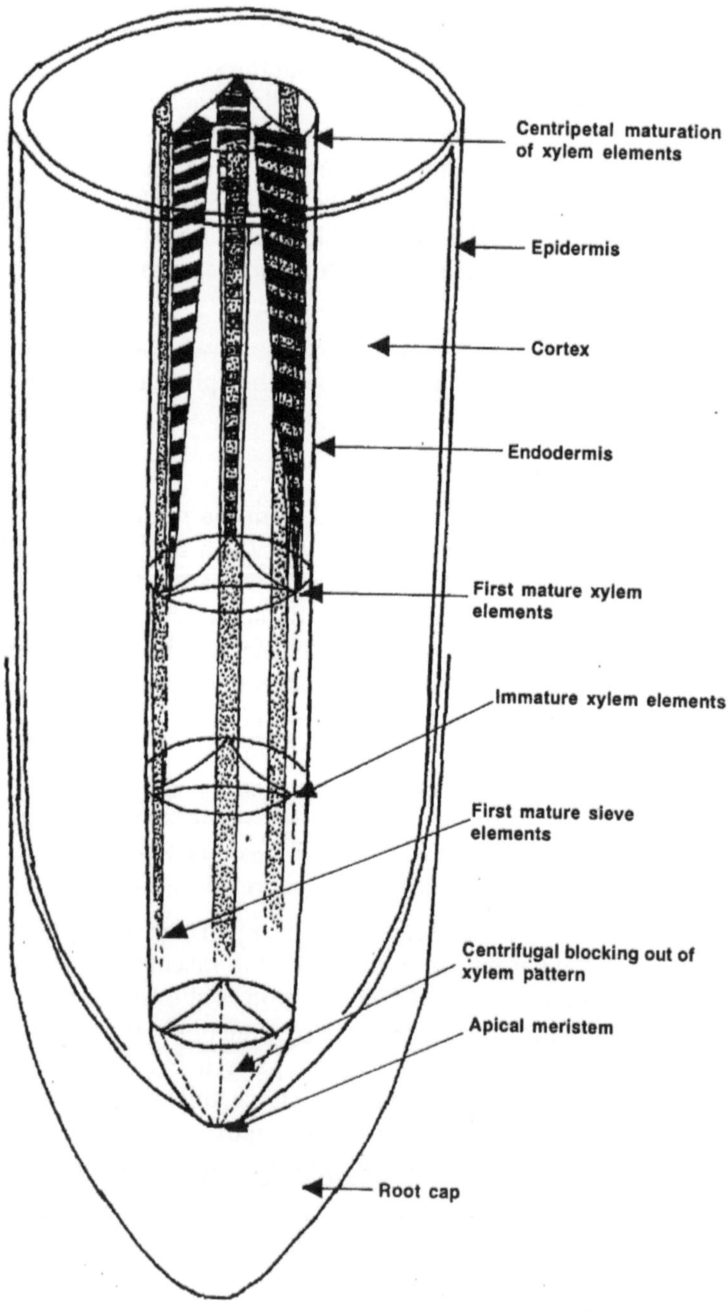

Figure 11.5. Diagram of Primary Vascular Differentiation in a Root of *Pisum sativum*.

Seedlings growth with limited nitrogen and high levels of sucrose initiate very few lateral roots. Sometimes, even the first division in the pericycle cells cannot be seen. Increasing nitrogen or decreasing sucrose restores initiation. High levels of salts such as NH_4NO_3 inhibit lateral root formation. Closer examination shows that lateral root initiation is unaffected, but lateral root primordia arrest take place. It appears that plants can regulate root system development in response to the environment both at the levels of initiation and at the level of development.

Development of Lateral Root

Once the lateral root primordium is formed, it is differentiated into a mature lateral root by a two stage process. First, the primordium emerges through the overlaying tissues by cell expansion. The increase in cell size is apparent in cells near the base of the primordium, while cell number remains relatively unchanged. Secondly, the new lateral root begins to elongate, and cell numbers increase at the root tip, as the lateral root tip is meristematic. This is similar to that of the primary root elongation by division of cells in the root apical meristem. Due to periclinal divisions in outer and inner layer of cells two derivative cell layers are formed. They are designated as OL_1, OL_2, IL_1 and IL_2. OL_1 by one more periclinal division constitutes the root cap and epidermis, OL_2 by periclinal division results in two layers of which the outer one constitutes the cortex and inner layer develops into the endodermis. IL_1 develops into pericycle and IL_2 into vasculature. In all the Angiosperms that have been studied, the endodermis contributes to the root primordium at least in the initial stage. It usually divides anticlinally producing a unistratose sheath around the elongating primordium. Apparently, whatever tissue is provided by the endodermis is just temporary. It is ruptured even before the root emerges from the cortex. Exceptionally, it has been reported that in *Zea mays,* the root cap of the new root is derived from the endodermis of the parent root. When the lateral root primordium has extended about 1/3 of the way through the cortex, the endodermal cells that cover its apex undergo periclinal division thereby producing a root cap meristem in *Zea mays.*

After completion of lateral root formation, it emerges out from the mother root. By that time, a well-defined apical meristem has been established and even a root cap may be present. The emergence of the root causes considerable damage to the cortex and epidermis and it results in an unprotected rupture that is a danger point for penetration by bacteria and fungi. A special corky tissue is formed and deposited around this to seal it off. As the lateral root begins to form its own endodermis, parenchyma cells at the point of origin of the root connects the endodermis of the parent root.

This method of endogenous origin for root is extremely beneficial, because if the soil becomes dry the epidermis and cortex of the older portion of the root may die. But, the endodermis, stele and other internal tissues remain alive. Lateral roots play an important role in water and nutrient uptake largely by increasing the root surface area.

Root Buds and Propagation

Some plants can produce new stems (called 'suckers') from the adventitious buds formed on their roots. The new seedlings emerge near a mother plant from below the ground; and these are carefully separated and transplanted in the rainy season. For example, *Millingonia hortensis* (the Great Indian Cork tree) and *Artocarpus heterophyllus* (Jack Fruit tree) are the trees that are widely cultivated by root cuttings in our country. In some plants (like Carrot and Sweet Potato), the buds are produced at the base of the old stem just above the tap root and help in vegetative propagation.

Formation of buds on roots is a familiar phenomenon which is utilized in propagation of plants by root cuttings. Root buds frequently arise endogenously as do lateral roots. The earliest stages of root and bud differentiation appeared to be identical, but later stages differed not only in rates of development but also in histologic organization and kind of participation of parent tissues in the growth of primordium. Root growth is regulated by the gravity perception mechanism (statoliths) and a light perception (phytochrome), bud growth is regulated by light received by the root during its occasional approach to the surface of the soil. Buds on roots are seen in *Murraya koenigii* and *Millingtonia hortensis*.

In vegetative propagation by roots, the root of the plant is taken and planted in the soil – by adventitious buds and from the reproductive roots. The roots of some plants develop new plants by producing the buds ('radical buds' or 'root buds'). Such buds grow as the leafy shoots above the ground (called the 'slips') and each slip gives rise to a new plant. The buds detach from the root and develop into new plants. The root tubers develop adventitious buds in Sweet Potato, *Dahlia, Asparagus* and Tapioca *etc.* Similarly, the tap roots of some plants develop adventitious buds to form the new plants (the shoots are produced by the roots of woody plants like *Dalbergia, Murraya* and *Millingtonia etc.*).

Contractile Roots or Pull Roots

A contractile root is a thickened, modified specialized root that develops from the base of a stem (the corm, bulb, rosette of other organ) which is designed to shrink vertically under conditions of seasonal drought that helps position the plant part at an appropriate level in the ground. The new branch develops at a higher level in the soil than the older one because the roots shorten and pull it down to a suitable level.

The anchorage of a plant body by means of roots may involve root contraction, a process that pulls the shoot closer to the ground or in bulbous plants deeper into the soil. The root contraction is widely distributed among Monocotyledons and herbaceous perennial Dicotyledons. The contractions usually occur in certain individual roots and are limited to parts of these roots. The shortening of the root depends mainly on the change in shape of the inner cortical cells, which at a certain stage of development expand radially and contract longitudinally. The root tissues not involved in this growth, that is central vascular tissues and the peripheral cortical tissues, become contorted and wrinkled.

These roots are very common in different families of Liliatae and Magnoliatae; and are also known from some species of Pteridophyta and Spermatophyta. The typical feature of these roots is the shrinkage at its proximal end and many of them swell and increase in their diameter before shrinkage. The roots show different zones (3 zones) during their development as in *Acidanthera bicolor*. The zone-I is pre-contractile (the root before swelling and shrinkage); Zone-II is the contractile stage (the proximal parts of the root are swollen); and Zone-III is the post-contractile stage (shows a wrinkled surface of the proximal root surface). The zones blend into each other as the swelling and shrinkage spread out towards the root tip and finally disappear. The contractile roots of *Oxalis bowjeana* and *Gladiolus* species are interpreted as transitory storage organs, while the species with real storage roots (*Hemerocallis fulva*) may also show contractile ability and these roots combine the functions of nutrition, contraction and storage. These roots are responsible for the movement of the underground part of the shoot (in a corm, a bulb, or a rhizome) and they must overcome the soil resistance which depend on the type of the soil, the moisture in the soil, and mechanical stability of the soil. The contraction of the root is mainly to improve the anchorage of the plant in the soil. The common examples are: *Ophioglossum multifidum* (Ophioglossaceae), *Cycas revoluta* (Cycadaceae), *Zamia* species (Cycadaceae), *Bryonia alba* (Cucurbitaceae), *Melilotus album* (Fabaceae), *Ficus benjamina* (Moraceae), *Oxalis hirta* (Oxalidceae), *Ranunculus bulbosus* (Ranunculaceae), *Agave americana* (Agavaceae), *Allium ursinum* (Alliaceae), *Narcissus pseudonarcissus* (Amaryllidaceae), *Philodendron bipinnatifidum* (Araceae), *Phoenix canariensis* (Araceae), *Asparagus officinalis* (Asparagaceae), *Muscari comosum* (Hyacinthaceae), *Lilium martagon* (Liliacae), *Musa ensete* (Musaceae), and *Cattleya crispa* (Orchidaceae) *etc.* Even the aerial roots of the trees are also contractile (*Coussapoa schottii*) and *Ficus benjamina*.

Trichoblast

A cell in on the exterior surface of a plant's root that is responsible for the production of the root hairs (of the root epidermis). The roots are covered in trichoblast cells to ensure healthy root growth and development.

Secondary Growth in Dicot Root

The secondary growth in the roots leads to increase in its thickness which happens by the addition of the vascular tissue. The process of secondary growth in the root takes place in the maturation zone soon after the cells stop elongating.

Most of the tap roots and main lateral roots of Dicotyledonous plants show secondary increase in thickness. As a rule, no secondary growth is present in Monocotyledonous roots. The primary xylem (with exarch protoxylem) and phloem tissues are arranged in separate radii in the dicotyledonous roots. Pith is usually absent. Parenchyma cells present beneath each phloem patch become meristematic and form cambial strips. The number of cambial strips is equal to the number of phloem patches present. For example, a tetrarch root is with four strips of cambium and a triarch root with three strips. Subsequently the pericycle cells outside the xylem ridges also become meristematic and join the cambial strips formed beneath the phloem groups. The cambial ring thus is initially in the form of a wavy ring, and starts producing new derivatives by periclinal divisions both inwards and outwards. The cambium also increases in circumference by anticlinal divisions. This wavy cambial ring passes internal to the primary phloem and external to the primary xylem groups. Secondary xylem produced by the cambium is more in quantity than the secondary phloem. After a while, the wavy outline of the cambium assumes circular shape.

As a consequence of secondary xylem formation, the primary xylem is pushed to the central part of the root. The primary phloem elements especially the sieve elements get crushed. At this juncture, the root is identified mainly by the radial position of the exarch primary xylem in the central region. Normally ray initials are differentiated in the cambial ring in the region just opposite to the primary xylem. They ultimately produce broad vascular rays running across the secondary vascular tissues in roots. In addition to these, narrower phloem rays are seen within the phloem tissue.

To compensate the ruptured epiblem, due to increased diameter of the root consequent upon secondary thickening, periderm is produced in the outer tissues of the root. Basically the periderm of the roots is similar to that of the stem. The pericyclic cells undergo periclinal and anticlinal divisions. The periclinal divisions cause an increase of the number of pericyclic layers in radial extent. The combined increase in thickness of the vascular tissues and of the pericycle forces the cortex outwards. The cortex does not undergo an increase in circumference but becomes ruptured and is cast off together with the epidermis and endodermis. A phellogen arises in the outer part of pericycle and forms phellem towards the outside. It may produce phelloderm towards the inside, but such phelloderm is difficult to distinguish from the pericycle which proliferated before the phellogen was initiated.

In perennial roots the activity of the vascular cambium continues through many years. The phellogen also continues its activity but may become replaced by phellogens arising at greater depths in the root. If such development occurs, the root, like a stem, has a rhytidome.

(Anomalous Secondary Growth in roots is described elsewhere in the book).

Chapter 12

The Stem

Introduction

A stem is one of the two main structural axes of a vascular plant, the other being the root. It supports the leaves, the flowers and fruits; it transports water and the dissolved substances between the root and the shoot in the xylem and phloem; stores the nutrients, and produces new living tissue. The stem is normally divided into nodes (these hold one or more number of leaves, as well as buds which grow into branches. Sometimes, adventitious roots are also produced from the nodes) and internodes (the distance between two successive nodes). The term shoot is often confused with stem – the shoot refers to the new fresh plant growth (both stems and leaves or flowers). Majority of the stems are aerial located above the soil surface) but in some plants it is underground.

The main functions of the stems are: to give the support for and the elevation of the leaves, flowers and the fruits. It keeps the leaves in the light and provides a place to keep the flowers and the fruits as well. It also helps in the transport of the fluids between the root and the shoot. It also helps in the transport of the nutrients. With the help of the meristems, it annually generates new living tissue. The xylem helps in the transport of water, while the phloem transports nutrients and food.

Primary Structure

The stem usually consists of three tissue systems, the dermal tissue system, the fundamental (ground tissue system) and the fascicular or vascular tissue system.

Epidermis

Externally, the stem is bounded by the epidermis which contains epidermal cells, guard cells, idioblasts and different types of trichomes. The stem epidermis commonly consists of one layer of cells and has a cuticle and cutinized walls. It is a living tissue capable of mitotic activity. The epidermis responds to stresses by tangential enlargement and radial divisions.

Cortex and Pith

In Dicotyledonous stems, the ground tissue remains differentiated into hypodermis, cortex and pith. Such a differentiation is not found among Monocotyledonous stem. The hypodermis in herbaceous stems is usually collenchymatous which may be in the form of a continuous ring (as in *Aristolochia*) or in the form of patches (in angular stems such as *Cucurbita* and *Peristrophe*). The chloroplasts may or may not be present in these cells. In many stems, as for instance *Pelargonium, Retama* and *Salicornia,* the parenchyma may have a photosynthetic function in addition to that of temporary storage of starch and other metabolites.

The cortex of stems contains parenchyma, usually with chloroplasts. The intercellular spaces are prominent but sometimes are largely restricted to the middle region of the cortex. In many aquatic Angiosperms, the cortex develops as an aerenchyma with a system of large intercellular spaces.

The stems of Conifers and Angiosperms commonly lack a morphologically differentiated endodermis. In young stems, the innermost layer(s) may contain abundant starch and thus be recognized as a 'starch sheath'. Some Dicotyledons, however, do develop casparian strips in the innermost cortical layer of the stem, and many lower vascular plants have a clearly differentiated endodermis in the stem.

The pith is commonly composed of parenchyma, which may contain chloroplasts. In many stems, the central part of the pith is destroyed during growth. Frequently this destruction occurs only in the internodes, whereas the nodes retain their pith (nodal diaphragms). The pith has prominent intercellular spaces, at least in the central part. The peripheral part may be distinct from the inner part having compactly arranged small cells and greater longevity. Since pith is also referred to as the medulla, the distinct peripheral zone of the pith is called peri-medullary zone or medullary sheath.

Both the cortex and pith may contain various idioblasts, including cells with crystals and other ergastic contents, and sclereides. If the plant has laticifers, they may be present in pith and cortex.

Vascular System

Internal to the cortex is the vascular system of the stem. In the Gymnosperms and most of the Dicotyledons, the vascular system consists of a continuous or a split cylinder which encloses the pith. In this cylinder, two types of vascular tissue can be distinguished, the phloem which is usually external and the xylem which is usually internal. In the case of the split cylinder, each strand is termed a vascular bundle. A vascular bundle in which the phloem is only external to the xylem is said to be a 'collateral bundle'. In some dicotyledonous families, *e.g.,*

Apocynaceae, Asclepiadaceae, Convolvulaceae, Cucurbitaceae, Solanaceae and certain tribes of Asteraceae, internal phloem is also present. This type of vascular bundle is called 'bicollateral' vascular bundle. A stele in which the primary vascular tissue is arranged in strands around the pith is called 'Eustele' *e.g.*, Dicotyledons, Gymnosperms.

The vascular bundles may be also 'concentric', in which either the phloem surrounds the xylem (amphicribral bundles) or the xylem surrounds the phloem (amphivasal bundles). The amphivasal bundles appear to be phylogenetically rather specialized and are found in certain positions in the stems of some Dicotyledons (medullary bundles in *Rheum, Rumex, Mesembryanthemum, Begonia*) and Monocotyledons (Araceae, Liliaceae, Juncaceae, Cyperaceae). Amphicribral bundles are most common in ferns but are found also in Angiosperms. Small bundles in flowers, fruits and ovules may be amphicribral.

Medullary and cortical bundles, present on the inside and outside of the stele respectively are found in certain plants. These bundles are associated with the stems of both anomalous and typical structure. 'Medullary bundles' occur in numerous Dicotyledonous families, *e.g.*, in the Amaranthaceae, Chenopodiaceae, Orobanchaceae, Berberidaceae and Cucurbitaceae. 'Cortical bundles' are less common and are known to occur, for example, in the Melastomataceae, Proteaceae, Araliaceae and Calycanthaceae. So-called "cortical bundles" are often leaf-traces which descend through the cortex for some distance before entering the stele, *e.g.*, as in *Begonia* and *Casuarina.*

In most Monocotyledons and in a few Dicotyledons, no distinct vascular cylinder exists. The primary vascular system consists of a large number of bundles which are scattered irregularly and it is impossible to distinguish clearly the boundary between the cortex, the vascular cylinder and the pith. A stele with the vascular bundles scattered within the ground tissue is called 'atactostele'.

Anatomy of the Node

In Angiosperms, the primary vascular cylinder is interrupted at each node by the exit of one or more bundles that enter the leaves. At each node, the portions of the vascular system are deflected into the leaf, attached at the node. The stelar bundles which are the continuation of bundles in the leaf bases are called 'leaf traces'. The leaf trace is a vascular bundle that connects the vascular system of the leaf with that of the stem. This is extended between the base of a leaf and the point where it is completely merged with other parts of the vascular system in the stem. One or more leaf traces may be associated with each leaf. There are 3 common types of nodes in the dicots. According to the number of leaf gaps per leaf, the node is termed unilacunar – the node with a single gap and a single trace to a leaf (Figure 12.1 A,D); trilacunar - the node with 3 gaps and 3 traces to a leaf, one median and two lateral (Figure 12.1B,C); or multilacunar – the node with several to many gaps and traces to a leaf (Figure 12.1 E). The traces also vary in length as measured by the number of internodes they traverse before they diverge towards the leaf (less than 1 internode in the Conifers, 1 to 35 in different species of Dicotyledons. Several Monocots possess the leaves with sheathing bases and nodes with a large number of leaf traces separately inserted around the stem.

The number of leaf traces and leaf gaps vary in different plants (Figure 12.1) and may vary in the same plant at different levels. In most of the vascular plants, only one trace is related to each gap. A deviating type of nodal structure is represented by the two-trace unilacunar condition (Figure 12.1F).

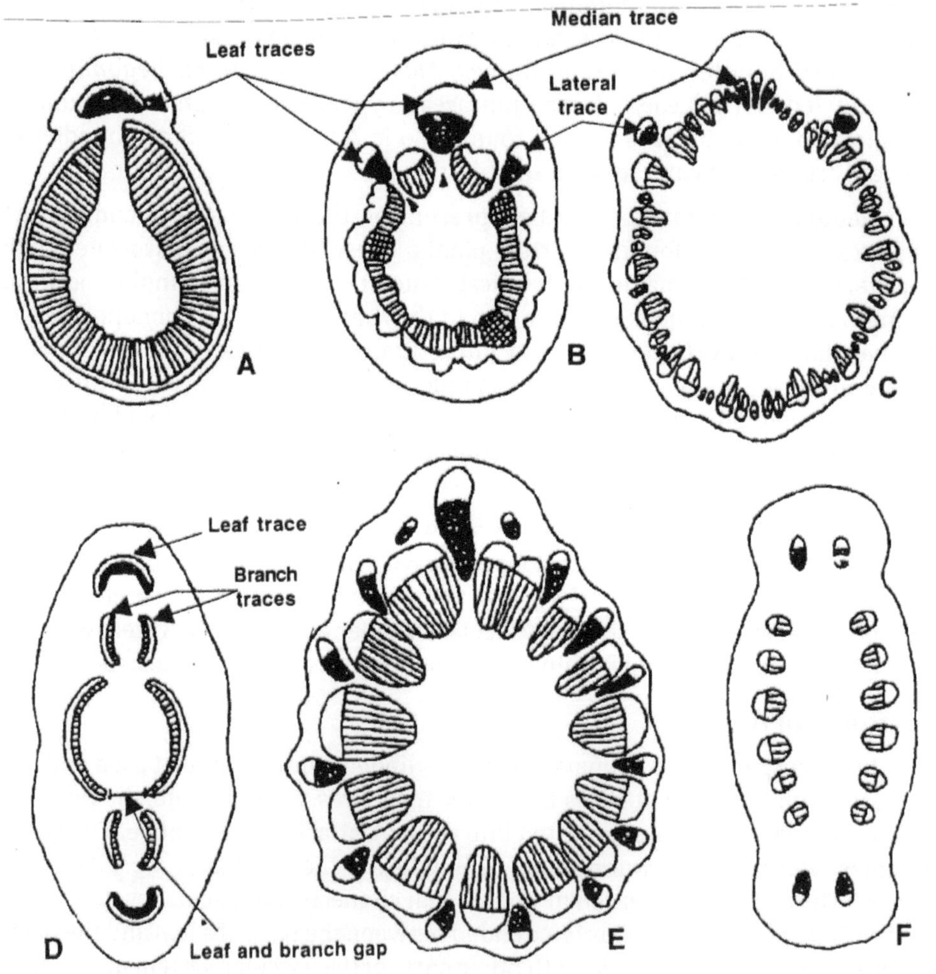

Figure 12.1: Cross Sections of Stems with different Types of Nodal Structure. Leaf traces are indicated by blackened xylem regions. A. *Spiraea*; B. *Salix*; C. *Brassica*; D. *Veronica*; E. *Rumex; F. Clerodendron.*

The anatomy of node is an important aspect of taxonomy. Sinnot (1914) concluded that the trilacunar node is the primitive type in the Angiosperms and that the unilacunar node developed, phylogenetically from it by the loss of two lateral gaps together with their respective traces or by approximation of the lateral traces to the median bundle to form a bundle composed of three traces, which is associated, therefore, with a single gap. The assumption that the trilacunar node is the primitive one among the Angiosperms has been refuted by later research on nodal anatomy in the Gymnosperms and Angiosperms (Bailey, 1956; Fahn and Bailey, 1957).

Branch Traces and Branch Gaps

The primary vascular supply to the lateral branches is also derived from the vascular system of the main axis, in the form of 2 bundles, and less often, one bundle, and these traces or strands are known as the 'branch traces' or 'ramular traces' (Eames and Mac Daniels. 1947). The Dicots and Gymnosperms have commonly 2 branch traces (connecting the vascular system of the branch to that of the main stem), while in Monocots, the connection to the axillary shoot with the main stem consists of many strands. The branch traces get extended within the main axis and the appendages are tied together by a primary vascular system. When the branch possesses 2 traces, the bundles unite within a short distance, forming a complete vascular cylinder; but when one trace occurs, this usually possesses the C.S of a typical horse-shoe shaped structure with the opening downward, and the vascular cylinder of the branch is formed by the closure of the opening as the branch traces pass out. In most of the vascular plants, the outward passage of a branch trace is associated with the formation of a break in the vascular cylinder around and above the point of departure of the trace. This opening is called as a 'branch gap' which always accompanies a branch trace. The branch gaps are present in all the vascular plants which possess a pith. But, in protosteles, the gaps do not occur because there is no pith. The branch gaps are commonly lower than the leaf gaps and extend for greater distances in the axis.

Branches develop from axillary buds and have vascular connections with main axis. These vascular connections are termed 'branch traces' (Figure 12.1 D). At the node, the branch traces are situated very close to the leaf traces that enter the leaf in whose axil the branch develops. The branch gap is situated above the leaf gap and together these two gaps appear as one. In Dicotyledons and Gymnosperms the vascular supply to the axillary branches usually consists of two traces (Figure 12.1 D). However, in some plants only one trace is present and in others more than two.

Leaf traces and stem bundles show structural differences. In transections of stems, the leaf traces are clearly circumscribed and have a larger amount of xylem than cauline bundles. As seen in longitudinal views, the leaf traces increase above their divergence from the stem bundles. Another notable feature of the leaf traces in members of diverse taxa is the occurrence of the vascular parenchyma cells differentiated as transfer cells with profusion of wall in growths.

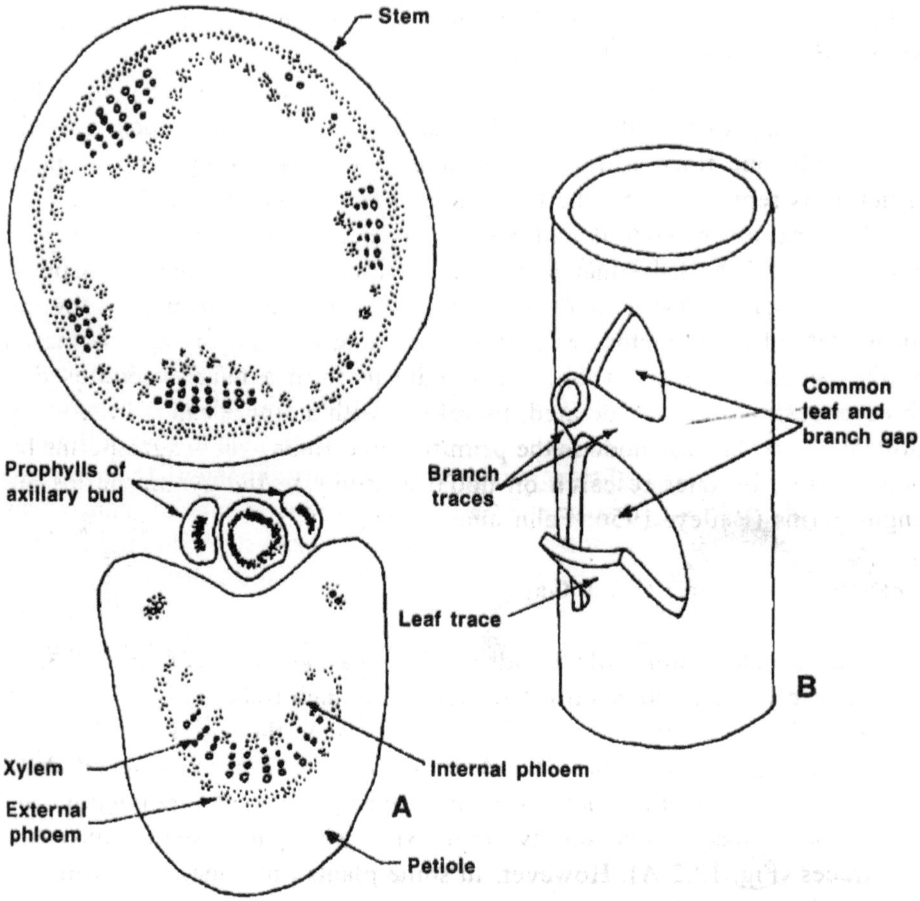

Figure 12.2: Relation between Stem, Leaf and Axillary Bud. A. Transection of shoot of *Solanum tuberosum*; B. Diagram showing relation of branch traces and a leaf trace to the vascular system in main stem.

Vascular System of Monocotyledons

The vascular system of Monocotyledons usually consists of bundles that are scattered throughout the ground tissue of the stem. Among the Dicotyledons, however, an arrangement of more or less scattered bundles does occur in a small number of plants, *e.g.*, in the Nymphaeaceae, many plants of the Ranunculaceae and the herbaceous genera of the Berberidaceae.

In the Poaceae, there are two basic types of arrangement of the vascular bundles: (1) That in which the vascular bundles are arranged in two circles, the outer circle consisting of thin bundles and the inner thick bundles, *e.g.*, *Triticum*, *Hordeum*, *Avena* and *Oryza*; and (2) That in which the vascular bundles are scattered throughout the entire cross section of the stem, *e.g. Sorghum, Saccharum* and *Zea*. The individual bundles of the Poaceae are collateral and are usually surrounded by sclerenchymatous sheath.

Types of Stele in the Stem

The pattern of arrangement of the primary vascular tissues differs in the various plant groups and sometimes even within the same species. It has been found that the different patterns illustrate various stages in the evolution of the primary vascular system. The pattern in each plant is determined in the shoot and root apices from which the vascular tissues differentiate according to its genotype.

In a vascular plant, the stele is the central part of the root or the stem, that contains the tissues derived from the procambium, which include the vascular tissue and in some cases, pith (the ground tissue).

According to older botanists, the vascular bundle is the fundamental unit in the vascular system of Pteridophytes and higher plants. The concept of stele was developed in the late 19th century, by P. E. L. Van Tieghem, and Douliot (1886), the two French botanists, as a model for understanding the relationship between the shoot and the root, and also for the discussion of the evolution of the vascular plant morphology (to interpret the body of vascular plant in a different way). At the beginning of the 21st century, the plant molecular biologists are trying to understand the genetics and the developmental pathways that govern the tissue patterns. But, physiologists are examining how the anatomy of different steles affects the function of the organs in a plant. As such, the fundamental parts of a shoot are the cortex and a central cylinder, called the stele. Hence, stele is defined as a central vascular cylinder, with or without pith and delimited by the endodermis of the cortex. The term, 'stele' has been derived from a Greek word, meaning a 'pillar'. van Tieghem and Douliot (1886) recognized only 3 types of steles, and they also thought that the monostelic shoot is rare compared to polystelic shoots. The stele of the stem remains connected with that of the leaf by a vascular connection called the leaf supply.

The stele of the sporophyte of vascular plants may be divided into two basic types: Protostele and Siphonostele.

1. Protostele

The earliest vascular plants had their stems with a central core of vascular tissue, which consisted of a cylindrical strand of xylem surrounded by a region of phloem. Around this vascular tissue there might have been an endodermis which regulated the flow of water into and out of the vascular system, and this type of arrangement has been referred to a 'protostele'. Jeffrey (1898) for the first time pointed out the stelar theory according to phylogeny. According to him, the most primitive and the simplest type of stele is the protostele. In this stele, the vascular tissue is a solid mass and the central core of xylem (central cylinder) is completely surrounded by the strand of phloem.

The protosteles are basically of 3 types: *Haplostele*, *Actinostele* and *Plectostele*.

(a) Haplostele

This is the simplest and most primitive type, in which the central solid smooth cylindrical core of xylem appears more or less circular in cross-section, and remains surrounded by a ring of phloem. An endodermis generally surrounds the stele. A 'centrarch' (protoxylem lies in the centre of a metaxylem cylinder) haplostele is seen prevalent in the members of the rhyniophyte grade, like *Rhynia*). It is also seen in *Selaginella.*

(b) Actinostele

This is a variation of the protostele (the core is lobed or fluted), and is a modification of the 'haplostele'. This is somewhat more advanced in having the central xylem core with radiating ribs in which the xylem is 'stellate' (hence the name) in cross section (*e.g., Psilotum, Lycopodium* and related genera). The actinosteles are typically exarch (the protoxylem is external to the metaxylem) and consists of several to many patches of the protoxylem at the tips of the lobes of the metyaxylem. The exarch protosteles are a defining characteristic of the lycophyte lineage.

(c) Plectostele

This is a protostele in which the plate-like regions of xylem appear in cross section surrounded by the phloem tissue. These discrete plates are interconnected in longitudinal section. This stele is the most advanced type of protostele in which the central core of xylem is divided into number of plates which are arranged parallel to each other. The phloem alternates with xylem, in which the longitudinal plates of some xylem are joined and others separate (*e.g., Lycopodium*). Some modern club mosses have the 'plectosteles' (derived from the actinostele) in their stems.

2. Siphonostele

The siphonosteles have a region of ground tissue called the pith internal to xylem. Hence, this is the modification of the protostele in which the stele is medullated (with a cylinder of pith within the xylem). The stele contains a tubular vascular region and a parenchymatous central region. The vascular strand comprises a cylinder surrounding the pith, and these often have interruptions in the vascular strand where the leaves (megaphylls) originate (referred to as the leaf gaps). Jeffrey (1898) interpreted that the vascular portion of the siphonostele possesses a parenchymatous region known as a gap immediately above the branch trace only or immediately above the leaf and branch traces. Both ontogenetically and phylogenetically the protostele is the more primitive, and it is thought that the siphonostele has developed phylogenetically from the protostele.

The types of siphonosteles can be 'ectophloic' (phloem being present only external to the xylem) siphonostele or can be 'amphiphloic' (with the phloem being present both external and internal to the xylem) siphonostele. Among the living plants, many species of ferns and some Asterid flowering plants have the latter type of stele.

(a) Ectophloic Siphonostele

In this type of siphonostele in that the pith is surrounded by concentric xylem cylinder and next to the xylem lies the concentric phloem cylinder externally.

(b) Amphiphloic Siphonostele

In this type of siphonostele, the pith is surrounded by the vascular tissue. The concentric inner phloem cylinder surrounds the central pith and next to the inner phloem lies the concentric xylem cylinder which is immediately surrounded by the outer phloem (*e.g.*, in *Marsilea*). Hence, the phloem surrounds the xylem externally and internally and the endodermis appears both on the outside and the inside of the vascular tissue on the borders of the cortex and pith respectively (*e.g.*, *Adiantum* and *Marsilea*).

On the basis of the branch and leaf gaps, Jeffrey (1910) distinguished two types of siphonosteles: 'cladosiphonic' (the leaf gaps are not found) siphonostele and 'phyllosiphonic' (in which both the leaf and branch gaps are present) siphonostele. The supporters of this theory believe that the inner endodermis and the parenchyma encircled by this have been originated from the cortex. According to Jeffrey and others, the siphonosteles with an internal endodermis are more primitive than those without an internal endodermis. The latter are believed to have been originated by disintegration of inner endodermis during the process of evolution. The siphonostele has been evolved from the protostele by a transformation of the inner vascular tissue into parenchyma, according to the theory proposed by Boodle (1901) and Gwynne Vaughan.

Siphonostele may consist of a continuous cylinder of vascular tissue or of a network of bundles. The latter type is the more advanced and the vascular tissues of this type appear in cross-section as a ring of separate bundles. The regions between the bundles are parenchymatous. Those parenchymatous regions that occur in the stele, above the positions where the leaf traces diverge from the stele to the leaves, are termed leaf gaps. The simplest form of the siphonostlele has no gaps (such as some species of *Selaginella*). But, among the simplest siphonostelic Pteropsida and Lycopsida, the successive leaf gaps in the stele do not overlap each other and are considerably apart from each other.

Such siphonosteles which lack the overlapping of gaps are called the 'solenosteles' according to Brebner (1902) and Gwynne-Vaughan (1901), which may be ectophloic or amphiphloic. But, some authors like Bower (1947), Wardlaw (1952) and Esau (1953), interpret the solenostele as an amphiphloic siphonostele in which the successive leaf gaps are of considerable distance, one from the other, is termed a 'solenostele'.

Other Types of Protostele

In addition to the three general types, several other types of protosteles can be distinguished. They are: Mixed Pith Stele, among which mainly are of the following types:

(a) Solenostele

An amphiphloic siphonostele is called a 'solenostele', if the cylinder of the vascular tissue contains no more than one leaf gap in any transverse section (*i.e.*, if has no overlapping leaf gaps). This type of stele is found primarily in the stems of ferns today.

(b) Mixed Pith Stele

In this type, the xylem elements (tracheids) are mixed with the parenchymatous cells of the pith, and this type of stele is found in primitive fossils and living ferns. Such steles are treated to be the transitional types in between the true protosteles on one hand and siphonosteles on the other (*e.g.*, in *Gleichenia* and *Osmunda*). The various types of protostele are characteristic of the Lycopsida (the lower Pteridophyta) and the siphonostele of the Pteropsida (the phylogenetically more advanced Pteridophyta and the Spermatophyta).

(c) Dictyostele

If multiple gaps in the vascular cylinder exist in any one transverse section, the successive gaps may overlap each other in the more advanced siphonosteles of Pteopsida. The numerous leaf gaps and leaf traces give a dictyostele the appearance of many isolated islands of xylem surrounded by phloem. Each of the apparently isolated units of a dictyostele can be called a meristele. Among living plants, this type of stele is found only in the stems of ferns. Such siphonosteles with overlapping gaps are called the "dictyosteles" (Brebner, 1902), and in such cases, the intervening portion of the vascular tissue between the lateral to such leaf gaps is known as a 'meristele' (protostelic type), and hence, the dictyostele with many meristeles appears like a cylindrical meshwork. An amphiphloic siphonostele with overlapping gaps *i.e.* that in which the lower part of one gap is parallel with the upper part of another gap, is termed dictyostele. In this case the bundles are interconnected to form a cylindrical network and each bundle is of concentric structure consisting of a central strand of xylem surrounded by phloem. Individually such bundles are termed meristeles. From the anatomical point of view these are *amphicribral* bundles).

(d) Eustele

The vascular arrangement of the most seed plant stems has been interpreted as a derived siphonostele and this modification is called as a 'eustele' (Brebner, 1902). In this type, the primary vascular tissue consists of vascular bundles, usually in one or two rings around the pith on its periphery. In such a case, the inter-fascicular areas and the leaf gaps are not distinguished from each other very clearly (*e.g., Equisetum*). In addition, this also appears in the roots of monocot flowering plants. The bundles can be collateral (with the phloem on only one side of the xylem developed from the ectophloic siphonostele) or bicollateral (with the phloem on both the sides of the xylem externally and internally, as seen in advanced Dicot families like Solanaceae, as a result of a secondary specialization and not a relic of the primitive structure characteristic of Filicinae). There is also a variant of eustele that is found in monocots like Maize and Rye, and this has numerous vascular bundles that are scattered in the stem, and is called an 'atactostele' (the characteristic of a monocot stem).

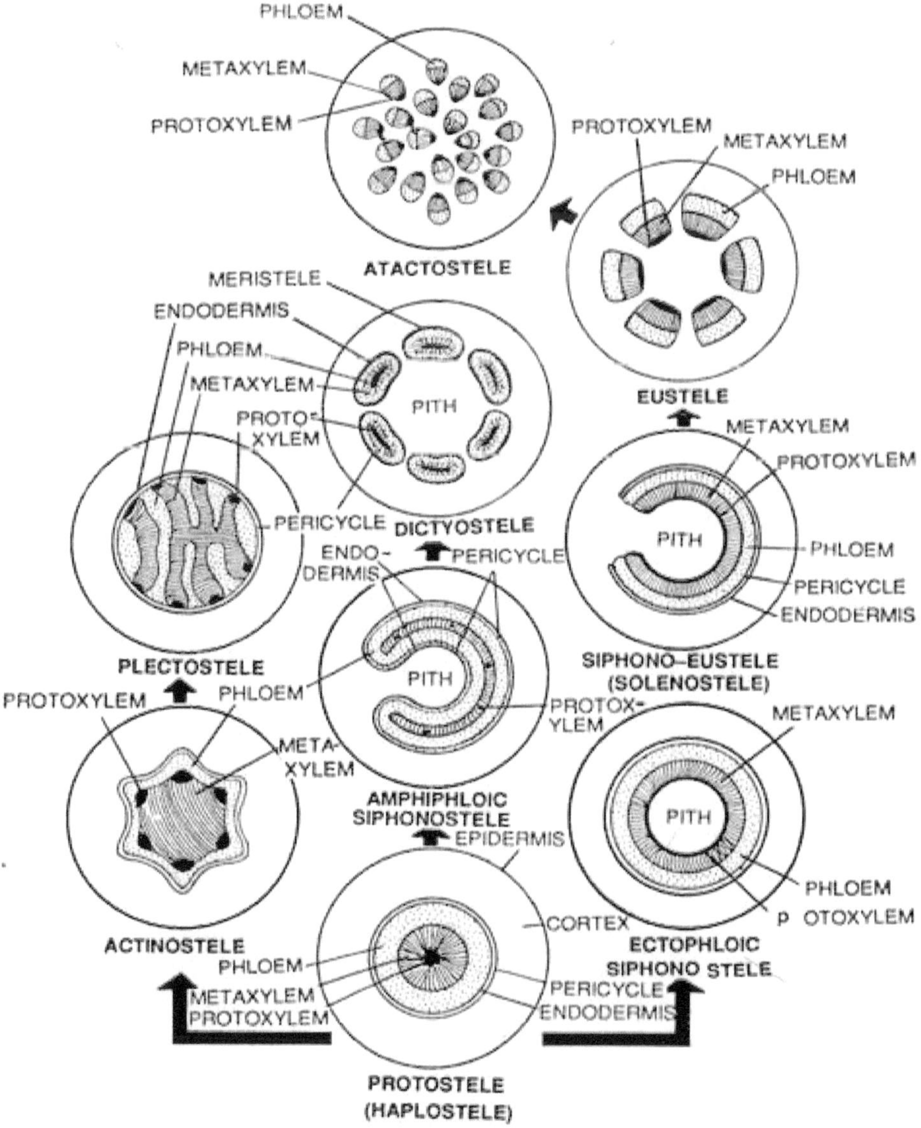

Figure 12.3: The Stelar System, Showing the Different Types of Steles Arranged in an Evolutionary Sequence.

The vascular system of eustelic plants composed of bundles that branch and anastamose is referred to as a 'closed system'. A vascular system characterized by bundles which branch, but still form independent, separate systems, is referred to as an 'open system'. In some plants, *e.g., Marattia, Pteridium*, and *Matonia*, two or more concentric cylinders of vascular tissue are present. Such a stele is termed a polycyclic stele. The individual cylinders in this case are interconnected. In rare cases, stems and roots contain more than one stele; such condition is termed 'polystelic'. Jeffrey (1902, 1910, and 1917) interpreted the evolution of the siphonostele from the protostele as follows. He supported that the parenchyma found internal to the phloem and xylem has been originated from the cortex. A different interpretation of the above nomenclature exists and has been summarized by Sporne (1962). According to this interpretation, the ectophloic siphonostele without leaf gaps, such as is found in the Pteridophyta, is considered as a protostele and is termed 'medullated protostele'. Sporne does not use the term siphonostele for the Pteridophyta.

(e) Polycyclic Stele

This type of stellar organization is the most complex one amongst all vascular cryptogams (Pteridophytes), and these steles are siphonostelic in nature. Each stele possesses an internal vascular system that is connected with an outer siphonostele (the connections are always found at the node). Typically, the polycyclic stele possesses two or more concentric rings of vascular tissue, which may be a solenostele or a dictyostele (2 concentric rings are found in *Pteridium aquilinum* and 3 in *Matonia pectinata*).

Primary Growth in Stems

The stems like the roots grow in their length by the division and elongation of the cells at their tips. The youngest cells (but not roots) are organized into two zones – the tunica (the divisions are perpendicular to the stem axis and give rise to a sheet of tissue several layers thick that cover the outside of the tip) and the corpus (the cells divide in all directions and give rise to an interior mass of cells). The cellular derivatives in both continue to divide and produce three recognizable primary meristems (transitional) – the protoderm, the ground meristem and the procambium – which elongate and differentiate to create three primary tissue systems – the dermal tissue system, the ground (or fundamental) tissue system and the vascular tissue system. The divisions of the apical meristem give rise to the leaf primordial close to the tip one after another, that the nodes and the internodes cannot be distinguished until the elongation and differentiation start. In the monocot shoots, at the base of the leaf primordial in the intermodal region, a zone of meristematic cells (the intercalary meristem) remains undifferentiated which retains the ability to divide throughout the life of the plant, which helps in the elongation of the monocot leaves from the base upward.

In contrast to the trees and shrubs, the herbaceous plants are composed of the primary tissues – and if the secondary growth does occur, the tissues are used for the rigidity but not conduction. The primary body of eudicots include two basic variations – a hollow cylinder of xylem cambium and phloem around central pith – while in others, a system of discrete vascular bundles (with xylem, cambium and phloem) arranged in a regular pattern between the pith and the cortex. In monocots, the primary body consists of bundles with no cambium which are scattered in an undifferentiated parenchyma, the ground issue. In many plants, most of the primary growth occurs primarily at the apical bud, rather than the axillary buds, and the influence of the apical bud on overall plant growth is called as the 'apical dominance', which prevents the growth of the axillary buds that form along the sides of the branches and the stems. Most of the coniferous trees exhibit a strong apical dominance producing a typical conical Christmas tree shape. If the apical bud is removed, then the axillary buds start forming the lateral branches, giving the plant an umbrella or a bushy shape.

The axis of the embryo in the seed consists of a hypocotyl and radicle. At the tip of the hypocotyl one or more cotyledons and the bud of the shoot, *i.e.*, the 'plumule', are found. At the tip of the radicle is the 'root cap'. Both shoot and root apical meristems arise during embryogenesis and act throughout the lifetime of the plant as source of new cells and as sites of organ formation. It is at the shoot apical meristem (SAM) that leaf, stem and floral primordia are formed.

Angiosperm shoot apical meristem (SAM) consists of a small number of morphologically undifferentiated, dividing cells located at the tips of shoots. The typical SAM is about 100 to 200 microns in diameter, although species with SAMs as small as 50 microns and as large as 900 microns have been reported. The morphology of SAM varies from flat (*e.g.*, sunflower) through mound shaped (*e.g.*, *Arabidopsis* vegetative SAM) to finger-shaped (*e.g.*, maize). The size and shape of the SAM vary within a plant during the plastochron (the period between the formation of successive leaves). Towards the end of the plastochron, just before the emergence of a morphologically distinct primordium, the SAM is at its largest, whereas at the beginning of a plastochron, just after a leaf primordium has been produced, the SAM is at its smallest. Presumably, near the end of the plastochron the SAM includes a leaf primordium that is not yet morphologically distinct from SAM.

One of the most obvious features of the Angiosperm SAM is its layered arrangement. The outermost region of the meristem consists of one or more sheets of cells that together comprise tunica. This sheet-like organization is a consequence of the regular orientation of cell divisions within the tunica. The tunica overlies a mass of cells called corpus (Figure 12.4 A, B). The irregular arrangement of cells within the corpus reflects the variable orientation of cell division in this region. However, in the basal/central region of the SAM, the rib meristem, composed of files of cells parallel to the long axis of the plant, can often be distinguished. Derivatives of the rib meristem give rise to the internal layers of the stem.

Cell layers within the SAM are designated as L1, L2, L3 *etc.*, where L1 refers to the most superficial layer and L2 to the next subjacent layer, and so on. Thus, the *Arabidopsis* SAM consists of three layers, L1 and L2 comprise tunica while L3 is the corpus. In maize, there are two layers in the SAM, the single-layered tunica (L1) and the corpus (L2). Despite the nearly universal occurrence of a stratified SAM, its development or function is not clearly unknown.

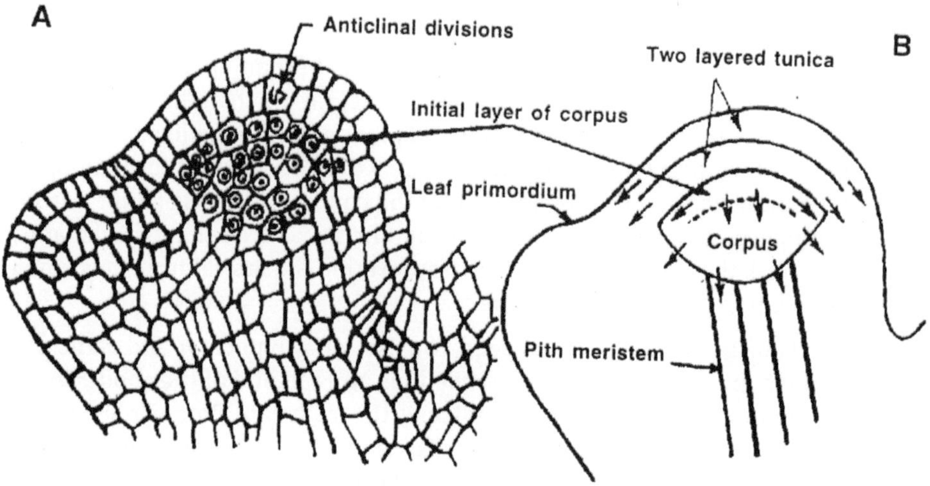

Figure 12.4. Shoot Apex of *Pisum*. Cellular details in (A) and interpretative diagram in (B) The pith meristem does not show typical rib-meristem form of growth.

The bud of the shoot usually consists of an axis, the epicotyl, containing a few internodes, which have not elongated, and some leaf primordia. With the germination of the seed, the embryo enlarges and starts to grow, the apical meristem of the young shoot adds further leaf primordia, and the internodes between the lower primordia which in the meantime have become distant from the apex, elongate. Buds develop in the axils of the developing leaves giving rise to a branched shoot. The details of shoot apex have been given earlier.

When the leaf primordium is initiated in the peripheral region of the shoot apex, localized cell divisions cause the formation of a protrusion (the so called "leaf buttress") on the side of the axis. In shoots with a helical leaf arrangement, the divisions alternate in different sectors around the circumference of the apical meristem and the resulting periodic enlargement of the apex, as seen from the above, is asymmetric. In shoots with a decussate leaf arrangement the enlargement is symmetrical because the intensified meristematic activity occurs simultaneously on opposite sides. Thus, the initiation of leaves causes periodic changes in the size and form of the shoot apex.

The differentiation of procambium is illustrated in Figure 12.5 in successive cross sections of a shoot of *Linum* (flax). No vascular differentiation is detectable above the level of leaf initiation (Figure 12.5 A). Farther below, where leaf primordia come into view, some vacuolation is present (Figure 12.5 B). Increased vacuolation in cortex and pith, delimits an approximately circular zone as seen in transections (Figure 12.5 C).

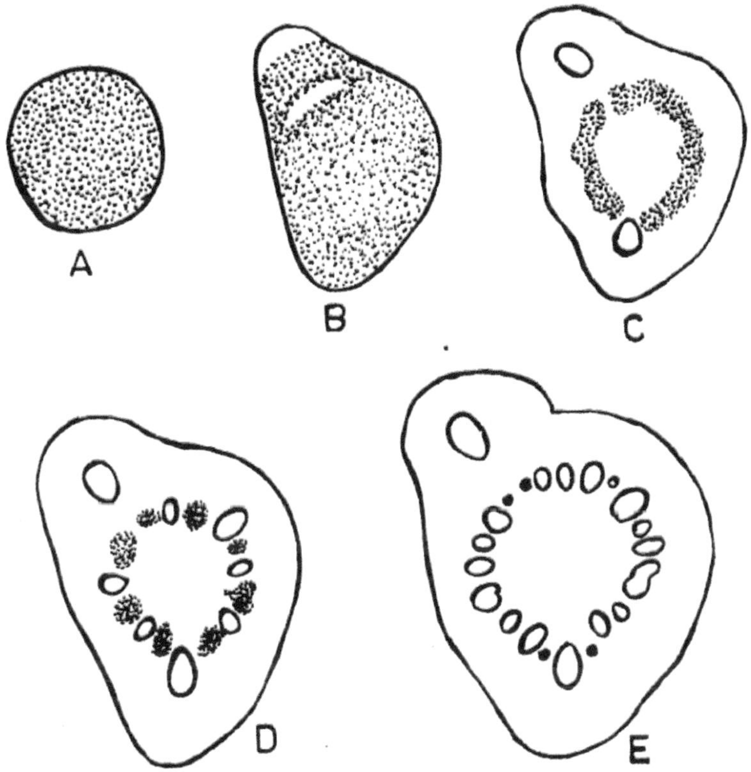

Figure 12.5: Cross Sections of Flax Stem taken at different Levels Beginning with Apical Meristem (A) and ending with node in which all primary tissues are delimited (E); Stages in differentiation of vascular system; vacuolation of cortex (starts in leaf primordium in B) and pith.

For descriptive purposes, this meristematic zone is called residual meristem. At successively later stages of development more and more cambial strands differentiate from the residual meristem (Figure 12.5 D). After all vascular bundles characteristic of the given stem level have differentiated, the remaining residual meristem differentiates into interfascicular parenchyma (Figure 12.5 E).

Branching

There are two types of branching patterns in plants: terminal and lateral. "Terminal branching" involves the separation of the apical meristem into 2, which then both continue growing as a shoot branch. This mechanism is common in lower plants but rare in angiosperms. How the single integrated growing centre is divided is not understood. Experimental manipulations such as bisecting the meristem can artificially induce terminal branching, but these do not appear to mimic any naturally occurring events. They do speak to the capacity of isolated meristem segments to reorganize into complete meristems.

"Lateral branching" in seed plants occurs mainly by growth of axillary buds. When leaf primordia are initiated, a small group of cells in their axils remain small and unvacuolated. These are referred to as "detached meristems" because they retain meristematic characteristics and do not differentiate as the cells around them do, but they become spatially separated from the apical meristem. As the shoot grows and the detached meristem attain a distance from the apex, they organize into a shoot meristem with several leaf primordia (a bud). At this point their growth is arrested because of "apical dominance" (the shoot apical meristem inhibits growth of the axillary meristems by producing the hormone auxin). In bushy plants the arrest is temporary until the main shoot has grown beyond the inhibitory range of the apex. In unbranched plants, the arrest is permanent unless the apical meristem is removed. Upon release from the apical inhibition, the lateral bud begins growth as a branch shoot.

Secondary Growth in Stems

The shoot grows in its length to give rise to the lateral organs (the branches, the leaves, the flowers, the fruits and the seeds), and at the same time, it also grows in its thickness, and this is accomplished by the primary and secondary growth. During primary growth, the stems increase in their length and thickness depending on the activity of the apical shoot meristem and the intercalary meristem; while during secondary growth, they increase in their diameter by the activity of the vascular cambium and the cork cambium. All plants perform primary growth; most of the monocots only show primary growth, while many Dicots and all Gymnosperms show both the primary and secondary growth.

The activity of the apical shoot meristem stops not far away from the shoot tip, and the growth in thickness is by increasing the cell size. The palm and other monocots may possess very thick stems by the addition of new vascular bundles in the apical tip of the stem, as well as by increasing the number of cells of parenchyma, termed as "anomalous growth", which is a consequence of the secondary growth meristem found more superficial than the vascular bundles, which gives rise to the parenchyma cells outward, and both parenchyma cells and vascular bundles inward. This is actually a non-well delimited region found in the periphery of the stem that contains the strands of pro-vasular tissue giving rise to vascular bundles and the leaf rays.

The process of secondary growth is controlled by the lateral meristems (the vascular cambium [in woody plants], and the cork cambium, and the process is similar in both the stems and the roots. The non-woody plants generally undergo the primary growth with hardly any secondary growth (increase in thickness). The wood (as a result of secondary growth) is noticeable in the woody plants (in dicots but very rare in monocots).

The following details are specific to secondary growth in stems (the details are somewhat different in the roots). The vascular cambium is located between the primary xylem (located towards the interior) and primary phloem (located towards the exterior) within the vascular bundle. The cells divide and form secondary xylem (tracheids and the vessel elements) to the inside and secondary phloem (sieve elements and companion cells) to the outside. The cells of the secondary xylem contain lignin which provides the hardness and strength. The xylem along with the pith forms the wood in a woody stem. In woody plants, the cork cambium is the outermost lateral meristem, and produces the cork cells, that contain a waxy substance which repels water. The phloem together with the cork cells forms the bark that protects the plant against any physical damage and helps to reduce water loss. The cork cambium also produces phelloderm, a layer of cells which grow inward from the cambium. The cork cambium, cork cells, and the phelloderm are termed collectively as periderm, which substitutes for the epidermis in mature plants. The action of both the vascular and the cork cambia result in secondary growth (or widening) of the stem of the plant.

Secondary growth in stem involves two processes – formation of secondary xylem and phloem and formation of periderm.

Formation of Secondary Xylem and Phloem

Secondary growth resulting from the activity of vascular cambium increases the amount of vascular tissues in stems. In Dicot stems, the vascular bundles are collateral, conjoint and open, with intra-fascicular or fascicular cambium. At the same time, some cells between the vascular bundles lying in line with fascicular cambium also become meristematic and get differentiated as inter-fascicular cambium. Thus, a complete ring of vascular cambium is formed.

The vascular cambium consists of two types of cells, fusiform initials and ray initials. Fusiform initials are many times longer than wide, while ray initials are nearly isodiametric. Fusiform initials give rise to secondary xylem and phloem while ray initials give rise to vascular rays. When the cambial initials produce secondary xylem and phloem cells they divide periclinally. At one time a derivative cell is produced toward the xylem, at another time toward the phloem although not necessarily in alternation. During the height of cambial activity, cell addition occurs so rapidly that elder cells are still meristematic when new cells are produced by the initials. Thus, a wide zone of more or less undifferentiated cells accumulates. Within this zone, the cambial zone, only one cell in a given radial file is considered to be the initial. The new cells formed towards the outer side facing the primary phloem get gradually modified into the elements of secondary phloem while those towards the inner side facing the primary xylem are modified into elements of secondary xylem. The ray initials of cambium cut off parenchymatous cells both outwards and inwards and thus form vascular rays.

Growth Layers – Annual Rings

In transverse section wood shows growth layers-"growth rings". Each growth layer is divisible into early wood and late wood. The early wood, also known as "spring wood", is less dense than the late wood, because wider cells with thinner walls predominate in the early wood, narrower cells with thicker walls in the late wood. The "late wood" forms a distinct boundary in a growth ring because of its sharp contrast to the early wood of the following season, but the change from early wood to late wood of the same growth layer is more or less gradual.

The relative amounts of early wood and late wood are affected by environmental conditions. Adverse growth conditions increase the relative amount of late wood in a Pine but decrease the relative amount of late wood in an Oak.

The successive annual rings formed year after year can be used in determining the age of a tree. The determination of the age of a tree by counting the annual rings is called "Dendrochronology".

Distribution of Vessels (Ring Porous Wood and Diffuse Porous wood)

The wood anatomist refers to a vessel in cross section as a pore. On the basis of distribution of pores in a growth layer two principal types of woods are recognised: 1. "Diffuse porous wood" with pores rather uniform in size and distribution throughout a growth ring, *e.g., Dalbergia, Shorea, Bombax, Artocarpus, Acacia, Pterocarpus, Terminalia, Mangifera, Fagus, Juglans, Liriodendron, Platanus, Populus, Pyrus etc.* and 2. "Ring porous wood" with pores distinctly large in the early wood than in the late wood, *e.g., Castanea, Morus, Quercus, Ulmus, Robinia, Tectona, Lagerstroemia, Toona, Tilia etc.*

Distribution of Axial Parenchyma

The distribution of the axial xylem parenchyma shows many intergrading patterns. The spatial relation to vessels, as seen in transections serves for its division into two main patterns; 1. "Apotracheal" - parenchyma not definitely associated with the vessels (Figure 12.6 A, B); 2. "Paratracheal" - parenchyma consistently associated with the vessels (Figure 12.6 C, D). The apotracheal parenchyma is further subdivided into: diffuse, single parenchyma cells or parenchyma strands scattered among fibres; apotracheal banded (Figure 12.6 B); boundary or marginal parenchyma, with single cells or a band at the end (terminal) or at the beginning (initial) of a growth layer (Figure 12.6 A). Diffuse apotracheal parenchyma may be sparse. The paratracheal parenchyma appears in the following forms: scanty vasicentric, forming complete sheaths around vessels; aliform vasicentric with wing-like tangential extensions (Figure 12.6 C) and confluent, coalesced aliform forming irregular tangential or diagonal bands (Figure 12.6 D). From the evolutionary aspect the apotracheal and diffuse patterns are primitive.

In a transverse section of a woody trunk, two regions, the outer light coloured and central dark coloured regions are seen. The light coloured region is called "sapwood" and dark coloured region is called "heartwood". The sapwood, also known as "albernum", consists of living and functional cells involved in conduction of water. Heartwood is also known as "duramen".

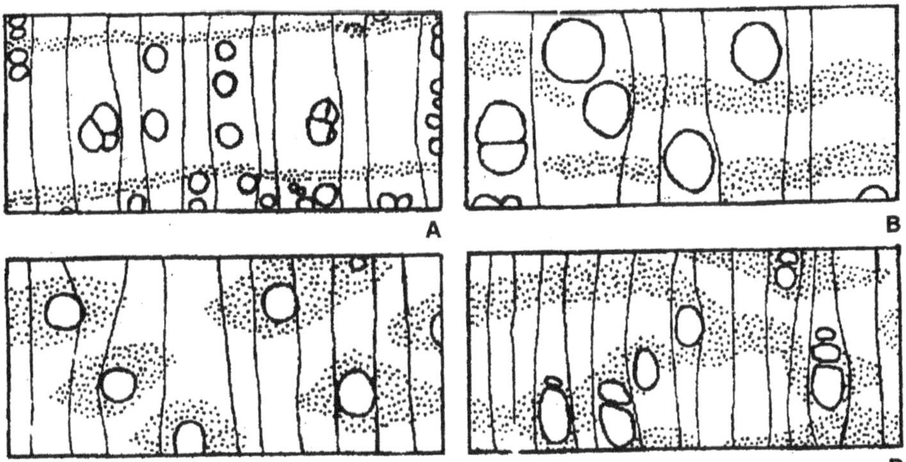

Figure 12.6: Distribution of Axial Parenchyma (Stipples) in Wood. A. Boundary apotracheal – *Michelia*; B. Banded apotracheal – *Saccopetalum*; C. Vasicentric aliform paratracheal – A leguminous species; D. Confluent paratracheal – *Terminalia.*

The earlier increments of secondary xylem gradually become nonfunctional in conduction and storage and it is known as "heartwood". Formation of heartwood involves removal of reserve materials or their conversion into heartwood substances and eventual death of protoplasts of the parenchymatic and other living cell types of the wood. Heartwood formation is explained as one of the processes enabling the plant to remove from the regions of growth those metabolic byproducts that may be inhibitory or even toxic to living cells. The translocation of these substances, possibly after their detoxification, occurs through the rays, in part towards the outer bark, in part towards the centre of the tree where the accumulation of excreta results in the death of cells. As the process continues the sapwood – heartwood boundary moves outward. The number of growth increments retaining the sapwood characteristics varies greatly among different species.

In certain cases, distinction between sapwood and heartwood is not sharp as in *Holoptelia integrifolia* and *Abies pindrow.*

The pigments found in heartwood in certain cases are of commercial importance *e.g.*, haematoxylin (*Haematoxylon campechiamum*), Brasilin (*Caesalpinia sappan*), Santalin (*Pterocarpus santalinus*).

Wood - Hardwood and Softwood

Woods are usually classified into two main groups, the softwoods and the hardwoods. The term "softwood" is applied to Gymnosperm wood, that of "hardwood" to Dicotyledon wood. The two kinds of wood show basic structural differences. The Gymnosperm wood is homogeneous in structure – with only tracheids and fibre-tracheids. The Dicotyledon wood is heterogeneous and contains vessels, tracheids, fibre-tracheids, libriform fibres and parenchyma.

On the basis of presence and absence of pores of commerce, *i.e.*, vessels, wood is classified into porous and non-porous wood. The wood in which vessels are not found is called "non-porous wood" such as that of Gymnosperms, whereas the Dicotyledonous wood which has vessels in xylem is known as "porous wood". The non-porous wood is technically called "softwood" and porous wood is called "hardwood". Softwood or hardwood terms are not at all related with softness or hardness of timber, *e.g.*, hardwood of *Bombax ceiba* (Angiosperm) is very soft and softwood of *Pinus* and *Cedrus* (Gymnosperms) is very hard. The wood of balsa (*Ochroma lagopus*), an Angiosperm is the lightest and still classified as hardwood.

Reaction Wood

The "reaction wood" is formed on the lower sides of branches and leaning and crooked stems of Conifer tree, and on the upper sides of similar structures in Dicotyledonous trees. This wood is called "compression wood" in Conifers and "tension wood" in Dicotyledons. This wood is called reaction wood because its development is assumed to result from the tendency of the branch or stem to counteract the force inducing the inclined position.

The reaction wood shows structural differences. The compression wood is denser and darker than the surrounding tissue. Its tracheids are shorter, rounded with heavily lignified than those in the normal wood. In Dicotyledons, the vessels of tension wood are reduced in number and width. The wood has also got characteristic presence of gelatinous fibres which have a thick highly refractive gelatinous layer in the cell wall.

Storied Wood and Non-storied Wood

In transverse sections the secondary xylem shows more or less orderly radial seriation of cells – a result of the origin of cells from tangentially dividing cambial cells. In the homogeneous Conifer wood this seriation is pronounced; in vessel-containing Dicotyledons it may be somewhat obscured by the ontogenetic enlargement of the vessel members and the consequent displacement of adjacent cells. Radial sections also reveal the radial seriation, and they indicate that the radial series of the axial system are superimposed one upon the other in horizontal layers, or tiers. The tangential sections are varied in their appearance in different woods. In some, the cells of one tier unevenly overlap those of another, in others the horizontal layers are clearly displayed in tangential sections. Thus, some woods are non-stratified or non-storied, in tangential sections (*Castanea, Fraxinus, Juglans, Quercus*), others stratified or storied (*Aesculus, Cryptocarya, Ficus, Tilia*, numerous Fabales). The storied condition is especially pronounced when height of the ray matches that of a horizontal layer of the axial system. In storied xylem the vessel members are usually short. From the evolutionary aspect the storied woods are more highly specialized than the non-storied.

Vascular Rays

A panel of tissue variable in height and width, formed by the ray initials in the vascular cambium and extending radially in secondary xylem and secondary phloem is known as the "vascular ray".

Vascular rays may be uniseriate (one celled in width) or multiseriate (two or more celled in width).

The radial parenchyma cells are divided into categories according to their form. The two most common types are the procumbent and the upright ray cells (Figure 12.7). A ray cell the longest diameter of which is oriented radially is procumbent; the cell elongated axially is upright. The two kinds of ray cells are often combined in the same ray, the upright cells appearing at the upper and lower margins of the ray. Rays composed of one kind of cell are called "homocellular" (Figure 12.7 C, D), those containing procumbent and upright cells "heterocellular" (Figure 12.7 A,B).

The rays of confiers are mostly one cell wide, occasionally biseriate and from one to twenty or even to fifty cells high. Presence of resin ducts makes the normally uniseriate appear multiseriate.

In contrast to the predominantly uniseriate rays of Conifers, those of Dicotyledons may be one to many cells wide, that is, they may be uniseriate or multiseriate and range in height from one to many cells. The multiseriate rays frequently have uniseriate margins.

The parenchyma cells of the vascular rays are involved in storage and translocation of oil and many other ergastic substances. Tanniferous compounds and crystals are common inclusions in vascular rays.

Tyloses

In many species parenchyma cells (axial or one in a ray) located next to the vessels form outgrowths through the pit cavities into the lumen of the vessels when the latter become inactive. Soon the wall of the parenchyma also gets thickened due to gradual deposition of lignin. The nucleus and part of the cytoplasm of the parenchyma cell commonly migrate into the tylosis. These outgrowths are called "tyloses". Tyloses block lumen of the vessels and reduce the permeability of the wood. Tyloses block the vessels during the formation of the heartwood and in the sapwood, in response to wounding and infection and diseases. Their development is possible only if the pit aperture on the vessel side is no less than 10 microns wide. Examples of woods with abundant development of tyloses are *Quercus, Robinia, Vitis, Juglans* and *Morus*. Tyloses occur in the primary xylem also and are seen in coleoptile and bean leaves. Tyloses also occur in certain herbaceous plants like *Cucurbita, Coleus* and *Canna.*

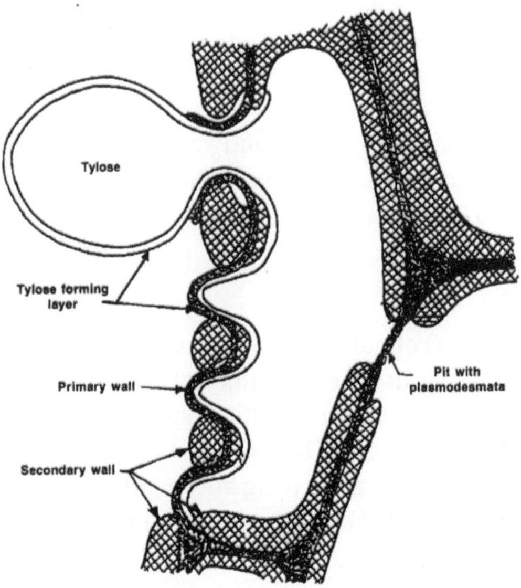

Figure 12.7 Diagram, of a ray cell that has formed a tylose

Formation of tyloses is important in the treatment of wood with preservatives and in its selection for tight cooperation. An outgrowth resembling a tylosis is known as "tylosoid". Parenchyma cells show outgrowths into sieve elements (phloem) and these are known as tylosoids. Similarly epithelial cells show out growths into intercellular resin ducts.

Resin Ducts and Gum Ducts

In most of the Gymnosperms, resin ducts are developed in the axial or ray systems. These ducts develop schiziogenously between resin producing parenchyma cells, which then form the epithelium of the duct. Sometimes the resin duct may become blocked by enlargement of the epithelial cells and these structures are called as the tylosoids. Differences exist in the thickness and lignification of cell wall of the epithelial cells in various Coniferous genera. In those genera, where the resin secreting cells have thick, lignified walls and in which these cells die after one season, relatively little resin is produced. In those plants where the epithelial cells are thin walled and function during several seasons, a large amount of resin is produced. Resin ducts with lignified epithelial cell walls are common in *Abies* and *Cedrus*, while thin walled, non-lignified secretory cells occur in *Pinus*.

Resin ducts appear as a normal feature of the wood in *Pinus, Picea, Larix* and *Psudotsuga*. However, these are also produced as a result of injuries, such as wounding, pressure and frost. In certain conifers like *Cupressus*, resin ducts are never developed in the secondary xylem.

Gum Ducts

Intercellular gum ducts occur in Dicot wood similar to the resin ducts of the Gymnosperms. They may arise normally in the wood or may be the result of various kinds of injury, *e.g.*, Simaroubaceae. These ducts contain various substances such as resins, oils, gums and mucilages. Normally formed gum ducts always occupy a characteristic position in the wood, while those of the traumatic origin may be variously situated. In a given plant the canals may be either horizontal or vertical, but only in a few cases are the two types present in a single plant.

Chapter 13

Anomalous Secondary Growth

Introduction

In plants, the primary growth occurs as a result of cell division at the tips of the stems and roots causing them to elongate, and to give rise to the primary tissue, while the secondary growth results from cell division in the cambia (or lateral meristems) which causes the stems and roots to thicken by the secondary tissues. This occurs in most seed plants while the monocots lack secondary growth (and if they do have it, it differs from the typical pattern of other seed plants). The formation of secondary vascular tissue from the cambium is a characteristic feature of dicots and Gymnosperms. In certain monocots, the vascular tissues also increase after the completion of the primary growth (because the cambium is of a different nature). This is rare in living Pteridophytes but occurs in *Isoetes* and *Botrychium*.

Secondary growth is the result of the activity of the two lateral meristems, the cork cambium and the vascular cambium in many vascular plants. The lateral meristems help in the increase the width of plant stem/root rather than its length. As long as these continue to produce the new cells, the stem/root will continue to grow in its diameter. This process produces the wood and shapes the plant into a tree (with a thick trunk) in woody plants. Because of such a growth the epidermis ruptures, and as this growth normally ruptures the epidermis of the stem/root, the plants with the secondary growth develop cork cambium, which gives rise to thick cork cells to protect the surface and also reduce water loss. After some years, the process may produce a layer of cork (as in Cork oak). The process of secondary growth also occurs in many non-woody plants (for example, tomato, potato tuber, tap root of carrot and tuberous root of sweet potato) as well as a few long-lived leaves. The palms (*Roystonea regia*) increase the diameter of their trunk by division and enlargement of the parenchyma, without a real secondary growth.

Abnormal Secondary Growth

The Dicotyledons and Gymnosperms exhibit continued accretion of secondary vascular tissues all through their life by the activity of the normal vascular cambium. This is usually described as *normal secondary growth*. In a number of Angiosperms, however, secondary growth takes place in a different way that is different from that of the normal type, and is then described as *anomalous secondary growth*. This is also termed *abnormal secondary growth* or *unusual secondary growth*.

This term *anomalous secondary growth* refers to a group of cambial conformations, cambial products, and cambial numbers which differ from the common normal condition, namely, a single cylindrical cambium which produces the phloem externally and xylem internally. This does not follow the pattern of a single vascular cambium (as in the ancestral lignophytes) producing xylem to the inside and phloem to the outside. In some dicots (like *Bougainvillea*) a series of cambia arise outside the oldest phloem during secondary growth. Monocots either have no secondary growth (as in the case of ancestors) or have abnormal secondary growth of some type (as in palms). The increase in the diameter of the trunk in palms due to division and enlargement of parenchyma is called as the "*primary gigantism*" because there is no production of the secondary xylem and phloem; or sometimes "*diffuse secondary growth*". In the stems of *Yucca* and *Dracaena* with anomalous secondary growth, a cambium forms but produces the vascular bundles and parenchyma internally and only parenchyma externally. Some monocot stems increase in their diameter due to the activity of a primary thickening meristem that is derived from the apical meristem.

Anomalous Secondary Growth – Types

According to Haberlandt (1914), from the physiological point of view the anomalous secondary growth may be of the "Non-adaptive" and "Adaptive" types.

Non-Adaptive type

The anomalous secondary growth exhibited by some herbaceous plants (*e.g. Amaranthus, Achyranthes, Boerhavia*) is not of any functional significance to these plants and may be described as of non-adaptive in nature.

Adaptive Type

In this type, the anomalous secondary growth is of definite functional advantage to the plant and hence may be described as adapive, *e.g., Bignonia, Bauhinia, Serjania*. The mechanical requirements of stem of woody climbers are quite different from those of the ordinary woody plants. The stem of a woody climber must be flexible so that it could be able to twist and twine around its support in an efficient way and at the same time it should be strong enough to support the weight of its own foliage. The above factors are responsible for the anomalous secondary growth in these plants. Further, a solid mass of vertical trunk does not serve the purpose of a woody climber. On the contrary, a woody climber requires a stem that is supple and flexible enough to enable it to survive around its support efficiently and with ease. As a result of anomalous secondary growth such a type of stem is produced in a woody climber.

Methods of Abnormal Growth

Anomalous secondary growth takes place through one of the following methods:

1. Abnormal activity (behaviour) of the Cambium
2. Abnormal position of the Cambium
3. Formation of more than one Cambial Ring
4. Formation of Inter-xylary (included) Phloem
5. Formation of Inter-xylary Cork

1 Abnormal Activity of the Cambium

In some cases, where the normal cambium starts cutting off the cells at many places irregularly, and produces much larger portions of xylem than that of phloem at some places, and at other places, more phloem than the xylem, a ridged and furrowed xylem cylinder is produced. This can be of a simple structure (as in some bignoniaceous genera) or very complex (as in species of *Bignonia*).

(a) *Bauhinia*

In *Bauhinia rubiginosa,* the cambium is more active at certain points but shows limited activity or no activity at other regions. In the regions of more activity, excessive secondary vascular tissue is produced, where as in the regions of less activity, little or no vascular tissue is formed (Figure 13.1 D). Such an abnormal growth pattern results in the formation of ridged vascular cylinder and the mature stem exhibits a ridged appearance externally. In other species of *Bauhinia,* the activity of the cambium is limited to the two opposite poles only, resulting in an elongated flattened strap-like structure (Figure 13.1 F).

(b) *Bignonia*

In *Bignonia* the output of secondary xylem is relatively less from the beginning at four points, which are arranged in the form of a cross. At these four points, however the secondary phloem tissues are more. Consequently, the stem appears to be provided with four longitudinal furrows. These grooves become constantly deeper, as the secondary growth proceeds. Later, the woody cylinders become split up into separate strips.

Representatives of Bignoniaceae illustrate yet another type of anomalous growth. After the usual type of cambial cylinder is formed at the end of primary growth, four strips of cambium cease to produce xylem but continue to give off derivatives to the phloem side. Thus, two kinds of cambium become established, one showing bidirectional activity, the other unidirectional activity. As the stem

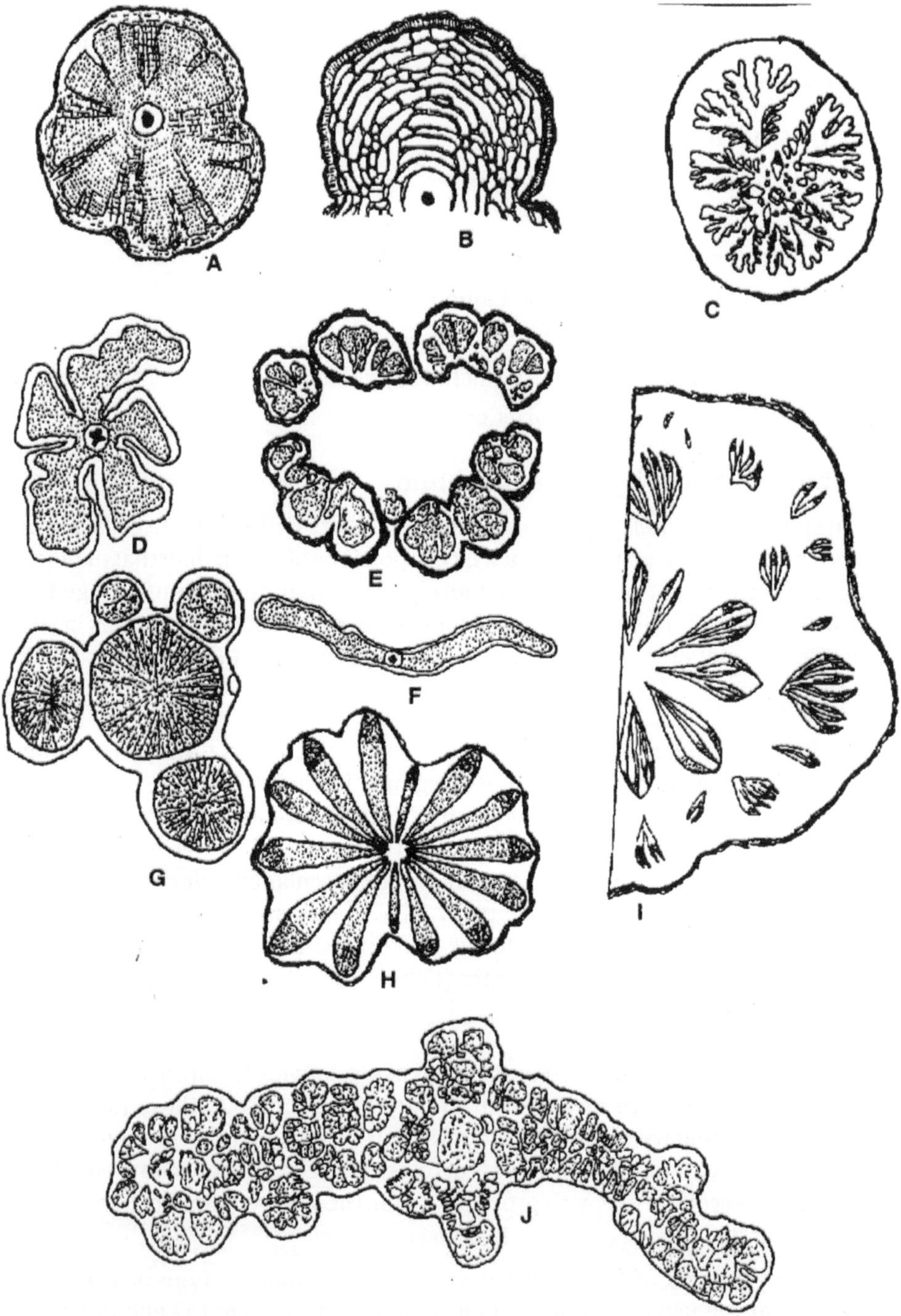

Figure 13.1: Anomalous Stem Structure – Liana Types. A. Bignoniaceous genus; B. *Securidaca lanceolata*: C. *Bignonia* sp.; D. *Bauhinia rubiginosa;* E. *Serjania ichthyoctona*; F. *Bauhinia* sp.; G. *Thionouia scandens;* H. *Begonia fruticosa*; I. *Piper fluminence*; J. *Bauhinia langsdorffiana.*

increases in circumference, additional strips of cambium change from bidirectional to unidirectional cell production. Some of the new phloem panels develop next to the earlier ones, others appear in new regions of the xylem. Each phloem panel is flanked by wide rays and retains its original tangential diameter, but the groups of panels increase in width towards outside through the addition of new panels to each group (Figure 13.1 A). The xylem is thus repeatedly fissured by the insertion of new phloem panels. In old stems, it is further broken up by dilation growth of xylem and pith parenchyma.

Where a cambium, normal in position and activity forms in some segments much larger proportions of xylem than of phloem, and in others more phloem than xylem, a ridged and furrowed xylem cylinder is formed. This may be simple in structure as in Figure 13.1A or very complex as in Figure 13.1 C. In a cross section of a stem, the cambium produces more xylem than phloem in certain places and more phloem than xylem in others, the xylem cylinder becomes ridged and sometimes even more complex structures may develop. Such a stem configuration occurs for instance, in *Passiflora glandulosa* (Figure 13.2 C).

The most interesting anatomical feature in *Bignonia* of Bignoniaceae is as follows:

A. Presence of Phloem Wedges in the Xylem

The young stems which exhibit such type of structure on attaining maturity are provided with a normal ring of vascular bundles. The vessels of the young stem are narrow in their diameter, and the wood produced in later stages contains wider vessels. At this stage, four furrows at four equidistant points appear in the xylem, extending almost to the pith, and the cambium is situated on the inside of the furrows. The cambium is not found on the radial surface. The phloem enhances in bulk and the tissues move towards the sides along the lateral surfaces of the furrows. Later, due to the development, the furrows become closed, and the four radial groups of phloem get united by the medullary ray tissue. In T. S, the narrow medullary ray can be seen traversing the phloem of furrows.

B. Presence of Fissured Xylem

The fissured xylem is only visible in fairly old stems. The wedges of the phloem are produced first followed by the xylem which becomes fissured by dilation and cell division in the wood parenchyma and the pith.

(c) *Urtica dioica* Stem

In young stem of *U. dioica*, a normal cambial ring is formed which behaves normally, and cuts off secondary phloem towards outside and secondary xylem towards inside. However, after sometime, the cambium begins to behave abnormally at certain places cutting of parenchymatous cells on the inner side, instead of secondary xylem. After cutting off a group of parenchymatous cells, the cambium again begins to behave normally to resume its normal activity, forming secondary xylem above the parenchymatous group. The process is repeated again and again, to result in the formation of islands of parenchyma which get embedded in the secondary xylem. The cambium forms vascular tissues only in the region of the vascular bundle:

(d) *Aristolochia*

During the secondary growth, the parenchymatous cells of the medullary rays adjacent to fascicular cambium become meristematic to form interfascicular strips of cambium in *Aristolochia* stem. The interfascicular cambium behaves abnormally by producing only the parenchyma cells. The fascicular cambium on the other hand, behaves normally, producing secondary xylem and phloem within the vascular bundles. Concomitantly, the vascular bundles gradually increase in their size but continue to maintain their individuality (Figure 13.2 A). Such a condition is also seen in *Begonia fruticosa* (Figure 13.1 H).

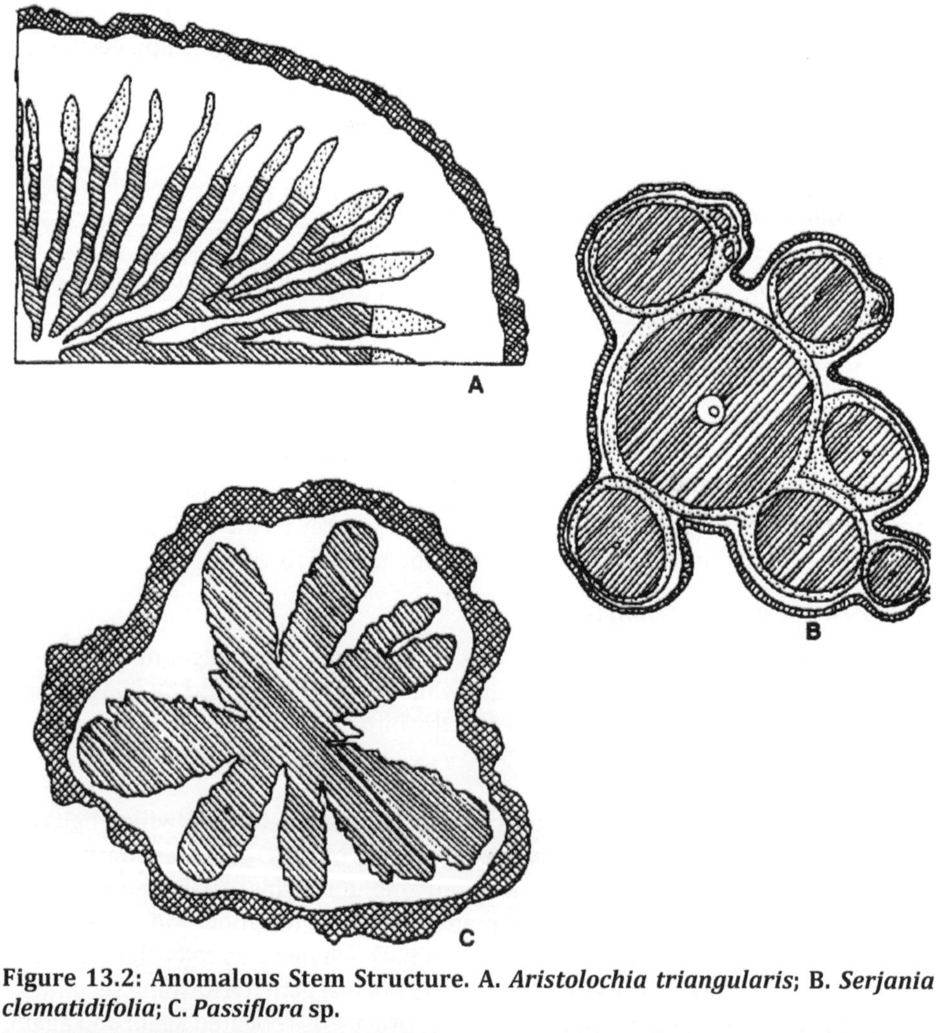

Figure 13.2: Anomalous Stem Structure. A. *Aristolochia triangularis*; B. *Serjania clematidifolia*; C. *Passiflora* sp.

(e) *Cucurbita*

In *Cucurbita,* the fascicular cambium produces xylem towards inside and phloem towards outside while the interfascicular cambium produces only parenchyma cells.

2. Abnormal or Anomalous Position of the Cambium

The stems of many unusual shapes or types are formed by the unusual position of the cambium. For example, in *Thinouia scandens,* while the stem is young, the cambium is thrown into folds or ridges, and the tips of the ridges are pinched off and after separation evolve steles in them. In *Serjania ichthyoctona,* the cambium appears originally in many separate strips, each of which surrounds portions, (even individual strands), of the primary xylem and phloem. Such a stem appears to be made up of several fused stems.

(*a*) *Thionouia scandens* (Woody climber of Bignoniaceae)

In young stems, the cambium is not circular but with ridges. Later, the cambial ridges get separated and act as individual cambia forming individual groups of vascular tissues. Consequently, the secondary growth becomes abnormal with separate steles with individual secondary phloem and secondary xylem resulting in a peculiar shape in the mature stem (Figure 13.1 G). In the older stems, the compound condition becomes more marked, because the parts are separated from each other as the outer layers of each strand die with the result of the evolution of the layers of periderm. Hence, there is formed a stem made up of strands lying together more or less like the strands of a rope.

(*b*) *Serjania* and *Paulinia* (Fam.: Sapindaceae)

In these tendril climbers, specific groups of vascular tissues are constricted off from the main vascular cylinder. These constricted off vascular tissues develop their individual cambial strips. With the initiation of secondary growth, each strip of cambium develops to form a complete and separate cambium ring, which functions normally forming secondary xylem on the inner side and the secondary phloem on the outer side. Thus, the mature stem consists of a number of discrete vascular cylinders each of which develops its own periderm. In the mature condition, the stem seems to be made up of several smaller stems appressed to each other giving the appearance of strands of rope (Figure 13.2 B). This apparently compound condition becomes more marked as the stem becomes older and the parts separate as the outer layers of each strand die because of the development of periderm layers (Figure 13.1 E).

(*c*) *Bauhinia langsdorffiana* (Fam.: Fabaceae)

In this plant, a similar structure as in *Serjania ichthyoctona* is produced. The change is brought about by the breaking into strips of the original cambium cylinder, and even of the vascular cylinder produced by this cambium by the proliferation of the xylem parenchyma. The parenchyma in the xylem and phloem increases excessively, finally rupturing the first formed original tissues and the cambium sheet which formed them. The parenchyma of secondary xylem and secondary phloem tissues undergo intercalary growth resulting in excessive amounts of parenchymatous tissues termed *dilation parenchyma* which makes the secondary tissues along with cambium split into many strands (Figure 13.1 J).

(d) *Bauhinia rubiginosa*

In this plant, there is restriction of the activity of the cambium to some regions which result in the formation of ridged stems. In other species, the straps of stems are generated because of the restricted activity of the cambium in certain regions, where the cambium is more active at the two opposite sides. Abnormal position of the cambium is also seen in *Securidaca* (Figure 13.1 B).

3. Formation of More than One Cambial Ring

In some Angiosperms (*Boerhavia diffusa, Bougainvillea, Mirabilis*), normal vascular cambium ceases to function after the production of limited quantity of vascular tissues. New cambia are, however, produced successively, in the pericycle and cortex. About 28 families of Dicotyledons, show formation of accessory cambia (Figure 13.1 I). The cambium in this zone behaves abnormally producing vascular tissues in the fascicular region and lignified conjunctive tissue in the interfascicular regions. Successive rings of cambia are formed which behave in a manner similar to the first ring, as a result, concentric layers of vascular bundles are formed embedded in conjunctive tissue.

Accessory Cambium Formation and its Activity and Medullary Bundles

In the stems of *Bougainvillea*, and other members of Nyctaginaceae (*viz. Boerhavia diffusa, Mirabilis, etc.*), many cambia arise successively in a centrifugal direction. Every cambium produces xylem and conjunctive tissue to the inside, and phloem and conjunctive tissue to the outside. The resulting tissue gives the appearance of concentric rings of vascular bundles embedded in conjunctive tissue.

The herbaceous and woody plants of Nyctaginaceae (*viz., Bougainvillea, Boerhavia, Mirabilis*) are noteworthy from the viewpoint of their anatomy that they contain the anomalous secondary growth in thickness of the axis. This type of secondary growth in thickness takes place by means of evolution of the successive rings of collateral vascular bundles. In herbaceous plants of this family, the bundles remain embedded in parenchymatous ground tissue, but in the woody species (*viz., Bougainvillea*), the ground tissue is somewhat prosenchymatous and lignified. Both kinds of conjunctive tissue are developed from successive cambia. In the woody species, there is no clear differentiation between the xylem and the conjunctive tissue and therefore sometimes in T. S. of the axis, the phloem in seen in the production of islands. In certain other cases, the strips resembling medullary rays are also visible in the conjunctive tissues. In *Bougainvillea glabra*, a robust herbaceous or slightly woody species, the inner bundles remain embedded in parenchyma and the outer bundles in the prosenchymatous ground tissue.

Sometimes, in certain species, the narrow radial strips resembling normal medullary rays are also visible traversing in the conjunctive tissue. In *Boerhavia*, thin-walled lignified groups of parenchyma are connected with the phloem; the phloem groups and adjoining ground parenchyma occasionally appear as concentric, annular or band-shaped strip, of tissue. In *Bougainvillea, Mirabilis, etc.*, innermost part of the conjunctive tissue, produced by the secondary meristem resembles true pith and thus causes the innermost secondary bundles to appear as if they are medullary in origin.

4. Formation of Inter-xylary Phloem or Included Phloem

Variations of another type in the activity of the cambium produce interxylary phloem (included phloem), and the phloem of this type is secondary phloem in the form of strands (patches) embedded in the secondary xylem. These patches are secondary in their origin. There appears to be two methods by which interxylary phloem comes to lie embedded in secondary xylem. Interxylary phloem is seen in about 25 families of Dicotyledons like Combretaceae, Malpighiaceae, Salvadoraceae, Asclepiadaceae, Acanthaceae, Gentianaceae *etc.* In *Salvadora*, this arises centripetally on the inner side of the normal cambium, and on maturation, the cellular disorganization takes place in the inter-xylary phloem. The inter-xylary phloem is also known as the "internal phloem", and this occurs in the form of strands or as a continuous band around the pith. The origin of this in most plants is primary. The cells are like those of the external phloem except that the fibres are few in number or absent, and the sieve tubes and companion cells exist in small groups surrounded by the parenchyma. The internal phloem develops after the evolution of the external primary phloem. The bundles are normally treated as bicollateral because of intra-xylary phloem (internal phloem).

(a) *Combretum* and *Entada*

In genera like *Combretum* and *Entada*, small segments of cambium are known to produce phloem cells towards the inside for a brief period, in place of the xylem cells which are normally produced. After a period of such activity, these cambium segments return to normal function and thus get buried in the inwardly formed phloem with xylem.

(b) *Strychnos*

In other forms such as *Strychnos*, the interxylary phloem strands are formed by the cambium towards the outside as a part of the normal external phloem, but these strands later become embedded in the xylem in the following manner. The small segments of the cambium cease to function, their cells becoming transformed into mature conducting tissue. The new segments of cambium then arise, as secondary meristem, in the phloem, some few rows of cells come out from the original cambium or in the pericycle. They unite with the edges of the segments of the general cambial cylinder – which have continued meanwhile their normal activity – as such they enclose a strand of phloem cells. This is repeated in other segments of the cambium so that the secondary xylem possesses numerous scattered strands of embedded phloem.

(c) *Leptadenia* and *Thunbergia*

In *Leptadenia* (Asclepiadaceae) and *Thunbergia* (Acanthaceae) also, phloem is formed not only towards the outside but from time to time also towards the inside.

Interxylary phloem plays a significant role in some xerophytic plants as it remains functionally active even during drought conditions when other tissues desiccate and become non-functional.

(d) *Salvadora persica*

The cambium functions normally cutting off secondary xylem on inner side and secondary phloem on outer side. After some time at certain places, the cambium begins to behave abnormally cutting parenchyma cells instead of xylem on the inner side. This is, however, a temporary change and soon the abnormally behaving cambium strip reverts back to its original function of cutting secondary xylem on the inner side. This behaviour is repeated several times, resulting in the formation of a number of patches of islands of thin walled cells. The parenchyma cells ultimately get converted to phloem parenchyma (peripheral cells) and sieve tubes and companion cells (central cells) giving rise to phloem patches termed as "included phloem" or "interxylary phloem".

5. Formation of Interxylary Cork

In several plants such as *Achillea fragrantissima, Artemisia tridentata* and *Epilobium angustifolium* interxylary cork is formed. The cambium ring cuts off secondary xylem and parenchyma towards inner side. Soon a layer of phellogen (cork cambium) develops in this parenchymatous zone between the secondary xylem, becomes active and cuts off cork cells which remain enclosed between two consecutive rings of secondary xylem, termed as "interxylary cork". In older stages, suberization of the medullary rays may occur or certain portions of the cambium may cease its activity resulting in splitting of the stem.

Anomalous Secondary Growth in Monocot Stems

In Monocotyledonous plants, secondary growth is usually absent as the vascular bundles are of closed type (cambium is absent). But, some members of Liliflorae like *Dracaena, Yucca, Aloe, Agave, Cordyline, Sansevieria, Lomandra, Xanthorrhoea* and *Kingia* exhibit secondary growth which is anomalous in origin as well as in function. The cambium arises in the parenchyma of cortex outside the primary vascular bundles. The activity of this cambium is unusual. It produces more parenchyma towards the inside and a small amount of perenchyama towards the outside. In the development of the vascular bundle, individual cells derived from the cambium divide longitudinally, and then two or three of the resulting cells repeat the longitudinal divisions. The products of the final divisions differentiate into vascular elements and associated sclerenchyma cells. Vertically, many tiers of cells combine to form a vascular bundle. The secondary vascular bundles may be collateral or amphivasal and occur in more or less definite radial files. The cells formed outwards by cambium are parenchymatous and constitute secondary cortex. They are thick-walled, lignified and are arranged in regular radial rows.

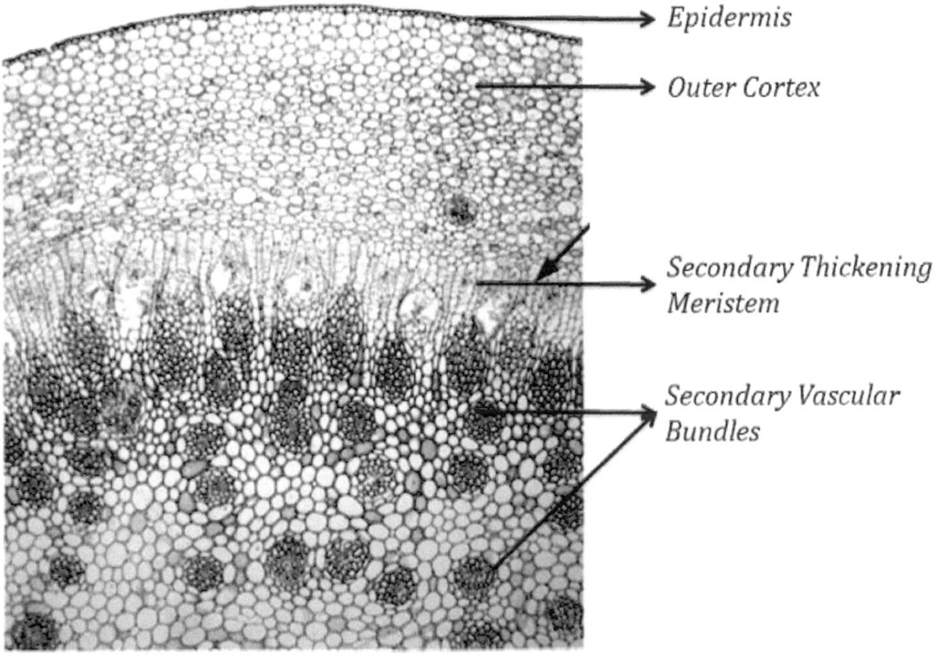

Figure 13.3: Anomalous Secondary Growth in *Dracaena*.

Formation of Storied Cork

The parenchyma of the cortex divides many times periclinally and the products of these divisions, get aligned in distinct radial rows (storied cambium) become suberized. The cork of this type is called the storied cork.

Anomalous Secondary Growth in Root

In a number of Dicot plants, the roots are modified to function as storage roots. In the roots of Apiaceae, as those of Fennel (*Foeniculum*) or carrot (*Daucus*), the secondary growth is of the ordinary type, but parenchyma predominates in the xylem and phloem; while in others like *Beta* and *Ipomoea batatas*, secondary growth in roots is usually of anomalous type as an adaptation to their storage function. Members of Apiaceae, Brassicaceae, Chenopodiaceae, Asteraceae, Convolvulaceae *etc.* show anomalous secondary growth in roots.

(a) *Beta vulgaris*

The secondary growth is initiated by the formation of a cambium in the interstitial parenchyma, except opposite the two protoxylem groups where it is from the pericycle. The cambium gives rise to a ring of essentially collateral vascular bundles separated by narrow bands of parenchyma. After some time, the cambium ceases to function and another cambium arises from the phloem parenchyma of the first ring of vascular bundles. This cambium produces a second ring of vascular bundles separated by large radial channels of storage parenchyma. Soon, a third ring of vascular bundles is produced from yet another cambium derived either from phloem or pericycle, and it also behaves in the same way as the earlier cambia. Thus, a number of accessory cambia are produced and contribute to the successive rings by the accessory cambia which consist of more of parenchyma and less of xylem elements.

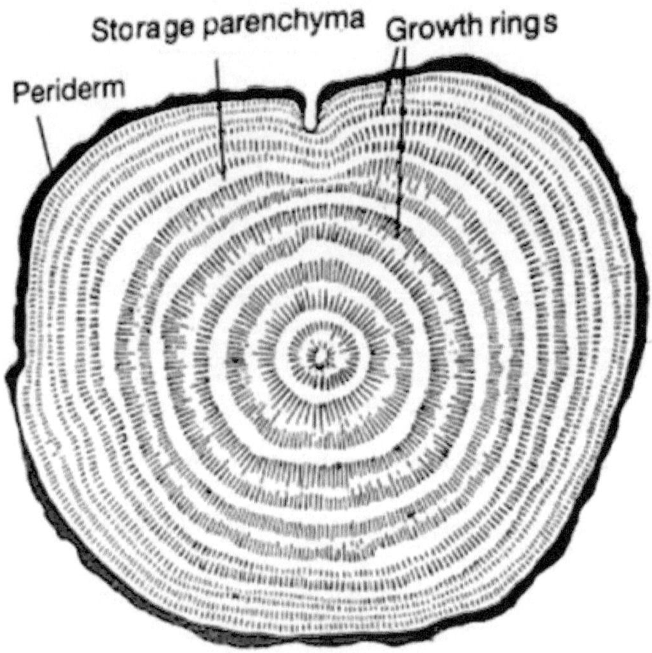

Figure 13.4: Anomalous Secondary Growth in *Beta vulgaris.*

(b) *Ipomoea batatas*

Another complex type of anomalous growth occurs in the fleshy adventitious roots of *Ipomoea batatas* (Sweet potato). Secondary growth of the storage root takes place by the formation of cambium ring which is normal in function initially. The xylem contains a large proportion of parenchyma. Cambia develop in the parenchyma around individual vessels or vessel groups to produce a few tracheary elements toward vessels, a few sieve tubes and laticifers away from the vessels and a considerable number of storage parenchyma cells in both the directions. Thus, the phloem elements appear within the part of the root that originally differentiated as xylem.

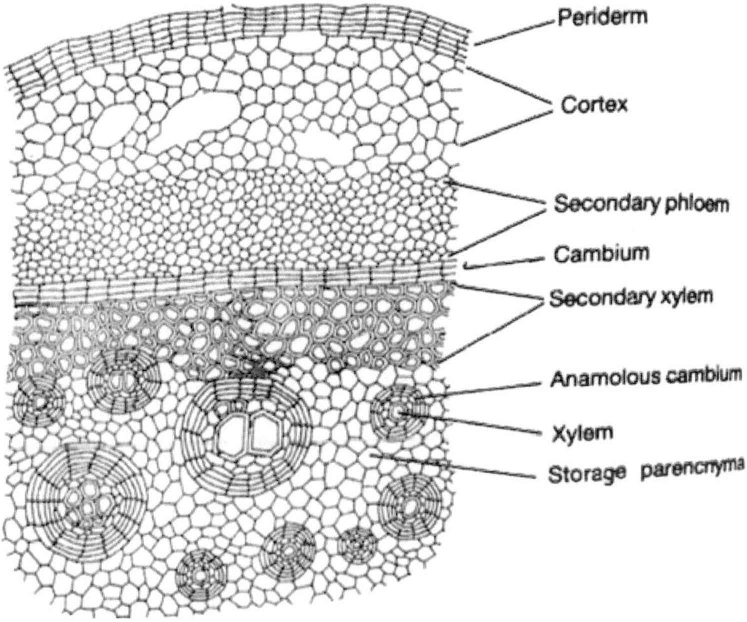

Figure 13.5: Anomalous Secondary Growth in Sweet Potato.

Chapter

14

The Leaf: Structure and Development

Introduction

A leaf (Plural: Leaves) is the main lateral appendage of the stem in vascular plants. This is usually borne above the ground and is specialized for photosynthesis. Both the leaves and the stem together constitute the shoot. Collectively, all the leaves on the plant are referred to as the "foliage".

The leaves appear in different shapes, sizes and textures. The broad and flat leaves with the complex venation are called as the "megaphylls", and the species that bear them are megaphyllous plants (flowering plants). In others, the leaves are simple (with only a single vein) called as the "microphylls", and the plants that bear them as microphyllous (Club mosses). Some others are called as "bulb scales" are associated with the underground stems and lie below the ground. The leaves are submerged in water in the aquatic plants; the leaves are often thick and juicy in succulent plants; some others are non-photosynthetic and may be dead at maturity, as in "cataphylls" and spines. In addition, there are several leaf-like structures in vascular plants that are not homologous with them, represented by flattened plant stems (phylloclades and cladodes); flattened leaf stems (phyllodes) which differ both in structure and origin. In non-vascular plants, some structures look like and function like the leaves (phyllids of mosses and liverworts).

The leaf is one of the three basic organs of higher plants along with the stem and root. It plays a vital role in Photosynthesis. The leaf is the most variable of all the plant organs. '*Phyllome*' is the collective term generally used for the various types of leaves appearing in plants. The phyllome of the spermatophyta is quite variable in its external and internal structure as well as in its function. The phyllomes are thus classified as: Foliage leaves, Cataphylls, Hypsophylls, Cotyledons *etc.* The foliage leaves are the main photosynthetic organs. The "cataphylls" are the scales which appear on buds and underground stems and function as protective agents as well as storage organs. The "hypsophylls" are the bracts present in the flowers and they also help flowers in protection. Rarely, they are coloured and function like petals. The 'Cotyledons' are the first leaves of the plant. The floral organs are also considered as the leaves.

Leaf - Morphology

This varies considerably from species to species. They may be isobilateral or dorsi-ventral. The former have both sides identical (dorsal and ventral surfaces are uniformly green) while the latter show different surfaces, dorsal surface being darker green while the ventral surface light green. Structurally, a complete leaf consists of a petiole (leaf stalk - if absent, the leaf is "apetiolate" or sessile - this mechanically links the leaf to the plant and transfers water and sugars to and from the leaf), a lamina (leaf blade), stipules (small leaf-like structures on either side of the base of the petiole - may be conspicuous as in beans and roses, or fall off early as in Moraceae or absent as in Magnoliaceae), and a sheath (lies at the base and either fully or partially clasps the stem above node as in grasses and Apiaceae). The leaf blade is the green, expanded, flat structure containing the chloroplasts. The adaxial (upper) angle between a leaf and the stem is known as the "axil" and this is the position of the bud. The veins (or the ribs) keep the lamina flat and expanded and give the proper support to it as an endoskeleton. The leaves have always a determinate growth and grow to a specific pattern and shape before the growth stops. The stems and the roots are non-determinate and continue to grow as long as they have the resources. The type of the leaf is characteristic of a species (monomorphic) though some species produce more than one type of leaf (dimorphic or polymorphic). The leaves are longest in *Raffia* palm, *R. regalis* and they grow up to 25 m (82 ft long and 3 m (9.8 ft) wide. The leaves are referred to as "prostrate", when they are basal and lie on the ground.

Leaf - Basic Types

The leaves that are shed annually are called "deciduous", while those that remain through the winter are called evergreen. The leaves that are attached to the stem by the stalks are "petiolate", while those that are attached directly to the stem without a stalk are called sessile (apetiolate). The leaves in the angiosperms are standard and include stipules, petiole and a lamina, while the leaves in ferns are called the "fronds"; needles (or awl-shaped or scale-like and evergreen in conifers; microphylls in lycophytes; sheathy in grasses and other monocots; and specialized as in pitcher plant.

The leaves are basically divided on the basis whether the stalk bears one or more leaflets (the division of the blade) into two types, namely: simple leaves and compound leaves. A simple leaf has an undivided lamina but it may be dissected into lobes. A compound leaf has a fully subdivided lamina and each leaflet (pinna) being separate along a main or secondary vein. The leaflets have petiolules and stipels (equivalent to petioles and stipules). The common stalk (the middle vein of a compound leaf or frond) is called the "rachis". The compound leaves are characteristic of Fabaceae. The central vein is called as the midrib. The different types of compound leaves are: Palmately compound (the leaflets radiate from the end of the petiole like the fingers of a palm of a hand – as in *Cannabis* and *Aesculus*); Pinnately compound (the leaflets are arranged along the mid vein); Odd pinnate (provided with a terminal leaflet - as in *Fraxinus*); Even pinnate (there is no terminal leaflet – as in *Swietenia*); Bipinnately compound (the leaves are twice divided and the leaflets are arranged along a secondary vein that is one of the several branches of the rachis. The leaflets are called "pinnules" – as in *Albizia*); Trifoliate or Trifoliolate (a pinnate leaf with three leaflets – as in *Trifolium, Laburnum, Toxicodendron etc.*); and Pinnatifid (the lamina is pinnately dissected to the central vein but with the leaflets not entirely separate – as in *Polypodium* and *Sorbus*).

Mostly, the leaves are dorsiventral and lateral, as in *Iris* or cylindrical as in *Sansevieria* or may be tubular as in some insectivorous plants. In certain plants like *Acacia*, the lamina of the leaves is reduced and the remaining parts form a flat, leaf like photosynthetic organ known as 'Phyllode'. In *Opuntia* species and *Muehlenbeckia*, the stems become flat and take up the photosynthetic function and are termed as 'Platyclades '. If the platyclades look like leaf as in *Ruscus*, they are called 'Phylloclades'.

Morphologically and anatomically the leaf is the most variable plant organ. The foliage leaf usually shows its specialization as a photosynthetic organ. If the base of either a sessile or a petiolate leaf encircles the stem, it is referred to as a "sheathing base". Sometimes, the sheathing part of the leaf is developed into a conspicuous leaf sheath. Plants with multilacunar nodes have sheathing bases. Outgrowths at the base of the leaf, the "stipules", are often present in the leaves associated with trilacunar nodes. The leaf surface which faces upwards and is closer to the internode above it is called as the "adaxial", ventral or upper surface and the opposite surface as the "abaxial", dorsal or lower surface. The morphology changes within a single plant and it may be "homoblasty" or "heteroblasty". The former is the characteristic in which a plant has small changes in the size, shape and growth habit of a leaf in between the juvenile and adult stages; while the former is the characteristic in which a plant has marked changes in the size, shape and growth habit of the leaves in between the juvenile and adult stages.

Leaf - Venation

Majority of the Dicot leaves have blades with pinnate venation (major veins diverge from one large mid vein and have smaller connecting networks between them). Less commonly, they have palmate venation (several large veins diverging from the petiole to the edges of the leaf). Rarely some may show parallel venation.

The monocots usually have narrow blades and usually have a parallel venation converging at the tips of the leaf blade. Some also show pinnate venation. These may be further reticulate (divergent and convergent) and parallel (divergent or convergent).

Leaf - Arrangement (Phyllotaxy)

The arrangement of the leaf on the stem (phyllotaxy) is of different types. It may be alternate (a single leaf or branch or flower attaches at each node, and these are alternate in their direction, along the stem); basal (the leaves arise from the base of the stem); cauline (the leaves arise from the aerial stem); opposite (two leaves, or branches or flowers attach at each node - paired or decussate – each successive pair rotates 90° along the stem); whorled or verticillate (three or more leaves or branches or flowers attach at each node and appear as whorls); rosulate (the leaves form a rosette) or in rows (2 - distichous, 3 - tristichous, 4 - tetrastichous, or more).

Histology of Angiosperm Leaef

Leaf tissues are composed of layers of plant cells. Different plant cell types form three main tissues found in leaves. Histologically, the leaf is composed of three types of tissue systems: Dermal tissue system, Ground tissue system and Vascular tissue system.

Epidermis

Epidermis forms the dermal tissue system of leaf. The epidermis constitutes the outermost layer of cells of the leaf. Functionally and morphologically the epidermal cells are variable. Both upper and lower epidermis is usually uniseriate composed of compactly arranged cells. However, in certain plants, a multiseriate epidermis may develop as a result of periclinal divisions of the protodermal cells, *e.g.*, Moraceae, certain species of Begoniaceae, Piperaceae, Chenopodiaceae. In *Ficus,* the upper epidermis is multiseriate whereas in *Nerium* both upper and lower ones are multiseriate. In *Begonia*, a succulent plant, the upper epidermis is three-layered while lower one is two-layered. The epidermis secretes a waxy coating called the cuticle (made of cutin) which is impermeable to water and water vapour and also forms a boundary that separates the inner tissues from the external environment. In some cases, the cuticle is thinner on the lower epidermis than on the upper epidermis. The epidermis also serves many functions – protection against water loss through transpiration; regulation of the gas exchange and secretion of metabolic compounds.

The epidermis includes many differentiated cell types, namely – epidermal cells, trichomes (epidermal hairs), the guard cells and subsidiary cells in the stomatal complex *etc.* The epidermal cells are numerous, largest and least specialized cells that form the epidermis and these are more elongated in the leaves of Monocots than that of Dicots.

The chloroplasts are generally absent in the cells of epidermis except the guard cells. The stomata are surrounded by the guard cells (containing the chloroplasts) and 2 to 4 subsidiary cells (without the chloroplasts) and form stomatal complex. The opening and closure of the stomatal aperture is regulated by the complex which regulates the gaseous exchange (gases and water vapour) between the plant and the outer environment. The guard cells control the size of the pores called the stomata (Singular: stoma) present in the epidermis. The stomata play an important role in allowing photosynthesis. These are numerous over the abaxial (lower) epidermis than the adaxial (upper) epidermis. The opening and the closure of the stomata allows the plants to either release or retain the gases including the water vapour, oxygen and carbon dioxide depending on their requirement.

Epidermal Cells

Various types of epidermal cells can be distinguished in different plants. These include the ordinary cells of the epidermis; single cells or groups of cells with special structure, form or content; cells connected with stomata; and epidermal appendages termed as trichomes.

The ordinary epidermal cells of the epidermis vary in shape, size and arrangement, but these are always closely attached to form a compact layer devoid of intercellular spaces. Many epidermal cells are tabular and in the leaf blade of Dicotyledons the anticlinal walls are mostly sinuous. In the leaves of many Monocotyledons, the epidermal cells are elongated. The external wall of the epidermal cells of certain leaves is raised in the form of papillae.

The epidermal cell wall differs in thickness. Some cells are thin walled, while in others the outer or the outer and inner periclinal walls are thicker than the anticlinal walls. Outside the outer cellulose wall there is usually a layer of pectin. Cutin is usually present in the outer walls of the epidermal cells and also constitutes a special layer, the cuticle, on the outer surface of the cell wall. The cuticle possesses in some genera like *Eucalyptus*, a complex and species specific ornamentation which can be used for diagnostic purposes. The cuticle is of varying thickness in different plants growing in dry habitats. The surface of the cuticle may be smooth, rough, ridged and furrowed. Lignified epidermal walls are found in the leaves of the Cycadaceae, needles of conifers, in the strips of epidermis above the bundles of sclerenchyma, in the leaves of Poaceae, Juncaceae and Cyperaceae, certain species of *Eucalyptus, Quercus, Laurus* and *Nerium*.

Bulliform Cells or Motor Cells

Some epidermal cells found in groups in the leaves of certain Monocots, especially grasses (maize, sorghum, sugarcane *etc.*) are comparatively larger, thin walled and highly vacuolated. These are called "Bulliform" (meaning bubble like) cells (Figure 14.3). These are commonly found in the upper surface of leaves, sometimes in lower surface also. Bulliform cells are mainly water containing cells, poor in solid and chlorophyll contents. Rarely tannins and crystals may be found. In cross sections, these appear as a fan-like band because the median cell being largest in size. There are three views regarding their function. (a) These are concerned with the unrolling of developing leaves by undergoing sudden and rapid expansion bringing about unrolling of lamina. (b) These are supposed to play an important role in hygroscopic opening and closing movements of mature leaves, due to changes in turgor. They are also called "motor cells" by some scientists. (c) These are concerned only with water storage and have no other functions.

Other Epidermal Cells

"Lithocysts" are other specialised epidermal cells containing cystoliths and are found in the members of Acanthaceae, Moraceae, Urticaceae and Cucurbitaceae. Secretory cells which contain the enzyme myrosin are known as the "myrosin cells" and occur sometimes in the epidermis of members of Brassicaceae. Large "idioblastic cells" containing mucilage occur in the epidermis of Lythraceae. Very large sac-like cells which intrude into the mesophyll are found in *Cleome aspera*. Secretory cavities may sometimes develop from epidermal cells, *e.g.*, in *Psoralea*. Long pipe-like cells filled with tannin substances appear in the epidermis of some species of *Saxifraga*. Silica cells may be present in epidermis as in certain grasses.

The continuity of the epidermis is interrupted by minute openings. These are intercellular spaces each of which is limited by two specialized cells termed as the "guard cells". These cells, together with the opening between them, constitute the stoma. In many plants subsidiary or accessary cells can be distinguished. These cells differ morphologically from the typical epidermal cells and they constitute two or more cells bordering the guard cells to which they are apparently functionally connected. The stoma together with the subsidiary cells, if present, is termed the "stomatal apparatus" or the "stomatal complex". The subsidiary cells usually develop from the protodermal cells adjacent to the stomatal mother cells, but they may also develop from the sister cells of the mother cell. Stomata are usually found on the aerial portions of the plant and especially on the leaves. If the stomata are present only on upper surface, the leaf is called epistomatic as in *Nymphaea*; if only on lower surface it is hypostomatic as in mango, oleander and banyan, and if on both the surfaces it is amphistomatic as in isobilateral leaves. Stomata are surrounded by kidney-shaped guard cells as in most dicots or dumb-bell shaped (double-headed) guard cells as in grasses. In the broad leaves of dicotyledons the stomata are scattered. In the narrow elongated leaves of Monocotyledons and Conifers, the stomata occur in rows parallel to the long axis of the leaf. The stomata may be on the same level as the other epidermal cells as in mesophytic leaves or they may be located above the surface of the epidermis (raised stomata) or may be sunken as in *Cycas, Pinus* and *Allium*. Several stomata may appear in a depression called "stomatal crypt" *e.g., Nerium*. The size of the aperture between the guard cells increases or decreases as a result of turgor changes in the guard cells. As a result of unequal thickness and consisting of the walls the guard cells undergo changes in form with the increase in volume. These changes result in the opening of the stoma. In most cases the thinnest wall is that closest to the subsidiary cells and is termed the back or dorsal wall.

Protruberances of the guard cell wall may be present above or both above and below, the stomatal aperture. In cross section these protruberances appear as horn-shaped ledges. The outer ledge delimits the front cavity above the aperture, and the inner ledge delimits the back cavity which abuts on the "sub-stomatal chamber".

The guard cells in Poaceae and Cyperaceae are different in form. They are elongated and dumb-bell shaped (Figure 14.9). The ends of these guard cells are expanded and thin walled, while the middle portions are elongated and thick walled and the cell lumen is narrow. As a result of turgor increase in this type of guard cell, the expanded ends swell and as a result push apart the middle elongated portions of the cells.

Below the stomata and directed inwards to the mesophyll are the large intercellular spaces which are termed as sub-stomatal chambers.

The details of trichomes are given elsewhere in this book.

Mesophyll

The "mesophyll" (Greek: Middle leaf) is the interior of the leaf between the upper and lower layers of the epidermis and is made of parenchyma (ground tissue) or chlorenchyma. This assimilatory tissue is the primary location of photosynthesis, generating assimilates (products of photosynthesis). In both in ferns and most flowering plants, in particular dicots, this is divided into two layers, namely, Palisade layer (Palisade parenchyma) and Spongy layer (Spongy parenchyma).

Palisade parenchyma

The palisade parenchyma consists of columnar cells that are elongated and perpendicular to the surface of the blade. This is one or two cells thick with intercellular spaces between them. The cells contain many chloroplasts than the spongy layer. These can take optimal advantage of light and the slight separation of the cells provides maximum absorption of carbon dioxide. The sun leaves have a multilayered palisade while the shade leaves or the older leaves nearer to the soil are single layered. The cells of the typical palisade parenchyma are elongated and in cross section of the leaf, they are rod-shaped and appear to be arranged in rows, while in a section parallel to the leaf surface, these cells are found to be round and separated or only slightly attached to one another. The amount of palisade varies in different plants which also correlate with external environment. The xerophytic plants have well developed palisade tissue. In plants of the temperate regions characterized by abundant available water in the soil (mesophytic habitat), the palisade parenchyma is usually located on the upper (adaxial) side of the blade, the spongy parenchyma on the lower (abaxial) side of the blade. A leaf with such structure is called "bifacial" or "dorsiventral". If the palisade tissue occurs on both sides of the leaf, as is common in the leaves of xerophytic habitats, the leaf is called "unifacial", "isobilateral" or "isolateral", *e.g.*, species of *Atriplex, Centaurea, Artemisia, Myoporum*. In some plants, however, the palisade parenchyma is found only on the abaxial side of the leaf, *e.g., Thymelaea*. In *Rosa* and *Lawsonia*, the palisade cells are arranged in a single layer while in *Musa, Mangifera* and *Ficus* the palisade parenchyma is two to three-layered. In *Polyalthia*, it is 2-layered. Many layered palisade parenchyma is seen in *Malus* and *Nerium*. In centric leaves such as *Allium* and *Juncus*, the palisade occurs all around the periphery of leaf. In a number of Dicotyledons such as *Caltha, Anemone* and *Aconitum*, and Monocotyledons such as *Bambusa* and *Arundinaria*, the palisade cells have peculiar folds projecting into the cell cavity and these are called "arm-palisade". In certain members of Piperaceae and Gesneriaceae, the palisade cells appear branched due to the presence of large lobes.

Spongy parenchyma

This lies beneath the palisade layer. The spongy parenchyma consists of cells of various shapes, frequently irregular, with branches extending from one cell to the other. The cells are not tightly packed and have large intercellular spaces between them. Limitation of such connections to the ends of the branches gives the spongy parenchyma an appearance of a three dimensional net, with the meshes enclosing the intercellular spaces. With regard to the connections between adjacent cells, the spongy parenchyma tissue has dominantly horizontal continuity parallel with the surface of the leaf, whereas the palisade parenchyma is continuous mainly in the direction perpendicular to the surface of the leaf. The stomatal apertures open into sub-stomatal chambers which are connected to the intercellular spaces between the spongy and palisade cells, so that oxygen, carbon dioxide and water vapour can diffuse into and out of the leaf so that the tissues of the leaf get access during respiration, photosynthesis and transpiration.

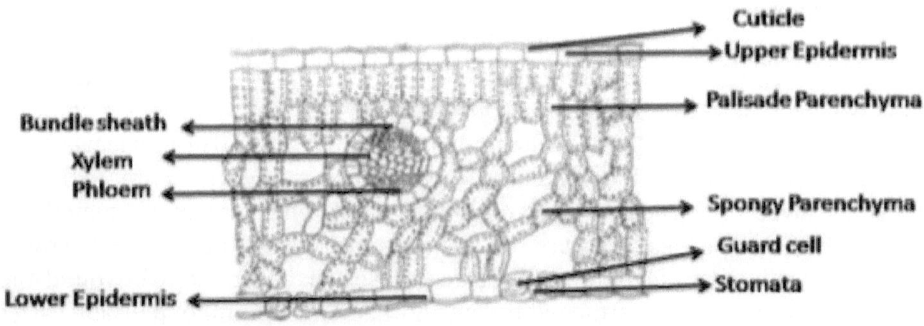

Figure 14.1. Internal Structure of Dicot Leaf.

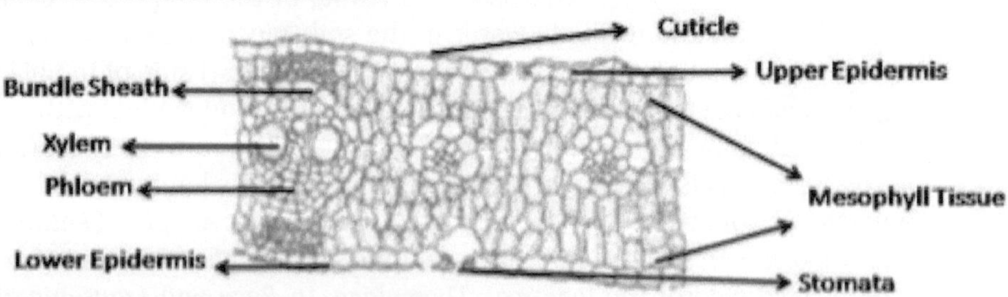

Figure 14.2. Internal Structure of Monocot Leaf.

In many Monocotyledonous leaves, such as grasses, mesophyll tissue is not differentiated into palisade and spongy parenchyma and there occur only one type of cells. These cells contain abundant chloroplasts and intercellular spaces. In *Eucalyptus* and *Atriplex*, the mesophyll is entirely composed of palisade cells. Because of the shape and arrangement of the palisade cells, the chloroplasts are so placed to enable the maximum utilization of light. Another factor which increases the photosynthetic efficiency is the presence of well-developed intercellular spaces present in the mesophyll which facilitates rapid gaseous exchange. The large surface areas of the mesophyll cells are brought into contact with air. Together these surface areas are termed as 'the internal surface area' of the leaf and are distinct from the 'external surface area' of the leaf. The ratio of the volume of the intercellular spaces to the total volume of the leaf varies between 77:1000 and 713:1000. The specialization of the palisade tissue that results in more efficient photosynthesis is brought about not only by the increased number of chloroplasts in the cells but also by the dimensions of its free surface area. The intercellular spaces of the mesophyll usually develop schizogenously, but in certain plants, the development may also be lysigenous (by the disintegration of group of cells), *e.g.*, water plants, marshy plants and banana leaf.

Other Structures in the Mesophyll

Sometimes, various types of idioblasts (isolated plant cells that differ from others and serve as storage of reserve materials, excretory materials, pigments and minerals; and also contain oil, latex, gum, resin, tannins and pigments *etc.*), are found in mesophyll region. Presence of lithocysts (the cells that contain the cystoliths) in *Ficus*, oil glands in *Citrus* and laticifers in *Calotropis* are few such examples. In certain plants, stellate cells and trichosclereids are present to provide mechanical strength.

Secretory structures, which participate in the secretion of water or other substances, are a common feature of leaves. Many of them are of epidermal nature. The substances produced may be excreted from the cells or they may be retained only to be released upon the disintegration of the cells.

Glands are present on many foliage leaves. These structures consist of a mass of dense parenchyma cells in which terminates a vascular bundle. The parenchyma is covered by a glandular epidermis, the cells of which is mostly elongated in a direction at right angles to the surface of the gland, and has dense cytoplasm and large nuclei. The glands in different organs or parts of an organ of the same species may secrete different substances. In *Rumex persica* and other related species the glands on the laminar teeth secrete a bitter resinous substance, while those on the petiole secrete nectar. Nectariferous glands are present on the petioles of many plants like *Passiflora, Ricinus* and *Impatiens.*

Another type of the secretory structure is the essential oil cavity which occurs characteristically in the mesophyll of the leaves of *Citrus* and the secretory cavity of *Gossypium*. These are examples of lysigenous cavities in which the secretions formed in the cells are released (the lysis). Secretory cavities occurring in mesophyll of *Eucalyptus* develop lysigenously (due to the disintegration of the cells) according to some authors and schizogenously (due to enlargement of the intercellular spaces) according to others. Other types of secretory structures common in leaves are the schizogenous resin ducts of the Asteraceae, Anacardiaceae and Coniferales; laticifers of *Euphorbia*, mucilage cavities of Sterculiaceae, Malvaceae, Moraceae *etc.*

In leaves, secretory substances may also be found in idoblasts. These secretory substances are classified according to the substances secreted. The secretory cells of the Lauraceae, Simaroubaceae and Onagraceae contain oils. Cells containing the enzyme myrosin are found in Brassicaceae, Capparaceae, Resedaceae, Tropaeolaceae and Moringaceae. Cells with a resinous content are found in the Meliaceae, Euphorbiaceae, Rutaceae and Rubiaceae. Tannin containing cells are common in Anacardiaceae, Annonaceae, Crassulaceae, Ericaceae, Euphorbiaceae, Buxaceae, Polygonaceae, Rosaceae, Tamaricaceae and Fabaceae. Cells with mucilaginous contents are found in Rhamnaceae, Malvaceae, Chenopodiaceae, Rubiaceae and many Monocotyledonous plants. Idioblasts with different types of crystals and cystoliths (*Ficus*) are present in the leaves of different species.

Special secretory structures are the "hydathodes" which secrete water in the form of drops from within the leaf. This phenomenon is called "guttation". Hydathodes secrete water which is brought to the surface by the terminal tracheids of the veins. The water passes through the intercellular spaces of the loosely packed parenchyma of the hydathode, devoid of chloroplasts known as "epithem". The intercellular spaces open to the exterior by special pores, which are of stomatal origin (open stoma) and which remain permanently open, *e.g.*, *Brassica*, plants belonging to Poaceae, Crassulaceae and Saxifragaceae.

Structure of Petiole

A similarity exists between the tissues of the petiole and those of the stem. The epidermis of the petiole is continuous with that of the stem. The parenchyma cells of the petiole, like those of the cortex, contain only few chloroplasts, compared to the cells of the lamina. The supporting tissues of the petiole are collenchyma and/ or sclerenchyma. The vascular bundles of the petiole may be collateral (*Ligustrum*), bicollateral (*Nerium*) or concentric (certain Pteridophytes and many Dicotyledons). In many species, the phloem is accompanied by groups of fibres. The arrangement of vascular tissues in the petiole differs in different plants. In a cross section of the petiole, it may appear as a continuous or an interrupted crescent (*Olea, Nicotiana, Nerium*) or as an entire or interrupted ring (species of *Ricinus, Quercus, Citrus*) or as a ring with additional internal and external bundles (*Vitis, Platanus, Robinia, Pelargonium*). An arrangement of many scattered bundles is seen in *Rumex* and many Monocotyledons. If there is a single collateral bundle in the petiole, the phloem is found on the abaxial side and if the bundles are arranged in a ring, the phloem is external to the xylem on the periphery of the ring.

In some plants, *e.g.*, *Mimosa* and *Albizia* of the Fabaceae the petioles or petiolules have pad like swellings termed pulvini (pulvinus). These are capable of bringing about movement of leaves or leaflets. The pulvinus contains a large amount of parenchyma and its surface is usually wrinkled. The closure and opening movements of the leaflets may be stimulated by environmental factors and are brought about by turgor changes in the motor cells of the pulvini.

Vascular System or Vascular Tissue

The veins represent the vascular tissue of the leaf and these are located in the spongy tissue of the mesophyll. The arrangement of the veins is called venation. It is reticulate in Dicots and parallel in Monocots. A vein is made up of a vascular bundle, and at the core the bundle are clusters of conducting cells, the xylem and phloem. Xylem brings water and the minerals from the roots into the leaf, and phloem moves sap with dissolved sucrose produced by photosynthesis in the leaf and out of the leaf.

The midrib in most Dicotyledons leaves have large bundle with adaxial xylem and abaxial phloem (both are embedded in a dense parenchyma, the sheath which includes some collenchyma). The bundle is conjoint, collateral and open. A thin strip of cambium is present between xylem and phloem. In *Liriodendron* and *Vitis,* the midrib region has several bundles arranged in a ring while in *Helianthus,* several but irregularly distributed bundles are found. The large and medium-sized veins contain vessels and sieve tubes. In the smallest veins, the tracheary elements are tracheids with annular and spiral wall thickenings. The larger veins in some Dicotyledons like *Ficus religiosa* show a limited amount of secondary growth as a result of activity of cambium. In Monocotyledonous leaves, the bundle is conjoint, collateral and closed.

In the small or minor veins, the parenchyma cells in contact with the sieve elements and tracheary elements constitute transfer cells. Some of these cells are believed to be concerned with short distance translocation between mesophyll and sieve elements and others with an exchange of solutes between the phloem and xylem. The specialized transfer cells develop wall protuberances which increase the internal cell surface.

Bundle Sheath

The large veins are surrounded by much parenchyma which is poor in chloroplasts. The smaller veins, also, are usually surrounded by a layer of tightly packed parenchyma cells; such a layer is termed the "bundle sheath". In Dicotyledons, the cells of the bundle sheath are usually elongated in a direction parallel to the vein. Sometimes, however, for example in *Atriplex halimus,* the cells are more or less cubical. The cells of the bundle sheath are thin-walled. They may contain as many chloroplasts as the mesophyll cell, or they may be devoid of them (Figure 14.3). Sometimes crystals may occur in the bundle sheath cells. Bundle sheaths extend to the ends of bundles so that the vascular tissue is rarely directly exposed to the intercellular spaces. An exception is found in those hydathodes in which the xylem elements at bundle end release water into the intercellular spaces. Bundle sheaths may be parenchymatic or sclerenchymatic and the two kinds of sheaths may be combined. Suberization of cell walls in bundle sheaths recorded in several species indicates that these cells might function as an endodermis.

In many Dicotyledonous leaves such as the species of *Styrax* and *Quercus*, the bundle sheaths are connected with the epidermis by panels of cells resembling the bundle-sheath cells. These panels are called "bundle sheath extensions". Bundle-sheath extensions have a conduction function in leaf, they conduct from the bundles to the epidermal cells. The latter are very closely connected laterally and therefore function to conduct in a plane parallel to the leaf surface. Water may also move from the bundle sheath and its extensions through the mesophyll to the cells lining substomatal chamber.

Bundle sheaths also occur in the Monocotyledons, especially in the grasses in which two types are distinguished, one- or two-layered. The outer sheath when two layers are present, or the single sheath where there is only one, generally consists of parenchymatous cells with thin walls. The cells of this sheath contain chloroplasts which are usually smaller than those of mesophyll. The cells of the parenchymatous sheath may store starch and form a "starch sheath". The cells of the inner sheath, the mestome sheath, are smaller in cross section; their walls are thickened and contain suberized lamellae. The mestome sheath is thus analogous to an endodermis. Single sheaths are generally characteristic of panicoid grasses (Figure 14.3A) and double sheaths (Figure 14.3B) of festucoid grasses.

Supporting Tissues of the Leaf

The epidermis itself, due to its compact structure, strength of the cuticle, occasional thickness of its walls or deposition of the silica, gives support to the lamina. Collenchyma is usually present close to the larger veins immediately below the epidermis and also on the leaf margins in the Dicotyledonous leaves. The bundle sheath extensions may also be collenchymatous. Additionally sclereids are also present in the mesophyll of many Dicotyledons. The large and medium sized veins are accompanied by groups of fibres in many plants. In Monocotyledonous leaves the vascular bundles are accompanied by many fibres. In Poaceae and many other Monocotyledons the fibres form girders on one or both sides of the bundles.

Leaf – Development

Leaf Determination

The primordium which gives rise to leaf, can also give rise to branch of the stem, the inflorescence, a flower or in some cases, even thorns. So the fate of the organ primordium is not specified until it has emerged. It is shown by interposing an impermeable barrier between the leaf primordium and shoot apical meristem. This results in the growth of fewer leaves and more lateral branches. It appears that there is an important stage immediately after the primordium emerges in which its fate is not decided. In presence of signals from SAM, the primordium

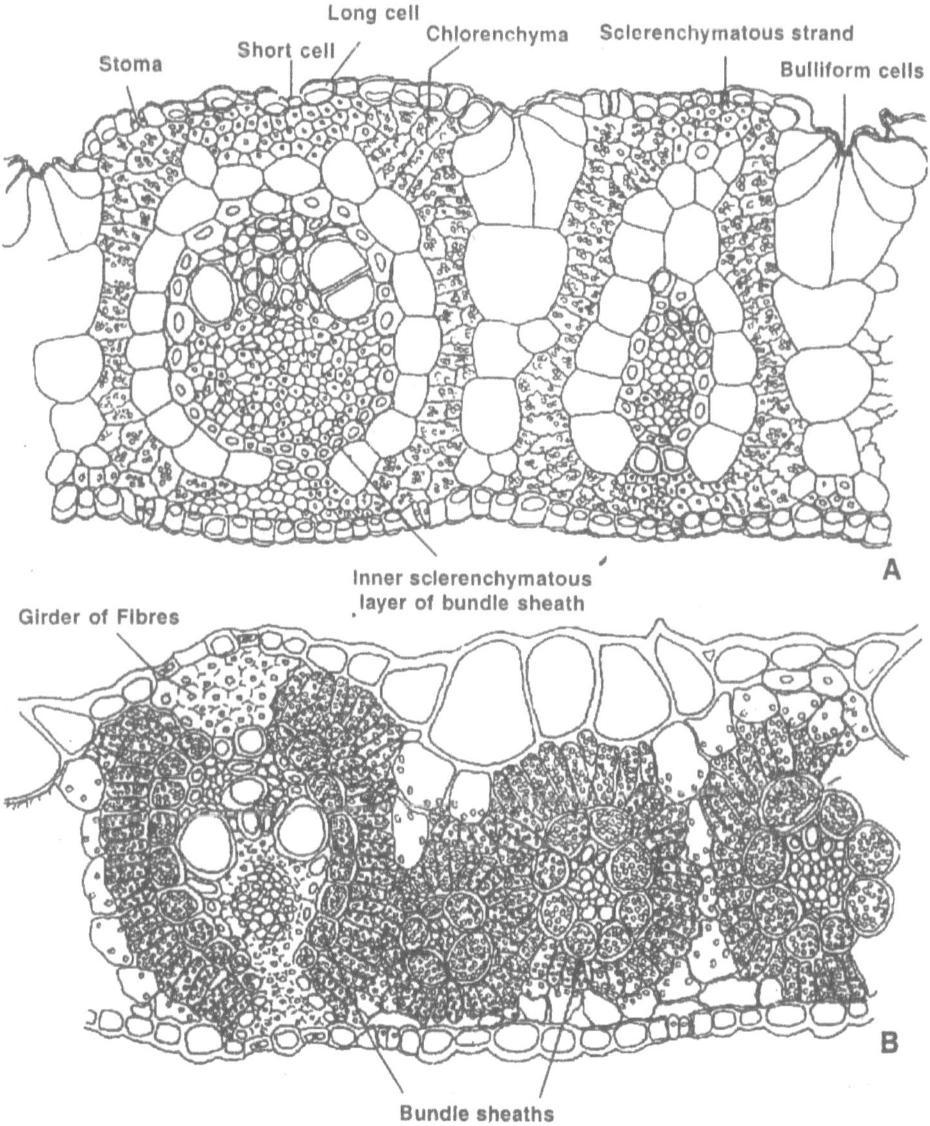

Figure 14.3. Portions of Cross-Sections of Grass Leaves. A. *Desmostachya bipinnata*; B. *Hyparrhenia hirta.*

gives rise to leaf, when the signals from SAM are absent (due to impermeable barrier) the primordium may develop into a shoot. The timing of the above process can be investigated by explaining the individual primordia at different timings after its emergence and see whether they are able to give rise leaf tissue in cultures. The results of this experiment showed that plants vary in the timing of leaf determination. But, in most plants it occurs soon after emergence of the primordium. By this, it is clear that leaf determination requires signals from SAM and these results in the polarization of dorsiventral axis. The dorsiventral polarity is seen in *phantastica* mutant. This mutant is identified in both *Antirrhinum* and *Arabidopsis*. The upper leaves in mutant plant are ventralized and form needle like structures, which resemble stem in their organization. The PHANTASTICA gene is very important component of dorsalizing signal in the beginning of leaf development.

Leaf Development

A leaf is initiated by cell divisions in the peripheral region of the shoot apex, more or less far below the distal zone of the apical meristem. However, the number of cell layers which divide and their position on the apex varies in different plants. In many grasses, the leaf initiation starts with periclinal divisions of the surface layer of the apex (outermost layer of tunica) and the cells of the layer immediately below it. So here, the main leaf primordium originates from the outermost layer of the shoot apex. In Dicots and *Tulipa* (Monocot), the first periclinal division occurs not in the surface layer of the apex, but one or more layers below it. Hence, the surface layer gives rise only to the protoderm of the young leaf. But, it is difficult to determine accurately the part played by each layer.

The periclinal divisions are soon followed by anticlinal divisions in the protoderm and the layers beneath it. In some taxa, the protoderm cells also divide periclinally. The initial divisions give rise to lateral protrusion, often called 'leaf buttress' on the shoot apex. This leaf buttress consists of a protoderm layer, an inner mass of ground meristem and a procambial strand which develops acropetally from the nearby procambium of the stem.

The leaves are partial shoots being derived from the leaf primordial of the shoot apex according to Agnes Arber's "partial-shoot theory". They are dorsi-ventrally flattened early in their development with both the dorsal and ventral surfaces. The compound leaves are closer to the shoots than the simple leaves. The studies of development have shown that the compound leaves (like the shoots) may branch in three dimensions. Eckardt and Baum (2010) on the basis of the molecular genetics concluded that "it is now generally accepted that the compound leaves express both leaf and shoot properties".

The leaf primordium extends upward from the initial protrusion as a bladeless conical or peg-like protuberance referred to as the leaf axis (Figure 14.4 A). The primordium is dorsiventral as it emerges. In subsequent stages, the primordium increases in length and width and by growing more actively on the abaxial side than on the adaxial side, develops a curvature toward the shoot apex.

The increase in width of the leaf includes the spread of the leaf base laterally so that the apical meristem is encircled to a greater or lesser extent depending on the degree to which the mature leaf ensheaths the stem. The encircling part of the leaf base may give rise to stipules or to a sheath as in the Monocotyledons.

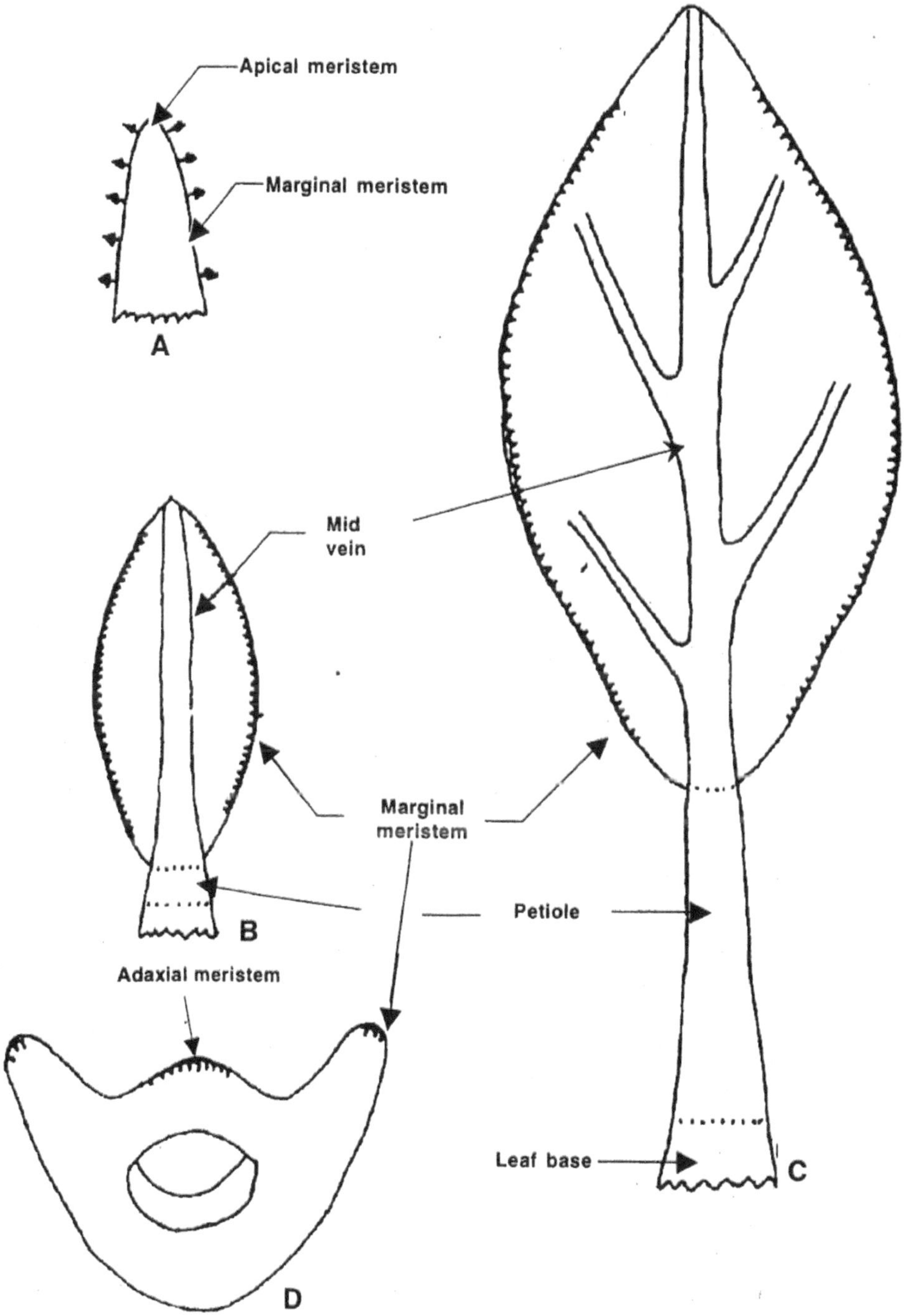

Figure 14.4. Diagrams Interpreting Growth of a Dicotyledon Leaf. A. Undifferentiated leaf primordium (leaf axis) without blade and petiole; Arrows indicate direction of growth; B, C. Two stages of growth of blade; D. Cross section of blade showing position of adaxial and marginal meristems.

The growth concerned with the lateral extension of the primordium becomes localized along two margins of the leaf axis and the primordium differentiates above the base into a midrib and two panels of the leaf blade (Figure 14.4B). Lateral veins become blocked out in the form of procambial strands as the blade continues to widen and, in petiolate leaves, a petiole develops between the blade and the leaf base (Figure 14.4C). The growing blade may spread sideways from the midrib or bend toward the adaxial side of the leaf (Figure 14.4D) so that it encloses the apical meristem and the younger leaf primordia. The midrib often increases in thickness on its adaxial side by orderly periclinal divisions within a region constituting the adaxial meristem. The petiole also increases in thickness by means of an adaxial meristem.

The early growth of a leaf is commonly divided into apical and marginal growth, the first concerned with the elongation of the primordium, the second with the lateral extension that produces the two panels of the blade. Thus, the growing primordium has an apical meristem at the tip and two marginal meristems along opposite sides of the axis (Figure 14.4). Each meristem is at least two cells in depth.

The new cells that are added to the different layers of the lamina originate from rows of marginal initials and submarginal initials. The marginal initials are the cells of the outer most layer on the margins of the young lamina. Generally in the Angiosperms these initials divide only anticlinally and so add new cells to the abaxial and adaxial protoderm. The submarginal initials undergo divisions in various planes. These cells give rise to new cells which are added to the inner layers of the young lamina. In Figures 14.5 and 14.6. it can be seen how, in different plants, these tissues develop from the cell layers of the young lamina.

In pinnately and palmately compound leaves lateral leaflets develop from the adaxial marginal meristem of the axis of the young leaf as two rows of papillae. In certain plants, the order of development of the leaflets is acropetal, *e.g., Carya*, but in many other plants the order is basipetal. The development of lobed leaf is a result of differential activity of the leaf margin which starts early in ontogenesis. In the regions of the future lobes meristematic activity is considerable, while in the regions of future sinuses there is little or no activity.

The derivatives of cells at the apex and along the margins of the leaf divide in their turn. These intercalary divisions, which are distributed at random, make the major contribution to the increase in size of the leaf. The intercalary meristematic activity ceases earlier at the apex than at the base so that the leaf tissues differentiate and mature in the basipetal direction. Moreover, cell divisions may be less numerous in the apical region than farther below, a difference that results in a greater width of the leaf toward the base. The basipetal order of maturation is particularly clear in the long narrow monocotyledon leaves in which growth becomes conspicuously localized at the bases of leaf blade and sheath zone is referred to as intercalary meristems.

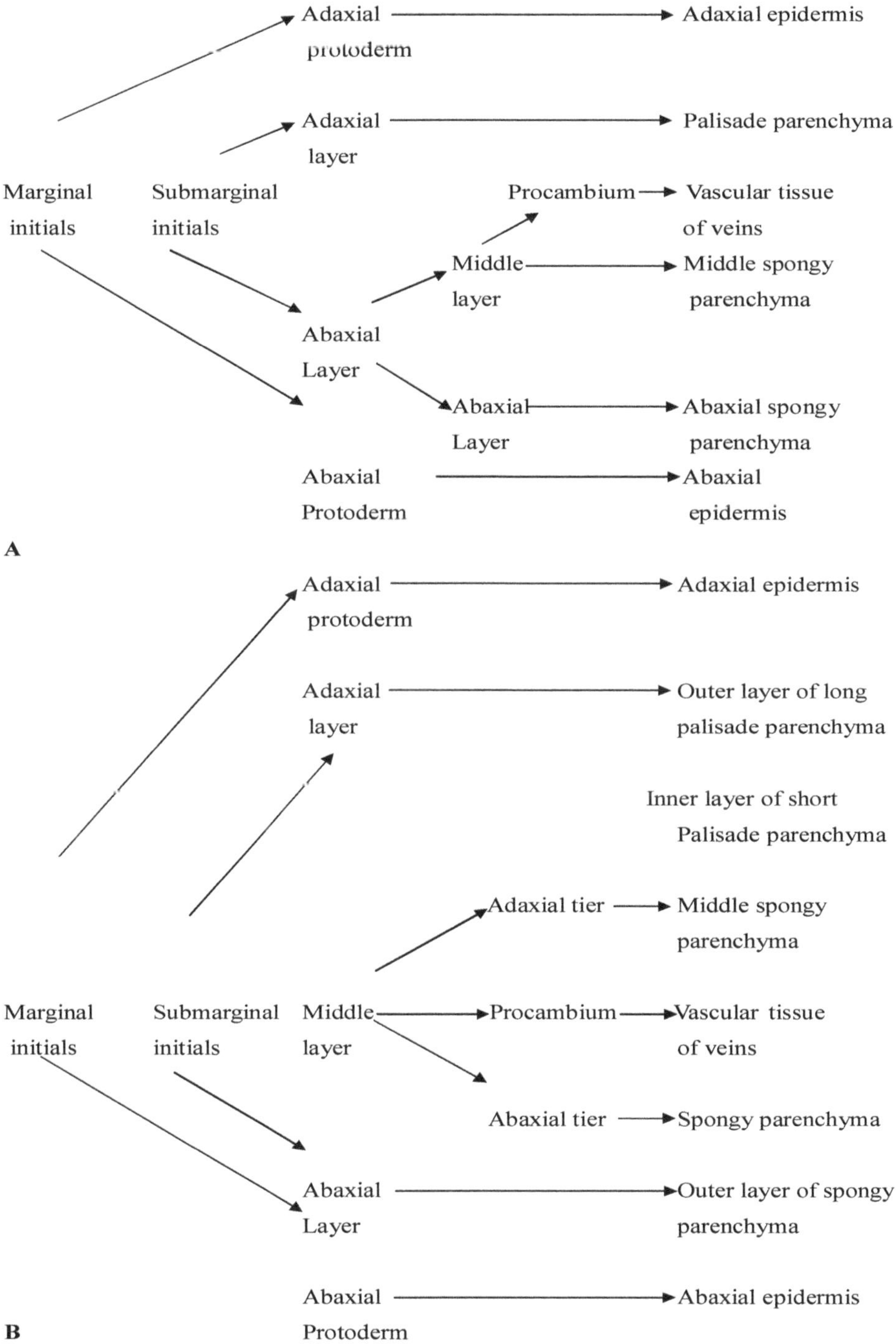

Figure 14.5: Diagrammatic Representation of Leaf Histogenesis. A. *Carya buckleyi*; B. *Pelargonium zonale*.

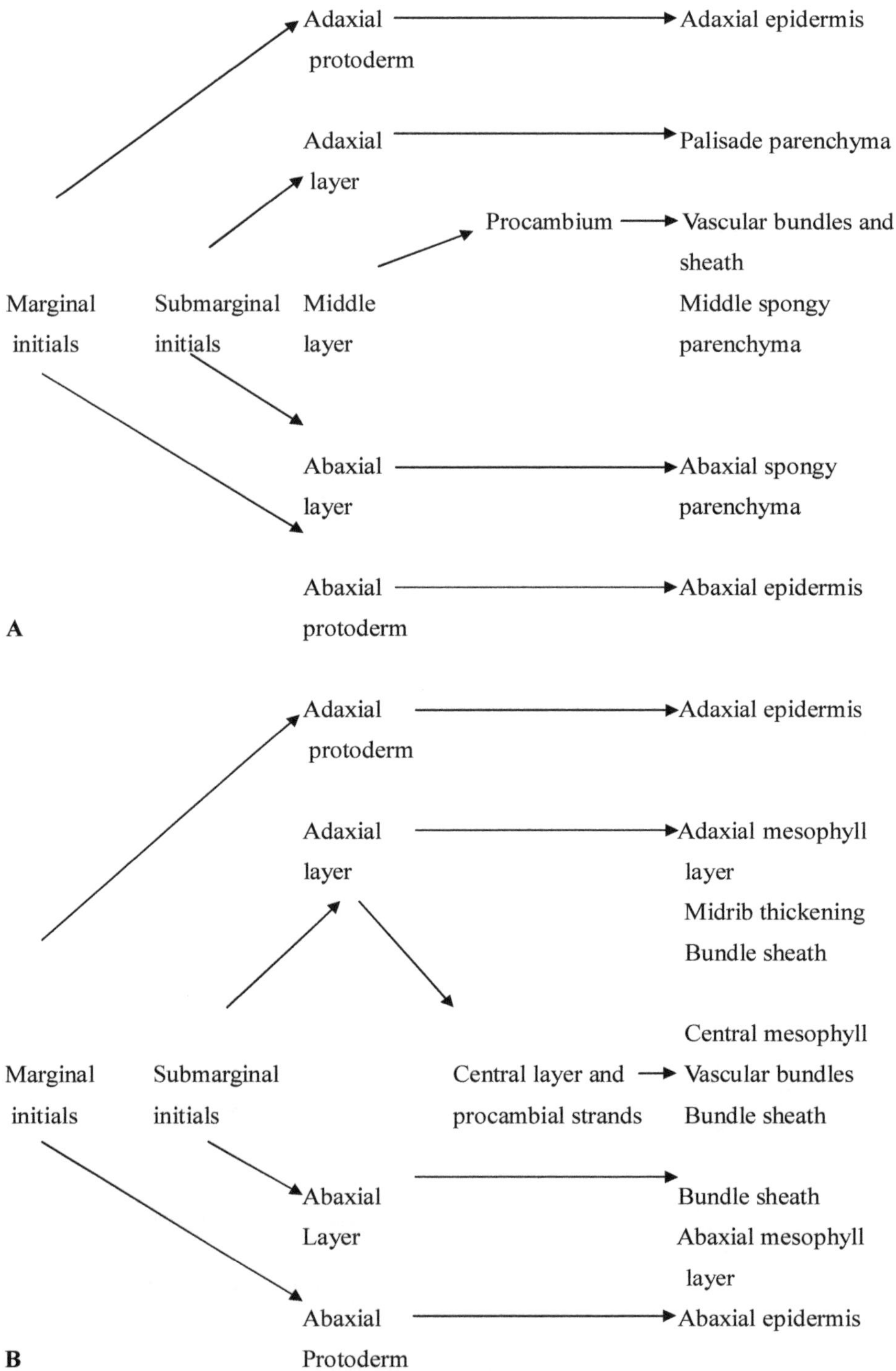

Figure 14.6: Diagrammatic Representation of Leaf Histogenesis. A. *Nicotiana tabacum*; B. *Oryza.*

Development of Stomata

Stomata are essential to the life of terrestrial plants. There is usually a stomata free zone surrounding each stoma in normal plants. This intervention of subsidiary (when they exist) or epidermal cells between stomata is thought to provide a pool of ions for the guard cells and optimize the access of the underlying mesophyll cells to the external environment. Two features that influence stomatal patterning are: internal anatomy (cell type, position of the vasculature), and contact between stomata. Little is known about that why stomata rarely develop over the vasculature. Certain underlying cell types seem to inhibit development of stomata over vasculature. Stomata also rarely develop side by side as there is a control over stomatal patterning.

Stomatal initials form by asymmetric cell division of precursor cells such that the smaller cell becomes a stomatal initial. The stomatal initial can produce both the guard cells and ordinary epidermal cells. But the mother cell of the guard cells always undergoes symmetrical division to produce two guard cells. Cell-to-cell communication takes place leading to the coordination of the plane of cell division as well as the regulation of precursor cell activity. Additionally, the timing of when a cell is competent to become a stomatal precursor and where such precursors can initiate are highly controlled.

In both Dicots and Monocots, the cell divisions leading to the production of stomata are highly asymmetric. In Monocots, the first asymmetric division results in the smaller, usually rectangular cell becoming guard mother cell, which divides symmetrically producing two guard cells. In Dicots, the smaller cell is usually triangular and continues to divide after the surrounding cells have ceased to do so. Eventually it converts into guard mother cell and divides symmetrically to produce two guard cells. Since Dicot stomatal initials continue to divide after the surrounding epidermal cells have ceased they have been termed "meristemoid cells" to emphasize their ability to continue cell division. In Dicots there are two types, primary meristemoids and secondary (satellite) meristemoids, the latter produced by an asymmetric division of a neighbouring cell to the stomata. Primary meristemoid positioning appears to be random, with the exception that it obeys the rule that two stomata should not be adjacent. Secondary meristemoids and stomatal initials in Monocots are placed in a highly regular fashion crucial to stomatal patterning. Primary meristemoids arise from protodermal cells. These cells undergo a series of asymmetric divisions that produce a small, triangular initial situated approximately in the middle of the future stomatal complex. Secondary (satellite) meristemoids arise once a primary meristemoid has completed development into a stoma.

Generally, the mother cell of the stomata divides symmetrically to form two cells that differentiate into the guard cells. At first these cells are small and have no special shape but, as they develop, they enlarge and become characteristically shaped. During their development the middle lamella between the two guard cells swells and becomes lens-shaped and finally it disintegrates to form the stomatal aperture. Stevens and Martin (1978) suggested that, pore formation is initiated enzymatically. The separation of the two guard cells is brought about by osmotic force derived from starch hydrolysis. The mature guard cells are sunken or rose relative to the level of ordinary cells of the epidermis. The sinking or rising are brought about during the maturation of the guard cells. The development of stomata in the leaf continues for a relatively long period during the growth of the leaf.

Timing of Stomatal Development

In *Arbidopsis* at least, stomata develop after trichomes have formed.

Cell-to-cell Communication during Stomatal Development

There is a wide range of signals traversing between cells thus coordinating stomatal development. Evidences for cell to cell communication are: (a) Inhibition of stomata formation above veins in the leaf, (b) Development of meristemoids besides another meristemoid or stomata is arrested, (c) The orientation of subsequent cell division in such a way to separate meristemoids when two happened to develop beside each other and (d) the initiation of subsidiary cell formation in Monocots.

Florin *et al.* (1931 and1933) proposed 2 types of stomatal patterns based on ontogeny (origin), namely:

(a) Haplocheilic Type

In this type, 2 guard cells originate from a common meristemoid which functions directly as a guard mother cell (GMC). The meristemoid does not divide prior to its transformation into GMC. GMC and the guard cells influence the neighbouring protodermal cells to divide and form the subsidiary cells (hence, these are "perigene"). *e.g. Ephedra*. This type of stoma is similar to perigenous type of Pant.

(b) Syndetocheilic

In this type, the stomatal meristemoid undergoes a number of divisions resulting in the formation of GMC and subsidiary cells that adjoin the guard cells (the guard and subsidiary cells are derived from the same mother cell). The subsidiary cells are "mesogene". This type of stoma is similar to mesogenous type of Pant. *e.g., Welwitschia*.

Florin and Pant recognized perigenous stomata as a distinct type but, Payne (1979) regarded that there exists no perigenous stoma, in which a protodermal cell directly serves as GMC (hence, it is mesogenous or mesoperigenous).

Considering the above facts, Payne proposed a classification on the basis of the manner of production and division of GMC, as follows:

Diameristic

In this type, GMC divides with a wall formed at right angles relative to the wall that cuts it off from its meristemoid. The meristemoid divides repeatedly resulting in 2 or more subsidiary cells which are parallel to each other, *e.g.* Mesogenous stoma of Lamiaceae, and mesoperigenous stomata of Monocots and Caryophyllaceae. At maturity, the stomatal complex consists of a pair of guard cells and a pair of subsidiary cells, the long axes of which are at right angles to each other.

Parameristic

In this type, GMC divides with a wall parallel to the wall that cuts it off from its meristemoid, which thus divides forming 2 cells. One of the daughter cells becomes the first lateral subsidiary cell, *e.g.* Mesogenous stoma of Rubiaceae, and mesoperigenous stoma of *Liriodendron*. The remaining daughter cell again divides in a way that the resulting cell wall curves to meet both ends of the cell wall that is formed in the first division. Among the 2 cells formed one becomes the second lateral subsidiary cell. Thus, a mature stoma exhibits a pair of guard cells enclosed by a lateral subsidiary cell, and the long axes of the guard cells and the subsidiary cells are parallel to each other.

Anomomeristic

In this type, the GMC divides with a wall formed at any angle relative to the wall that cuts it off from the meristemoid, which divides to form the GMC, *e.g.* Mesoperigenous stoma of *Ranunculus*. The GMC in turn divides to form a pair of guard cells the partition wall of which has no definite orientation in relation to the wall that formed the GMC.

Pant (1965) recognized three categories of stomata on the basis of development.

Mesogenous Type

A stomatal complex is said to be mesogenous where all the subsidiary cells or a single ring-like subsidiary cell have a common origin with the guard cells, that is, the same meristemoid gives rise to the subsidiary cells and the guard cells.

Guard cells and neighbouring cells, which may or may not differentiate as subsidiary cells have common origin, *e.g.*, Rubiaceae, Brassicaceae. The initial of the stomatal complex is formed by an asymmetric division of a protodermal cell. Two divisions result in the partitioning of the initial cell into a mother cell of the guard cells and two subsidiary cells. One more division in the mother cell of the guard cells leads to the formation of guard cells (Figure 14.7).

Perigenous Type

A stomatal complex is said to be perigenous where the subsidiary cells and the guard cells do not have a common origin. The subsidiary cells originate from the cells that lie around the meristemoid. This divides symmetrically to form 2 daughter guard cells (as in *Psilotum, Selaginella, Isoetes, Ginkgo*, and Lemnaceae *etc.*). Neighbouring or subsidiary cells do not have common origin with the guard cells, *e.g.*, Cucurbitaceae and Nymphaeaceae, Poaceae (Figure 14.9). In grass leaves and some other monocotyledons, the first step in the differentiation process is the symmetric division of certain protodermal cells. Before the cell divides, the nucleus

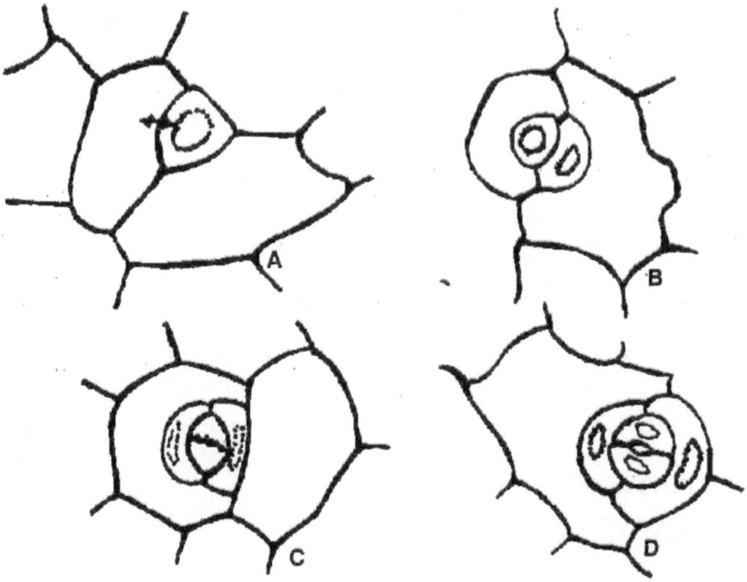

Figure 14.7: Development of Stoma with Mesogenous Subsidiary Cells in a Leaf of *Thunbergia erecta*.

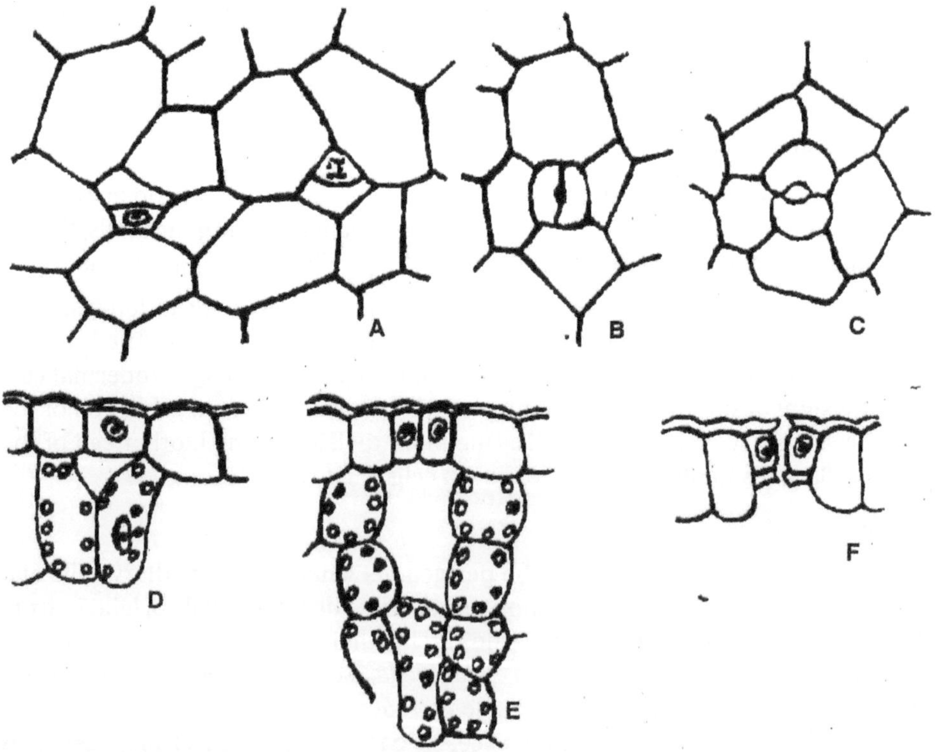

Figure 14.8: Development of Mesoperigenous Stomata in *Beta vulgaris*. A-C. Surface view; D-F. Sectional views.

becomes displaced in one direction and the vacuoles occupy the other end of the cell. The nucleus then divides. One of the daughter nuclei which is closer to the vacuoles becomes larger and more weakly staining than the other at the denser end of the dividing cell. The smaller distal nucleus divides later to form the two guard cells, but before doing so it induces neighbouring epidermal cells to divide (Figure 14.9A, B). In these cells also there is denser cytoplasm adjacent to the guard mother cell, and the induced divisions are also asymmetrical. The daughter nucleus in the centre of the cell becomes larger and weaker staining than that near the guard mother cell, which becomes the nucleus of a subsidiary cell. After the formation of the subsidiary cells the guard mother cell divides to form the guard cells (Figure 14.9C-G).

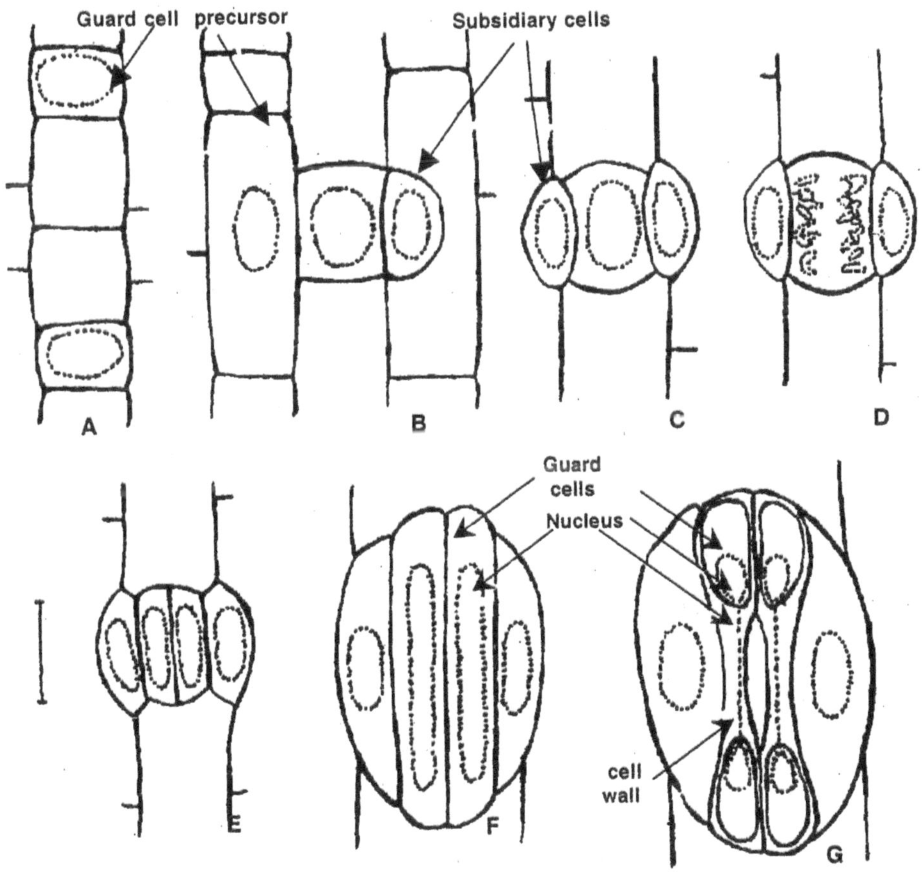

Figure 14.9: Development of Perigenous Stomata in *Avena sativa.*

Mesoperigenous Type

A stomatal complex is said to mesoperigenous where one stomatal meristemoid gives rise to one mesogene subsidiary cell and gcmc (guard cell mother cell) that divbides symmetrically into 2 guard cells. The rest of the subsidiary cells are perigene and originate from the cells lying around the meristemoid.

At least one neighbouring or subsidiary cell and stomata are derived from the same mother cell (Figure 14.8), *e.g.*, Ranunculaceae and Caryophyllaceae. As a rule the precursor cell undergoes an asymmetric division to produce a smaller stomatal initial and larger ordinary epidermal cell. The smaller cell undergoes a division giving rise to a subsidiary cell and an initial cell. Then the initial cell undergoes a division at right angles to the subsidiary cell wall that is already formed and produces another subsidiary cell. Finally, the initial divides to produce third subsidiary cell and mother cell of the guard cells. Then the mother cell of the guard cells divides symmetrically to produce two guard cells. As they develop, they enlarge and become characteristically shaped (Figure 14.8).

Genetic Analysis of Stomatal Development

There are two mutants identified that violate the rule that two stomates should not be in contact. The four lips (flp) and too many mouths (tmm) mutations show two different developmental mechanisms, tmm mutants have greater numbers of stomata in rosette leaves, cotyledons and abaxial sepal surface. Other tissues of tmm plants (inflorescence, stem, abaxial sepal, epidermis *etc.*) entirely lack stomata where as wild type plants possess them. The flower pedicel shows an abnormal stomatal frequency from none proximal to the stem to abnormally greater numbers near the flower base. Hence, TMM is thought to control the entry of protodermal cells into the stomatal developmental process and appears to have opposite effects on meristemoids depending on the tissue (compare rosette leaves with sepal epidermis). The tmm mutation appears to increase the formation of satellite meristemoids as well as to alter the polarity of cell divisions such that meristemoids can now form next to guard cells. The flp mutation can produce unpaired guard cells as well as result in stomata twinning. Analysis of stomata; development in flp mutants suggests that stomata develop normally until the guard mother cell stage. At this stage the guard mother cell undergoes symmetric division and the resulting daughter cells divide symmetrically again to produce two contiguous stomates. Hence, flp may control guard mother cell identity, or regulate the number of symmetric divisions a guard mother cell can undergo.

Development of Trichome

A trichome has been defined as any protrusion from the epidermis. Accordingly conical-papillate petal cells, both glandular and non-glandular hairs, and surface glands are all classified as trichomes. Although some plants have single type of trichome, but often different types of trichomes are reported on the same plant. For example, tomato has six different types of trichomes. Trichomes play a role in protecting the plant from predators and the disease-causing organisms they carry. Trichomes mature basipetally.

Trichomes are differentiated from single cells originating in the epidermis. Although trichomes occur on different parts of the plant, the leaf trichomes are special because they branch characteristically and are evenly distributed across the leaf's surface. The majority of flowering plants produce multicellular trichomes. Although these trichomes must undergo processes of commitment, expansion and morphogenesis, which may, in principle, be similar to those of unicellular trichomes, they must also do this in the context of continuing cell division. Multicellularity allows trichomes to function as miniature organs, with differentiation within the trichome with regard to cell function. Many multicellular trichomes are glandular, developing a terminal gland, which may secrete a variety of compounds, such as alkaloids to poisonous substances. Tobacco leaves develop two types of multicellular trichome, described as short and long-stalked. The short-stalked trichomes have an eight-celled glandular head, and secrete nicotine, while the long stalked trichomes have a gland that produce diterpenoids.

Ontogeny of Trichomes

Trichomes develop from single epidermal cells in young leaf primordia and have a remarkable cell morphogenesis. The trichome cell differentiate while epidermal pavement cells are still dividing and it synthesizes more DNA at the same rate as they do, but cell division is arrested. After three rounds of endoreduplication of nucleus, the cell expands outwards from the leaf surface as a single cone. The enlarging cell grows outwardly from the leaf surface, so its nucleus is pushed into a subapical location. When the volume of the trichome cell has increased tenfold, it branches. The nucleus enlarges again by endoreduplication, while the two newly formed branches grow at their tips. One of the branches elongates to form the main branch, then branches again. The nucleus then relocates within the cell so that it lies between the two branching points. The mature trichome rests on a small podium formed by a circle of 8-10 epidermal cells, around a layer of subepidermal cells. The epidermal podium cells differ in shape and size from other cells in the epidermis.

Trichome development is linked to leaf development. Trichomes emerge from the leaf primordium a distance of approximately four cells apart. As they grow and branch, the epidermal cells between them divide. In the end, approximately 30 epidermal cells separate neighbouring trichome cells. The branching of the trichomes is oriented according to the longitudinal axis of the leaf primordium. The first-formed branch points towards the base of the leaf. The latter-branching main branch grows towards the leaf tip. The trichomes of a leaf primordium are developed to different degrees: at the leaf tip are trichomes that are already mature, while at the base some trichomes have yet to branch, and in the leaf centre there are trichomes that have one branch.

The Pattern of Trichome Development

The pattern of trichome development in the *Arabidopsis* leaf epidermis is not random. There are two models to explain how trichome spacing is determined. The first suggests that precursor cells produce inhibitory signals that prevent neighbouring cells to develop into trichomes. The second suggests that, once a precursor cell is induced to become trichome it undergoes several divisions that separate it from other trichomes by a group of cells, all derived from the same precursor daughter cell. A model based on genetic analysis has been proposed to explain how events proceed to commit individual cell further to trichome differentiation while inhibiting neighbouring cells.

Differentiation of Mesophyll

The mesophyll differentiatiates from the derivatives of the marginal meristem after these derivatives have undergone the intercalary growth. The establishment of the characteristic differences between the palisade and the spongy parenchyma results from an unequal growth in the various layers of the leaf: there are differences in duration of cell division and cell enlargement and in the direction of cell expansion in the epidermis and in the various layers of the mesophyll. Cell division commonly ceases first in the upper epidermis and continues longest in the future palisade tissue.

The difference in amount of cell division between the upper epidermis and the palisade parenchyma may be illustrated by the numerical relationships between cells of the two tissues in young and old apple leaves. The ratios between the numbers and diameters of the epidermal and the palisade cells were 1:1 in young leaf. In the mature leaf the diameters of the epidermal cells are 3 to 4 times larger than those of the palisade cells, and there were 8 to 10 palisade cells to one epidermal cell. In *Xanthium* leaf, the upper epidermis and the palisade parenchyma differ in duration of cellular enlargement and rates of expansion. In the epidermis this rate is high in the horizontal plane but low in the vertical. The opposite relation characterizes the palisade tissue. While the division and the enlargement of the palisade cells remain correlated with the growth of the epidermal cells, the palisade cells are closely packed. When the growth of the two kinds of cells becomes differential, intercellular spaces develop in the palisade tissue. The palisade cells divide mainly by anticlinal walls. The palisade cells also elongate at right angles to the surface, and during the formation of intercellular spaces they separate from each other along the anticlinal walls. All these phenomena are responsible for the characteristic appearance of the palisade, that is, of a tissue composed of orderly rows of elongated cells largely separated from each other along their anticlinal walls. The developmental relation between the lower epidermis and the spongy mesophyll in bifacial leaves appears to be somewhat variable. This epidermis may cease cell division before the spongy mesophyll but continue longer with cell enlargement or it may show cell division after this activity has stopped in spongy tissue. In both situations the epidermis grows in surface without formation of intercellular spaces, while the spongy parenchyma grows in a tangential plane by some enlargement of cells and by a loosening of contacts. Among the various tissue elements of the leaf, the epidermal hairs, the stomata, and the large veins complete their differentiation before the mesophyll. The stomata develop concomitantly with or after the development of the intercellular spaces in mesophyll.

Development of Vascular Tissues

Vascular development in a leaf of a Dicotyledon begins with the differentiation of procambium in the future midvein. The procambium differentiates acropetally in continuity with the leaf trace procambium in the axis. With the commencement of the development of the lamina the procambial strands of the large lateral veins and, later of the smaller veins begin to form gradually. The procambial strands of the small veins, which form in a basipetal direction develop mainly during the intercalary growth of the lamina.

In the Monocotyledons the procambium of the larger veins differentiates acropetally, that of the smaller veins alternating with the larger and of the transverse interconnections progresses basipetally.

Structure of Modified Leaves

1. Xeromorphic Foliage Leaves

The term xeromorphic is used to indicate plants that have anatomy and growth form which are adapted to desert conditions, particularly where the water is scarce. In addition, the desert soils are typically alkaline and are rich in minerals and sometimes toxic too. They are poor in organic materials which help to hold water in the soil. Temperatures vary between extremely hot and severely cold on both daily and on annual basis. The plants meet these challenges in various ways. They germinate, grow, set seed and die within few weeks and complete their life cycle. So their adaptation to xeric conditions is their ability to complete their life cycle in a short span, when the desert is temporarily mesic after rains.

Plants that occur in xeric microhabitats are perennial and show well developed xeromorphic characters, such plants are called xerophytes. These xeromorphic characters are: thick walled epidermis and hypodermis covered by a dense, waxy cuticle, trichomes are abundant. These characters are advantageous in many ways, *i.e.*, all these characters are effective in blocking excess sunlight. They are also good deterrents against insect feeding and egg laying. Salt glands are also present in many cases and they also help in protecting against feeding and sunlight as well as keeping salts accumulating in the protoplasm. Some of these characters are directly induced by the environment. Plants of *Prosopis* growing outdoors have cuticle ten times thicker than the plants that are grown in doors.

The surface area of the xeric leaves is greatly reduced for decreasing the transpirational surface area. Leaves are smaller, more cylindrical and are more succulent. To further reduce water loss, the stomata are sunken and are restricted to well protected crypts, *e.g.*, *Nerium, Ficus.*

Inside the leaf too, the surface area is reduced, as the cells are packed together and are smaller. The spongy mesophyll may altogether lost, with only palisade parenchyma remaining. The vascular bundle sheaths are sclerenchymatous. Water storage cells are present as large mucilage cells, *e.g.*, Cacti.

2. Submerged Foliage Leaves

A small number of Angiosperms and a few Gymnosperms have migrated back into aquatic habitats. These plants are quite variable ranging from the plants growing in marshy swampy areas to growing as floating unanchored plants and plants rooted to the bottom of the stream, pond or ocean, its leaves are fully submerged and plants are called hydrophytes, *e.g., Ceratophyllum, Vallisneria, Sagittaria, Nymphaea.* As the water provides a uniform habitat, the anatomical features of water plants are less varied than the xerophytes. The factors influencing the water plants are temperature, air and the concentration and composition of salts in the water. The most striking features of the water plants are, the reduction of supporting and protective tissues, decrease of vascular tissues especially xylem and the presence of air chambers.

The epidermis of water plants does not have a protective function, it plays an important part in the uptake of nutrients from the water and in gas exchange. The cuticle and the cell walls are very thin and the epidermal cells contain the chloroplasts. Immersed leaves are generally devoid of stomata or may possess vestigial stomata. They are present on the upper surface of the floating leaves. In the epidermis of the immersed surface of leaves of some water plants, small groups of cells appear which stain deeply with certain dyes. These cell groups are termed as 'hydrophytes' are thought to be structures which facilitate water and salt transport into and out of the plants.

In many water plants like *Ceratophyllum, Myriophyllum* and *Utricularia*, the leaves are divided into narrow cylindrical lobes which greatly increase the area of contact with the water. In *Ranunculus aquatilis*, the floating leaves are entire while the immersed leaves are divided into narrow lobes. In the leaves and stems of hydrophytes air chambers are present which are filled with gases. These chambers are intercellular spaces regular in shape and pass through the entire leaf. The air chambers are seperated from one another by thin partitions and allow only the passage of gases but not water. Immersed hydrophytes contain very little or no sclerenchyma. However, strips of sclerenchyma are present along the leaf margins.

3. Halophytic Leaves

Halophytes grow in saline soil of water where there is abundance of salt in the soil. Typical examples of halophytes are *Chenopodium, Acanthus.*

Halophytes growing in marshy places near the sea-shore collectively form special vegetation called mangroves. Anatomically mangroves show several xerophytic characters. The epidermis is thick-walled and strongly cutinized. Sunken stomata are found in the epidermis. Mesophyll cells are usually compactly arranged.

Kranz Anatomy

Haberlandt (1918) distinguished in the leaf cross-section of certain plants a layer of radially oriented mesophyll cells which surround the vascular bundles

(Figure 14.3B). He called this layer "Kranz" (wreath) and the type of leaf anatomy Kranz type.

The C_3 and C_4 plants show anatomical and ultrastructural differences with a high degree of consistency. In C_3 plants this sheath has few organelles and rather small choloroplasts so that at low magnifications cells appear empty and clear in striking contrast to the chloroplast-rich mesophyll. In C_4 plants, the sheath has a high content of organelles especially mitochondria and its chloroplasts are deep green and large, commonly larger than the mesophyll chloroplasts. The sheath cell chloroplasts may be localized next to the outer tangential wall or on the opposite side of the cell. The walls of the sheath are somewhat thickened. When the vascular bundles of C_4 plants are enzymically isolated from the mesophyll the sheath is found to be tightly adhering to the vascular tissue, with its cells arranged in double helix. An extreme case of this ultrastructural dimorphism is seen in sugarcane where the bundle sheath chloroplasts lack grana. In other C_4 plants, grana are found in all the chloroplasts. Numerous large mitochondria are present in bundle sheath cells. Microbodies are present in bundle sheath cells than in the mesophyll cells. Another characteristic feature of C_4 chloroplasts is the occurrence of the peripheral reticulum (PR). The PR is generally more developed in mesophyll chloroplasts than the bundle sheath chloroplasts. The PR is involved in transport of materials between the chloroplasts and cytoplasm. It appears in the peripheral stroma as a series of anastomosing tubules contiguous with the chloroplast envelope. C_4 photosynthesis has been recorded in Aizoaceae, Molluginaceae, Amaranthaceae, Portulacaceae, Nyctaginaceae, Zygophyllaceae, Asteraceae, Cyperaceae and Poaceae.

Modified Leaves

a. Cataphylls or Cataphylla or Cataphyll Leaves or Scale Leaves

The cataphylls or "bud scales" or scale leaves function as a means of protecting the apical or axillary shoot tips during the times of dormancy and also in the underground stems. They are either dry or papery or fleshy and are small and reduced. Cataphylls include bracts, bracteoles, and bud scales as well as scale leaves (small leaves that resemble the scales). Examples include bud scales, bulb scales, corm scales, rhizome scales, scaly bracts and spines *etc.* Such leaves are also found on the aerial parts of Bamboo, *Casuarina, Asparagus etc.*, or as bud scales. The bud scales represent the different parts of a leaf axis. Annual plants do not have terminal resting buds and hence lack bud scales. Some of the tropical plants like *Hevea braziliensis, Mangifera indica* have resting buds with bud scales or cataphylls. Many temperate, perennial species have a means of protecting their apical regions from damage. There are several dangers to the meristems and leaf primordia within a bud. The most important among them is the desiccation and insect predation. Winter winds absorb a great deal of moisture and the immature vascular tissues are not able to conduct water rapidly during winter. Further, the young stems and leaves are the richest sources of proteins, a favourable source of food for insects. To combat these dangers, the bud scales are small, sessile or have a very short petiole. The longitudinal growth of the bud scales stops earlier than the foliage leaf of the same plant and the marginal growth starts sooner. The marginal growth of the bud is also less than that of the foliage leaf. But, in conifers with needle leaves, the cataphylls are broader than the needles. The epidermis is thick walled or it forms a cork cambium and produces a thin, protective, corky bark. The cork is a good water retention device and is also indigestible by the insects. Resin, oils and waxes also are secreted. Secretion and cork formation occur only in the exposed portions of bud scales. Stomata are rare. The vascular supply is simple, because very little conduction is needed. The mesophyll is only spongy but palisade is also present in green bud scales. Secretory cells and cavities occur in those that produce pungent oils.

b. Hypsophylls or Bract Leaves

Hypsophylls or floral bracts are special leaves which bear the floral buds in their axils. These protect the developing inflorescences and their function is similar to cataphylls and is active during the formation of inflorescences. The environmental conditions that they shield against are milder than those faced by cataphylls. Hence, the hypsophylls are not as resistant as cataphylls. Mostly they are green and photosynthetic. Anti-insect mechanisms are present in the hypsophylls, when insects are more abundant.

c. Cotyledons or Cotyledonary Leaves

These are the first leaves on the shoot. The node on the tigellum where the cotyledons are attached in a Dicot seed represents the first node of the shoot. The cotyledons may also bear the buds in their axils (Potato). The cotyledons are fleshy and store the food material in Pea or Bean. In castor, the cotyledons are thin with vein markings. These turn green like foliage leaves after germination. Sometimes, the cotyledons are lobed (*Tilia* and *Geranium*). In Monocots, the scutellum represents the cotyledon and its main function is the suction of the food material from the endosperm.

d. Prophylls

The first few leaves of a branch often differ from the other leaves and these are called as the prophylls. Usually there is one prophyll in Monocots and two prophylls in Dicots. The prophylls are represented by spines in wood-apple (*Aegle marmelos*). Similarly, in lemons and oranges (species of *Citrus*), the prophylls are represented by the spines. In many flowering branches, the prophylls are the first few leaves which are very small and do not subtend the floral buds. In winter buds, the outermost bud scales may be prophylls. Similarly, the tendril of *Cucurbita* also is a modified prophyll.

e. Floral Leaves and Sporophylls

These are specialised leaves (such as sepals, petals, stamens and carpels). Of these, the last two (stamens ands carpels) are also termed as sporophylls in common with the lower plants as they bear spores in Angiosperms. The sporophylls are the leaves that bear the sporangia, either megasporophylls (megasporangia) or micropsporophylls (microsporangia). These vary greatly in their appearance and structure, and may or may not look like sterile leaves. These include: *Alaria esculenta* (a brown alga) shows the sporophylls atached near the base; the lycophytes show the sporophylls aggregated into strobili (*Selaginella* and *Lycopodium*) or distributed among the sterile leaves (*Huperzia*); the ferns produce the sporophylls that appear very different from the sterile fronds, that are non-photosynthetic and lack typical pinnae (*Onoclea*); *Ginkgo* produces the microsporophylls aggregated into a pollen strobilus and the ovules are borne on the sporophylls; *Cycads* produce the microsporophylls in pollen strobili and the ovules are produced on cone scales; while some do not produce the sporophylls (for example, *Psilotum* produces the sporangia fused in a synangium or the terminus of the stem); *Equisetum* always produces strobili; and Gnetophytes produce both compound pollen and seed strobili.

f. Foliage Leaves

These represent the common and normal green leaves of the plant. These are the sites of photosynthesis. These differ from the scale leaves, floral leaves, which are also the structures present on the plant. The scale leaves refer to the bracts while the floral leaves are the accessory whorls of the flower, namely, the sepals and the petals.

Chapter 15

The Flower

Sometimes known as a "bloom" or "blossom", the flower is the reproductive structure in flowering plants (of the division Magnoliophyta, called also as angiosperms). The biological function of a flower is to facilitate reproduction, by providing a mechanism for the union of sperm with the egg.

The flower is the characteristic reproductive structure of angiosperms. The term – "flower" – applies when the part or all of the reproductive structure is distinctive in its colour and form. The flowers present an endless variety of combinations in their range of colour, size, form and anatomy. In their size, the flowers range from minute blossoms to giant blooms. In some plants, each flower is large and showy and is produced singly (as in Poppy, Magnolia, Tulip and Petunia) while in others, the individual flowers may be very small and are borne in a distinctive cluster called the inflorescence (as in Sunflower, Snapdragon and Lilac). Irrespective of their variety, all have a common function, the reproduction of the species through the seed production. The largest flower found on earth, *Rafflesia arnoldii*, weighs 15 pounds and can grow up to 3 feet; and the smallest is the *Wolffia*, is of the size of a rice grain.

Form, Types and Parts of the Flower

The essential parts of a flower are: the vegetative part (parts of perianth – collectively sepals and petals) and reproductive or sexual parts (stamens and carpels). Each flower consists of a floral axis upon which the rest of the floral organs (or the essential organs like stamens and pistils, and the accessory organs like sepals and petals) are borne. The latter serve to attract the pollinators and also to protect the essential organs. The part of the axis that represents the internode terminated by the flower is termed the "pedicel" (the "peduncle" – axis of the inflorescence). The distal end of the pedicel is swollen to various extents and this portion is termed the floral receptacle or the "thalamus". This is a greatly modified stem, and the part of the flower to which the stalk is attached, small, central, and usually contracted so that the parts of the flower are attached crowded together on the tip. The parts of the flower are usually arranged in whorls (or cycles) and may also be disposed spirally if the axis is elongate.

A typical flower has four distinct whorls of flower parts - an outer calyx consisting of sepals - within it is the corolla consisting of petals in the second whorl - followed by the "androecium" (from Greek: *andros* and *oikia* - meaning man's house) of a group of stamens in the third whorl - and in the centre is the "gynoecium" (from Greek: *gynaikos* and *oikia* - meaning woman's house) consisting of pistil(s) made of carpels, forming the central and innermost whorl. The relationship of the whorls to the gynoecium on the receptacle is - "hypogynous" (beneath a superior ovary) - "perigynous" (surrounding a superior ovary) - or "epigynous" (above an inferior ovary).

1. The outer most organs of the flower are the sepals (calyx) which are usually green, small, leaf-like parts that often resemble reduced leaves, and are found lowest (at the base of the petals) on the receptacle. The calyx mainly protects the flower before it opens (in the bud stage).
2. On the inside of the sepals is the corolla (petals) which are generally showy and are often brightly coloured. These two types of organs when they are indistinguishable (as in Lilies and Tulips) together refer to the "perianth" (the floral envelope made of tepals); however, sometimes one of them may be lacking. The corolla helps to attract the pollinators such as butterflies, birds, bees and wasps.
3. Within the perianth, two kinds of reproductive organs are found: Externally the stamens (which together form the androecium - the male parts of the flower - each of which consists of a long, slender and supporting filament attached to an anther at the tip, which is the head of a stamen, in which the pollen is produced and transferred to the pistil).
4. Internally the carpels (which form the gynoecium - the female parts of the flower - comprises one or more number of pistils, each consisting of an ovary at the base, which encloses the potential seeds, the ovules (the egg cells contained in the ovary that forms the fruit after fertilization and the ovules transform into the seeds). The ovary is provided with an upright long, slender stalk or an extension, the style, to take the pollen to the ovary through the pollen tube, for fertilization). On the top of the style rests the stigma, the receptive surface (which being sticky catches the pollen during pollination). The pistil may be simple (made up of a single carpel - the modified leaf that bears the ovule) or compound (formed of several carpels joined together).

A flower is said to be "complete", when it possesses all the 4 floral whorls – sepals, petals, stamens and the pistils. It is "incomplete", when it lacks one or more parts. If both the stamens and the pistils are found in a flower, it is called "perfect or bisexual or hermaphrodite" (contain both the male and female organs) and others which lack one or both the sex organs are called are "imperfect, or unisexual" (contains one of the two sex organs – a flower that lacks the stamens is called "pistillate" or female flower; while that lacks the pistils is called "staminate" or male flower). When the same plant bears the unisexual flowers of both the sexes (the male and the female), it is said to be "monoecious" (for *e.g.*, tuberous Begonia, Hazel, Oak and Corn); when the male and the female flowers are borne on different plants, it is said to be "dioecious" (as in Date, Holly, Cottonwood and Willow); and when the male, the female and the bisexual flowers are on the same plant, it is called "polygamous".

A flower is called a "regular or actinomorphic" (when it is radially symmetrical as in Rose and *Petunia*); and is "irregular or zygomorphic" (when it is bilaterally symmetrical as in Orchids and Snapdragons).

Development of the Flower (Floral Initiation)

The molecular control of floral organs and determination of their identity appears to be well understood in some species. It is known that 3-gene activities (called A, B and C-gene functions) interact in a combined manner to determine the developmental identities of the organ primordia within the floral meristem through the study of simple models. Only A-genes are expressed in the first whorl which leads to the formation of the sepals. Both A- and B-genes are expressed leading to the formation of the petals in the second whorl. In the third whorl, B- and C-genes interact with each other to form the stamens while C-genes alone give rise to carpels in the centre of the flower. This model is based on the studies of the mutants in *Arabidopsis thaliana*, Snapdragon and *Antirrhinum majus*. When there is a loss for example of B-gene function, the mutant flowers are produced as usual with sepals in the first whorl, but also in the second whorl instead of normal petal formation. The lack of B-gene function in the third whorl, but the presence of C-function mimics the fourth whorl, which leads to the formation of the carpels in the third whorl. Most of the genes in this model, belong to the MADS-box genes and these are the transcription factors that regulate the expression of the genes specific for each floral organ.

The identity of the floral organ primordia is controlled by 3 classes of "homeotic selector genes" (A, B and C) with overlapping areas of activity – and the stamens develop in the third whorl, where both B and C genes are expressed. It has been discovered in *Arabidopsis*, that the additional transcription factors, encoded by SEP (SEPALLATA) genes are required to confer the full activity on the homeotic genes in the stamens, as well as the sepals and the carpels. This model of ABC (proposed by Coenand Meyerowitz, 1991) was based on the genetic studies in *Arabidopsis thaliana* and *Antirrhinum majus*. The B class genes are AP3 (APETALA3) and PI

(PISTILLATA) in *Arabidopsis* while in *Antirrhinum* they are DEF (DEFICIENS) and GLO (GLOBOSA); whereas the C genes are referred to as AG (AGAMOUS) and PLENA respectively. The mutations in any one of these genes will result in homeotic conversion of the organs of the third whorl to a different type. For example, the loss of B function will result in the transformation of stamens to carpels, and the loss of C function results in the conversion of the stamens to petals, while the loss of both B and C transform stamens into sepals. According to the observations of Pelaz *et al.* (2000), the single gene mutations of SEP genes have very subtle phenotypes in SEP1, SEP2 and SEP3 triple mutants, and all the floral organs will resemble the sepals, which suggest that B and C function have been abolished. The signals that determine the number of floral organ primordia in each whorl are unknown according to Irish (1999), but the number of stamens can be perturbed by different mechanisms, as such extra stamens develop in CLAVATA1 mutants, which result in enlarged meristems (Clark *et al.,* 1993), and in superman mutants which have an expanded BC domain (Bowman *et al.,* 1992). The arrest of stamen primordial is a feature of normal development in some species by contrast. For example, in the radially asymmetrical flowers of *Antirrhinum,* 5 stamens initiate, but the dorsal stamen arrests early in its development in response to the activity of the CYCLOIDEA gene according to the observation of Luo *et al.* (1995). But, in some species that produce the unisexual flowers like maize and white campion, the stamens in the female flowers initiate but later either arrest or abort in their development (Tanurdzik and Banks, 2004). All the B class, C class and SEP proteins belong to the MADS family of transcription factors which bind a target DNA sequence (the CArG box) as homodimers or heterodimers. *In vitro,* the B class proteins, DEF and GLO (in *Antirrhinum*) or AP3w and PI (in *Arabidopsis*) bind CArg box sequences only as heterodimers (Jack, 2001; Thiessen, 2001).

The floral initiation programme is activated in many plants by signals from the environment, especially day length, which converts the leaf producing SAM into a flower producing inflorescence meristem. Floral organ formation and pattern occurs in the determinate floral meristem, which arises from the indeterminate inflorescence meristem during *floral initiation programme* (FLIP). The programme begins with the expression of three genes - LEAFY, APETALA-1 and CAULIFLOWER, which specify the floral meristem by activating the floral patterning genes. This is a critical phase in flower development. The floral meristem undergoes a series of developmental changes that eventually give rise to the four basic whorls of the flower – sepals, petals, stamens and carpels. Each of these structures is derived sequentially from a whorl that develops from the floral meristem.

Anatomy of Sepals and Petals

The external structure of sepals and petals may be leaf-like. However, the internal structure only of green sepals resembles that of the foliage leaves and the internal structure of coloured sepals and petals is distinctly different.

In the perianth, the vascular system is poorly developed and the veins usually lack sclerenchyma. Usually the epidermal cell walls are thin. In many flowers, the anticlinal walls of these cells are folded and undulated and are arranged so that the projections of the cells dove-tail one into the other, a feature that strengthens epidermis. At the base of petals, as well as along the length of the veins, the anticlinal walls are usually straight even when they are wavy in the other epidermal cells. The outer walls of the epidermal cells commonly have papillae, which make the petals glisten. More papillae are present on the adaxial than the abaxial epidermis and they are not developed at the base of the petals. The epidermis of both sepals and petals may have stomata and trichomes. The stomata on the petals either resemble those on the foliage leaves or are incompletely differentiated. Stomata, if present, are scanty and non-functional. Sometimes, the intercellular spaces covered by cuticle are formed in the epidermis of petals. The thickness of the cuticle varies in different plants and it may be variously sculptured. Various pigments are found in the epidermal cells of the petals and sepals. The green sepals contain chloroplasts but rarely show differentiation into palisade and spongy parenchyma. The colour of petals, results from the pigments in chromoplasts (carotenoids) and in the cell sap (flavonoids, mainly anthocyanins) and from various modifying conditions (such as, the acidity of cell sap). The crystal containing cells, laticifers, tannin cells and other idioblasts may be present in sepals and petals.

Stamen – Structure and Function

Stamens (*stamen* is the Latin word meaning "thread" – originally the thread of the warp, in weaving) are the male reproductive organs of flowering plants and are represented by the third whorl of the flower and all are collectively referred to as the androecium (from Ancient Greek – *andros* meaning man, and *oikos* meaning home, room or chamber). They consist of an anther (derived from French – *anthere,* and from Classical Latin - *anthera* meaning medicine extracted from the flower and in turn from Ancient Greek - *anthera* or *antheros* meaning flowery (from the flower), the site of pollen development, and in most species a stalk-like filament (derived from Latin –*filum* meaning the thread), which transmits water and nutrients to the anther and positions it to aid pollen dispersal. Within the anther, male sporogenous cells differentiate and undergo meiosis to produce microspores, which give rise to pollen grains, whereas other cell types contribute to pollen maturation, protection, or release. Pollen development involves an array of extraordinary events, including cell division and differentiation independent of a conventional meristem, the transition from sporophytic to gametophytic generation, and modifications of cell division to produce structures that are unusual in plant development, including coenocytic tissues (the tapetum and the microsporocyte mass), and subsequently free cells (microspores) that give rise to self-contained units for genome dispersal (pollen grains).

Stamens – Arrangement, Cycly, Position and Number

The arrangement refers to the placement of the stamens relative to one another. There are two basic arrangements – "Spiral" (all the stamens are arranged in a spiral), and "Whorled" (all the stamens are arranged in one or more discrete whorls or series).

Additional arrangements account for the relative lengths of the stamens to one another, and these are three types – "Didymous" (stamens in two equal pairs) - "Didynamous" (stamens in two unequal pairs as in the members of Bignoniacese, Lamiaceae and Scrophulariaceae *etc.*) - and "Tetradynamous" (with stamens in two groups of 4 long and 2 short which are typical in Brassicaceae).

Cycly refers to the number of whorls (or series) of stamens present in whorls to begin with. These are of three types namely, "Uniseriate" (with a single whorl of stamens) - "Biseriate" (with two whorls of stamens) - and "Multiseriate" (if additional whorls are present, such as triseriate (in 3 whorls), tetraseriate (in 4 whorls), pentaseriate (in 5 whorls), hexaseriate (in 6 whorls) *etc.*

The position of the stamens is the placement relative to other floral parts, the sepals and the petals. This is of various types such as – "Antisepalous" or "Antesepalous" (in which the point of stamen attachment is in line with [or opposite] the sepals, lobes of the calyx or the outer whorl of tepals) – "Alternisepalous" (the stamens positioned between the petals or corolla lobes). These are very common in the taxa with uniseriate stamens – "Antipetalous" or "Antepetalous" (position is one in which the point of attachment is in line with [or opposite] the petals, lobes of the corolla or inner whorl of tepals) – and "alternipetalous" (the stamens are positioned between the lobes of the calyx or the sepals). This is relatively rare and serves as a diagnostic character for specific groups of plants like Primulaceae and Rhamnaceae.

In other positions, cycly and the number are taken into account, and they are – "Haplostemonous" (when the stamens are uniseriate, equal in their number to the petals, and lie opposite the sepals [antisepalous] – "Obhaplostemonous" (when the stamens are uniseriate, equal to the number of the petals, and lie opposite the petals [antipetalous] - "Diplostemonous" (when the stamens are biseriate and the outer whorl lies opposite the sepals while the inner whorl lies opposite the petals) - "Obdiplostemonous" (when the stamens are biseriate and the outer whorl lies opposite the petals while the inner whorl lies opposite the sepals) – and "Polystemonous" (this term is often used for an androecium with numerous stamens, usually many, more than the number of the petals). The haplostemonous condition is much more common among the taxa with a uniseriate stamen cycly; while the diplostemonous condition is more common among the taxa with a biseriate stamen cycly; and the obdiplostemonous stamens are relatively rare and are diagnostic for some members of Crassulaceae.

In Angiosperms, the most common type of stamen, consists of an anther divided into four microsporangia and borne on a thin single-veined stalk or filament. The anther generally contains four microsporangia which are paired in two lobes. The two lobes are separated by a zone of sterile tissue, termed the "connective", through which a vascular bundle passes.

In majority of Angiosperms, the anthers are "tetrasporangiate". But in Malvaceae, Adoxaceae, Circaesteraceae, Epacridaceae and Restionaceae, the anthers are "bisporangiate". In Bixaceae, the anthers are 8-sporangiate. Both bisporangiate and tetrasporangiate anthers are seen in the families Asteraceae, Araceae, Asclepiadaceae, Celastraceae, Lauraceae, Lemnaceae, Moringaceae and Piperaceae. Only one microsporangium per anther has been recorded in *Arceuthobium minutissimum*.

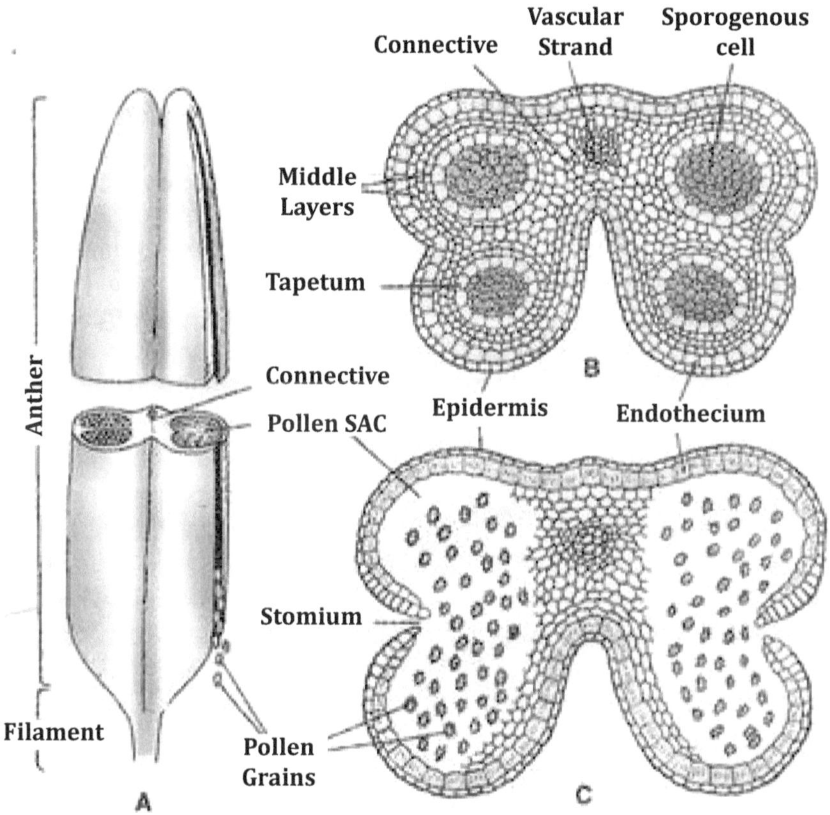

Figure 15.1: Structure of the Stamen and Anther (C.S).

The young anther in cross section consists of an oval mass of meristematic cells surrounded by an epidermis. When a four-lobed contour is formed, archesporium gets differentiated hypodermally at the four corners of the anther. The archesporial cells are larger in size, slightly radially elongated and have prominent nuclei. The archesporial cells undergo periclinal divisions forming primary parietal layer and a primary sporogenous layer. The primary parietal layer undergoes periclinal and anticlinal divisions forming the anther wall.

The anther wall, at the pollen mother cell stage comprises of the epidermis, endothecium, middle layer(s) and tapetum (Figure 15.1B). The epidermis is single-layered and is the external covering of the sporangium. The epidermis at first keeps pace with the developing internal tissues by anticlinal divisions, but soon its cells become radially flattened and stretched tangentially. The epidermis performs its usual protective function. The layer of cells lying immediately beneath the epidermis is the endothecium. Its cells enlarge and develop fibrous thickenings on their inner tangential walls at maturity. These bands extend radially outwards and upwards terminating near the outer tangential wall. The presence of fibrous bands, differential expansion of the outer and inner tangential walls and hygroscopic nature of the endothecial cells help in the dehiscence of anthers at maturity. In majority of the angiosperms, there is a single middle layer but in some families middle layers are 2 to many. The cells of the middle layer ultimately become flattened and crushed at the time of meiotic divisions in the pollen mother cells. The tapetum is composed of single layer of cells which are full of dense cytoplasm and prominent nuclei. In Angiosperms, the anther tapetum is of two types, "glandular" (or secretory) and "amoeboid" (or periplasmodial). In glandular type, tapetal cells remain *in situ* till the maturation of pollen grains and provide nutrition. In periplasmodial tapetum, the walls of the tapetal cells break down; protoplasts enlarge, enter the locule and fuse to form a coenocytic periplasmodium. The primary sporogenous cells undergo mitotic divisions and form pollen mother cells. The pollen mother cells undergo meiotic divisions resulting in microspore tetrads. The microspores separate from the tetrad and develop into pollen grains.

The filament is relatively simple in structure. It consists of parenchyma with well developed vacuoles, small intercellular spaces, and often with pigments in the cell sap. Parenchyma surrounds the vascular bundle which may be amphicribral. The cutinized epidermis may have trichomes, and on both the filament and anther, stomata may be present. The vascular bundle traverses the entire filament and ends blindly either at the base of the anther or in the connective tissue located between the two anther halves.

The primitive dicotyledonous taxa have stamens which are leaf-like (Laminar stamens). A very primitive type of stamen is found in the genus *Degeneria*, where there is no filament, and the anther connective can be distinguished as the stamen is broad and leaf-like, and has three vascular bundles. The four microsporangia are deeply sunk into the abaxial side of the stamen. The microsporangia are found between the lateral and median bundles. Similar stamens are found in other Ranalean genera *e.g., Austrobaileya, Himantandra* and certain genera of the Magnoliaceae. In the Magnoliaceae, intermediate stages may be found from broad stamens with three bundles and laminal microsporangia, *i.e.*, microsporangia and distinct filaments and anthers.

Carpel

In primitive Angiosperms, the carpels appear as folded style-less structures with the stigmatic tissue covering the unsealed margins. In *Degeneria,* the carpel is a folded organ of which the margins flare outwards and remain unfused for a long time. This type of carpel is termed "conduplicate". Swamy observed an excessive development of hairs on the flared margins and extending into the locule, even beyond the attachment of the ovules. The cleft between the closely appressed margins is filled with interlocking hairs. On the inner side of the carpel the trichomes are small and papilla-like. According to Swamy, all the hairs together form the surface area of the stigma. When pollination takes place, the pollen grains become attached to the hairy, outwardly flared margins of the open carpel. Swamy also observed that the germination tubes of the pollen grains grow into the locule of the carpel between these hairs and that they never penetrate the carpel tissue. In the early stage of carpel development, the two rows of ovules are very remote from the true margins. The vascular supply to the ovules may be derived from the ventral or dorsal bundles or both.

In *Drimys* also, the carpels are folded in half and they have widened margins with stigma-like characteristics. Bailey and Swamy (1951) found that the young carpel has a stalk, and the blade-like portion of the carpel encloses two rows of ovules which are situated on the inner adaxial surface a great distance from the margins. When such a mature carpel is cleared and unfolded, it can be seen that the ovules, which are situated between dorsal and ventral bundles, are supplied with traces from both these bundles. Here also the pollen grains reach the hairy margins of the carpels and the pollen tubes penetrate the ovules by growing between the hairs on the inner surface of the carpels. A structure somewhat similar to this has been found in the genus *Cananga* of the Annonaceae (Periasamy and Swamy, 1956).

According to Bailey, from this type of conduplicate carpel with flared margins, the present carpel has developed by closure and the concentration of the stigmatic margins to the upper part of the carpel only. According to Bailey, the loss of the marginal portions outside of the ovules has resulted in the impression that the carpel margins are involuted and that the placentae are marginal. During the course of evolution, the number of ovules was reduced, and they remained only on the lower portion of the carpel while the upper part underwent differentiation to form style and stigma.

The carpel is classically interpreted as the basic unit of the gynoecium. A flower may have one carpel or more than one. If two or more carpels are present, they may be free from one another ("apocarpous" gynoecium) or they may be united ("syncarpous" gynoecium). An old established term used with regard to the gynoecium is "pistil". It refers to a single carpel in an apocarpous gynoecium (simple pistil) or to an entire syncarpous gynoecium (compound pistil).

As a foliar structure, the carpel is described as folded lengthwise in such a way that the adaxial surface is enclosed and the margins are more or less completely united. According to the prevalent view, the folding is conduplicate with margins remaining flat (Figure 15.2 A). Evolutionary changes in this type of carpel are thought to have led to a reduction of the marginal areas (Figure 15. 2B). An older concept implies that the folding includes an involution of the margins so that the surface is lined with abaxial surface (Figure 15.2 C).

The union of carpels in a syncarpous gynoecium follows two basic plans: the carpels are joined either in a folded condition, abaxial surface to abaxial surface

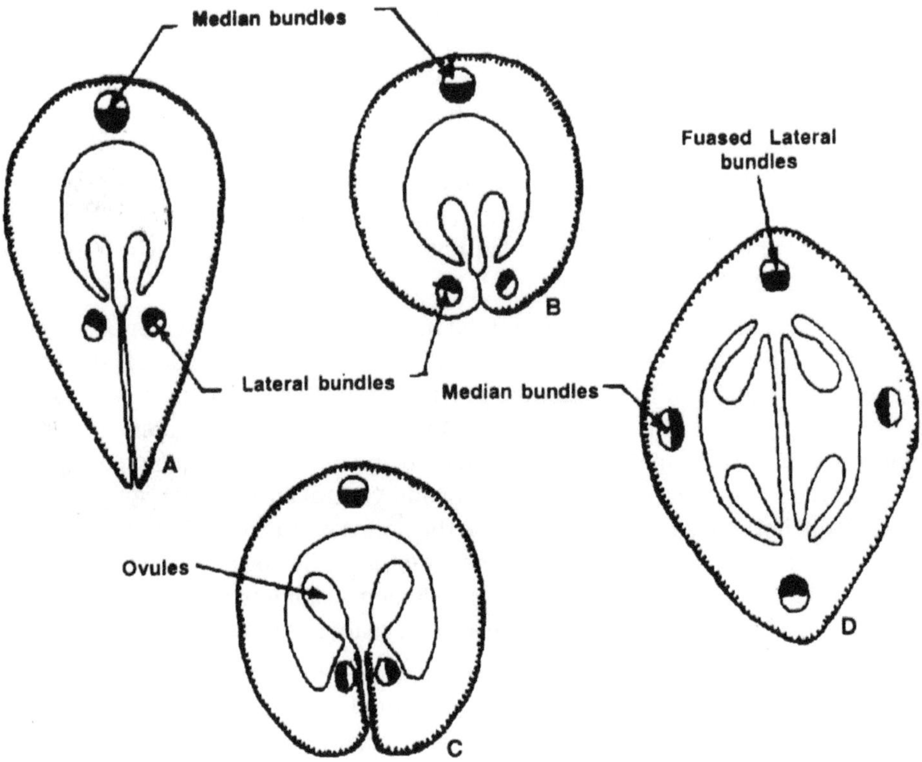

Figure 15.2: Diagrams of Cross Sections of Ovaries Composed of One (A-C) and Two (D) Carpels.

(Figure 15.3 C) or in an unfolded or partially folded condition, margin to margin (Figure 15.3 D). The first type of arrangement results in a bilocular or multilocular gynoecium (Figure 15.3 A, C). The second type is a gynoecium with one locule (Figure 15.3 D). Secondary modification may bring about variations in the basic plans. One example of such modification is found in the Brassicaceae in which two carpels are united by their margins but the gynoecium is not unilocular because a partition is formed by an outgrowth from the placentae at the carpel margins (Figure 15.3 D).

The carpel of an apocarpous gynoecium or the entire syncarpous gynoecium is commonly differentiated into the lower fertile part, the ovary and the upper sterile part, the style. A more or less extensive peripheral part of the style is differentiated as a stigma, usually in the upper region of the style.

Ovary

Structurally ovary consists of ovary wall, the locule (cavity) or locules, and the partitions in multilocular ovary. The ovules are borne on the ovary wall located on its inner (adaxial) side. The mode of arrangement of the ovules attached to the

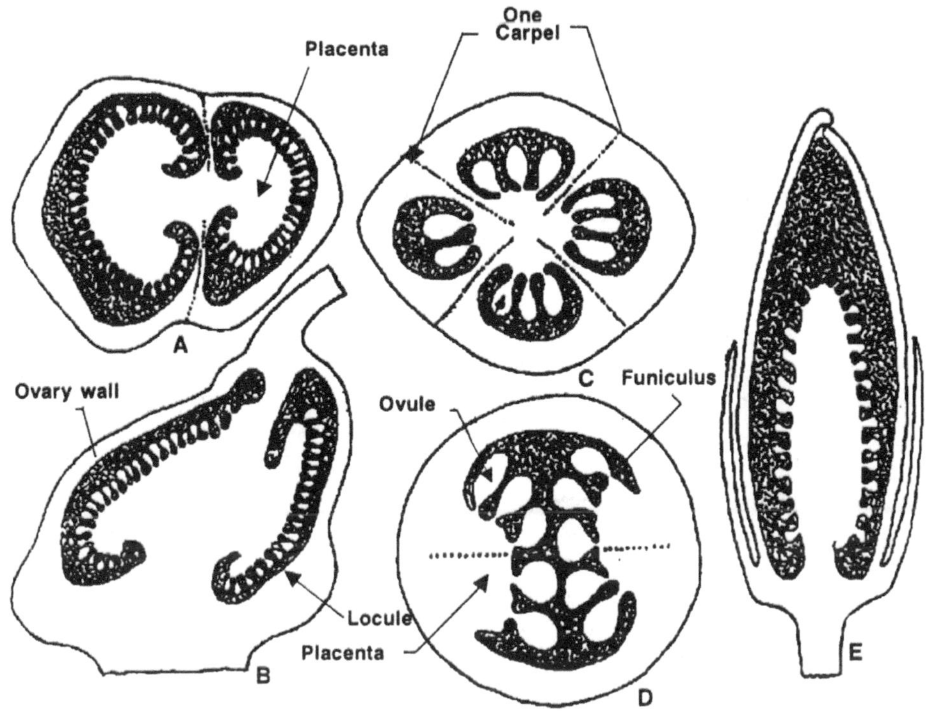

Figure 15.3: Placentation in Ovaries in Cross (A, C, D) and Longitudinal (B, E) Sections. A, B. *Antirrhinum*; C. *Fuchsia*; D. *Ribes*; E. *Dodecatheon*.

walls of the ovary constitutes the placentation. Each carpel has two placentae. In a given carpel, the placentae occur either close to the margin or at some distance from them. On the basis of this character, one can distinguish between the marginal and laminar placentations. The latter is considered to be primitive. A placentum may be a conspicuous outgrowth, sometimes almost occluding the lumen of the ovarian cavity.

Morphology and Function of the Stigma and Style

Stigma

The stigma (Plural: stigmas or stigmata) is the receptive tip of the carpel (or of several carpels) in the gynoecium, and it consists of specialized cells. The stigma forms the distal portion of the style (or stylodia), and is composed of stigmatic papillae, the cells that are receptive to pollen. These cells are restricted to the apex of the style or may cover a wider surface (in wind pollinated species).

Shape of the Stigma

The stigma is often split into lobes, for example, 3-lobed (trifid), 5-lobed (pentafid), many lobed (multifid), or may resemble the head of a pin (capitate) or may be pointed (punctiform) or may be feathery. The shape of the stigma varies considerably.

The morphology of the stigma and style show great diversity and to some extent depends on the type of pollination. The function of the stigma is to accept the pollen and permit its germination. The stigma also provides nutrients to the pollen and directs the pollen tube growth. Both the pollen coating and stigma coating are involved in recognition of and receiving signals.

On the basis of the morphology of the stigmatic surface, the amount of secretion, and the nature of the surface cells, a general classification is suggested by Heslop-Harrison and Shivanna (1977). They have divided thee stigmas into two categories namely – wet stigmas (with copious fluid secretion), *e.g.*, *Petunia* and dry stigmas (limited secretions). In the latter, the receptive cells are dispersed on multiseriate branches (*Zea, Sorghum, Hordeum etc.*) or concentrated in distinct ridges or zones. However, in the former, the receptive surface may have papillae (Liliaceae, Orchidaceae *etc.*) or non-papillate (*Dracaena, Alpinia etc.*). Trinucelate pollen is mostly associated with the dry stigmas, whereas binucleate pollen is seen in both the cases.

The stigma receives the pollen and it is here that it germinates. The stigma may be long and slender to globe-shaped to feathery. The pollen may be captured from the air ("anemophily" – pollination by wind-borne pollen); or from insects and other animals which often visit the flowers attracted by their colours *etc.* (biotic pollination or "zoophily"); or surrounding water in rare cases ("hydrophily" – epihydrophily [pollination at the surface as in *Vallisneria*] or hypohydrophily – pollination under water as in *Zostera*]).

Structural differences have been noted in the stigmatic papillae of floral morphs in many heterostylous plants. Ghosh and Shivanna reported that in *Linum grandiflorum,* the stigma of the "pin morph" (long-styled) is the 'dry' type and smooth and the papilla is covered by a distinct cuticle-pellicle layer. The stigma of the "thrum morph" (short-styled), on the other hand, is of the wet type, warty and uneven and the cuticle-pellicle layer is disrupted.

Dumas (1978) classified the stigmas based on the function and nature of secreted products. Stigmas have an abundant secretion of lipophilic or polysaccharide substances or in some cases, less abundant secretions with polyphenols, tannins, flavonoids *etc.* Some others have no liquid secretion at all, but stigmas are covered with waxes or proteins. This classification can also be combined with the stylar structure. Hollow styles of Monocots produce mainly polysaccharides whereas solid styles of Dicots, produce a lipophytic exudates. Some families like Orchidaceae, Solanaceae, Scrophulariaceae excrete lipopolysaccharide substances.

Nature of Stigmatic Covering

The receptive surface is formed by the surface cells which are glandular in nature. This is followed by a transition zone of stigmatoid tissue without glandular cells. The stigmatoid tissue is also known as the neck of the "transmitting tissue".

The dry stigma has cuticle covered with a pellicle. The cuticle has some discontinuities for the water. Both the pellicle and the surface of the stigma remain hydrated. The pellicle is an extracellular product originated from the protein of the stigmatic cells.

The exudates of the wet stigmas mainly consist of amino acids, lipids, antioxidants, proteins and alkaloids. During pollination, the exudate serves as nutritive for birds, bees, beetles, bats, moths *etc.* The exudates also have boron and serve as a nutrient for pollen. The exudates of wet stigmas vary in quality and composition and can be considered as a complex fluid. In *Trifolium*, it contains glucose, sucrose and traces of galactose, arabinose, some proteins and esterases. The exudates functions in the hydration of the pollen grain, stimulation or inhibition in pollen germination as well as in nutrition.

Fresh exudates of *Phaseolus* contain lipids, proteins, amino acids, sugars (namely glucose), reducing acids and phenols. Alkaloids are absent. Sugars stimulate pollen germination; lipids prevent desiccation and also attract insects. Proteins function as protection against chewing insects and pathogens. Thus, the nature of the stigmatic covering is very complex and it is not possible to make any generalizations.

Morphology and Function of the Style

In plants, style is a long, slender stalk that connects the stigma and the ovary. The former is at the top of the style and is a sticky platform where the pollen is deposited; while the ovary is at the bottom and houses the ovules which contain eggs and supporting cells for reproduction. These three are collectively referred to as the pistil. When there is more than one carpel, each carpel may have a separate style-like structure (called as stylodium) or share a common style. In Iridaceae (as in *Iris*), the style gets divided into 3-petal-like (petaloid) style-like branches (sometimes called as stylodia), almost to the base of the style (called "tribrachiate"), and these are the flaps of the tissue, that run from the perianth tube above the sepal. The stigma is represented by a rim (or the edge) on the underside of the branch, near the end lobes. The stylar branches are also seen on *Dietes, Pardanthopsis* and most species of *Moraea*. In *Crocus*, the 3 divided branches of the style create a tube. The stylar branch spreads out in *Hesperantha*. The style sometimes may also be lobed rather than branched. In *Gladiolus*, the style is bilobate (bilobed). In *Freesia, Romulea*, and *Watsonia*, the style is bifurcated (2-lobed) and the branches are recurved.

The style is very important during the process of fertilization because it is the location where the pollen tube forms, and is also involved in stopping incompatible pollen from penetrating the ovary. When the pollen tube begins to extend within the style, genetic information gets exchanged between the pollen and the plant. At this juncture, the plant also creates a toxin, and if the pollen is incompatible and from a different species or too closely related, the toxin is released to stop the growth of the pollen tube. So, the style is the location where the compatibility is checked to determine whether it can participate in fertilization or not.

The morphology and anatomy of the style are variable. The style is a narrow, upward extension of the ovary, connecting it to the stigmatic papillae, and if the style is absent, the stigma is referred to as sessile. In general, the styles are tube-like, either long or short. The style can be open or closed. Three different types of styles are seen in angiosperms, namely open, solid and half-closed.

1. *Open Type (Hollow Style)*

These are seen in majority of the syncarpous monocots and some eudicots. The style can be open (containing few or no cells in the central portion) with a central canal which may be filled with mucilage. This type is more common in monocotyledons, *e.g., Lilium, Amaryllis, Gladiolus.* In the hollow styles, there may be one or more canals; the number of canals normally correspond to the number of carpels. The stylar canal is lined with canal cells. The pollen tubes creep on the surface of these canal cells, which are secretory in nature. The canal cells show dense cytoplasmic contents and are rich in cell organelles. The secretion from the canal cells contains proteins, carbohydrates, lipids and also shows activity for esterases and acid phosphatases. It has been suggested that in *Linum,* the parenchyma cells underneath the canal cells synthesize some mucilaginous substances and transport them to the canal cells, which in turn, secrete them into stylar canal. After pollination, the canal is filled with secretion which serves as nutrition for the growing pollen tubes. In the open style, the canal cells provide elements for pollen tube growth and functionally similar to the intercellular substances of the solid style.

2. *Solid Ttyle (Closed Type)*

Alternatively, the styles may be closed completely filled with tissue. These are seen in syncarpous eudicots and grasses and contain specialized secretory transmitting tissue which links the stigma to the centre of the ovary. This forms a nutrtient rich tract for the growth of the pollen tube. This type of style is mostly seen in dicots, *e.g., Datura, Gossypium etc.* It has a solid core of transmitting tissue. The transmitting tissue of the solid style has either loosely arranged or it may be more compact. This depends on the amount and quality of the intracellular substance. In general, the cells of the transmitting tissue show a thick cell wall with a small lumen, or a thin cell wall with large amount of intracellular substance. The cells of the transmitting tissue contain numerous mitochondria, plastids, rough endoplasmic reticulum, dictyosomes and ribosomes. The transmitting tissue also contains, carbohydrates and isoenzymes such as peroxidase and acid phosphatase. The function of the intercellular substance is to aid in the pollen tube growth. In *Oryza sativa,* the solid style is devoid of any transmitting tissue.

3. *Half-closed Type*

This is present in a few taxa in which the transmitting tissue can be limited to one side of the stylar canal. It is seen in family Cactaceae.

Style - Attachment to the ovary and Position of the Style

This may be terminal (apical), sub-apical (sub-terminal), lateral, gynobasic or sub-gynobasic. The apical position refers to the style attached at the apex of the ovary which is the commonest position in plants. In sub-apical pattern, the style arises to the side slightly below the apex. A lateral style arises from the side if the ovary (as in Rosaceae). The gynobasic style arises from the base of the ovary (or between the lobes of the ovary) which is characteristic of the family Boraginaceae; while the sub-gynobasic styles occur in *Allium* species. Due to some self-incompatibility reactions in the style, the full growth of the pollen tubes is prevented so that they fail to reach the ovules. In species like *Gasteria*, the pollen tube is directed to the micropyle by the style.

Vascular System of the flower

The position of placentae in an ovary is related to the method of union of carpels. The placentae of marginally joined carpels are said to be inserted on the ovary wall and the placentation is called parietal. If the carpels are joined in a folded condition, the ovary is bilocular or multilocular and the placentae occur in the centre of the ovary where the carpel margins meet. This kind of placentation is called axile. Various modifications of these two basic placentations occur in different taxa. The partitions in a multilocular ovary may disappear in terms of either phylogeny or ontogeny so that a free central placentation results. The placenta in a unilocular ovary may occur at the base and the placentation is then called basal.

In relatively unspecialized flowers, the vascular organization is readily interpreted as being comparable to that of vegetative shoot. The number of traces of the different floral parts varies in the same flower. The usual pattern is as follows: Each sepal has as many traces as the foliage leaf of the same plant. Each petal in a dicotyledon has one trace; each tepal in a monocotyledon, one to many. Within the sepal and petal, the vascular bundles form more or less complex systems resembling those in the foliage leaves. The specialized type of stamen has one trace, which is continued as a single bundle in the filament as well as the anther. The carpel has three traces and their prolongations within the carpel may have branches. The ovules are supplied by branches from carpellary bundles, usually the two lateral ones. The carpellary vascular bundles are continuous through the style. The vascular bundles of floral parts may be fused when they are united.

In some epigynous flowers, certain bundles show an inverted orientation of xylem and phloem. Such arrangement is interpreted as an indication that the ovary is embedded in receptacular (axial) tissue, whereas the absence of inverted bundles is regarded as evidence that the extra-carpellary tissue is appendicular in nature (floral tube).

Chapter 16

The Fruit

In plant science, a fruit is the fleshy or dry, ripened ovary (seed-bearing structure) in the flowering plants (called the angiosperms) formed after flowering, enclosing the seed(s). The seeds are the means by which the flowering plants get disseminated. Edible fruits propagate with the movement of humans and animals in a symbiotic relationship for dispersal of seeds and nutrition. In common usage, the "fruit" normally means - the fleshy seed-associated structure of a plant that is sweet or sour, when raw (like apples, bananas, grapes, lemons, oranges and strawberries *etc.*). In botanical usage, "fruit" is a mature ovary and its associated parts, and usually contains the seed(s) which are the ovules developed after fertilization. The main purpose of the fruit is the protection and dissemination of the seed. For a botanist, a fruit is a reproductive structure of a flowering plant which develops from the ovary and accessory tissue which surrounds and protects the seed. The fruits are important in seed dispersal.

Structure of the Fruit

The classification systems for the mature fruits consider the number of the carpels which constitute the original ovary; dehiscence versus indehiscence; and dryness versus fleshiness. The properties of the ripe ovary wall (or the epicarp) are also important. The pericarp is often edible and formed from the ovary as well as surrounding parts around the seeds. In some, cases, other tissues contribute to form the edible portion.

Development of the Fruit

A fruit is a developed and ripened ovary or ovaries together with adjacent floral organs and other plant parts. Such organs may be tepals (*Morus*), the receptacle

(*Fragaria*), bracts (*Ananas*), and the floral tube, which is formed by the floral organs together with receptacle (*Pyrus malus*) or enlarged axes of the inflorescence (*Ficus*).

The body of the fruit developed from the ovary wall, which surrounds and encloses the seed is known as the pericarp. When this is not homogenous histologically, distinct outer, median and inner parts are evident, and these parts are known as exocarp (epicarp), mesocarp and endocarp respectively. In many fruits the pericarp is more or less nearly homogenous and not separable into parts; in other fruits only exocarp and endocarp are evident. Strictly speaking, the 'Pericarp" refers only to the modified ovary wall, but in general usage it is applied to the outer tissues of fruits regardless of their morphological nature.

A fruit is the result of the maturation of one or more number of flowers and the gynoecium of the flower(s) forms all or a part of the fruit. The ovary (the ovaries) has one or more number of ovules inside where the mega-gametophyte contains the egg cell. These ovules will become the seeds after double fertilization (or triple fusion). The ovules are fertilized in a process that commences with pollination, and involves the movement of the pollen from the anthers (of the stamens) to the stigma (of the gynoecium). Following pollination, the pollen tube grows through the stigma into the ovary to the ovules. Two sperms are transferred from the pollen to the mega-gametophyte, and within, one of the sperms unites with the egg to form a zygote (fertilization), while the second sperm enters the central cell to form the endosperm mother cell (triple fusion) of double fertilization. The zygote will give rise to the embryo of the seed, while Endosperm Mother Cell will give rise to endosperm, a nutritive tissue for the developing embryo. The ovules develop into seeds and the ovary begins to ripen into ovary wall. The pericarp may be fleshy (in berries and drupes) or a hard covering (as in nuts). The simple fruits in some are derived from an inferior ovary, and other parts (sepals, petals and the stamens) fuse with it and ripen. In other types of fruits, all the outer whorls (sepals, petals, stamens and stigma fall off).

The general mode of development of the fruit is of 3 types, namely: Apocarpous fruits (the fruits develop from a single flower with one or more separate carpels (simplest fruits) – Syncarpous fruits (the fruits develop from a single gynoecium with 2 or more carpels are fused together – and Multiple fruits – which form from many different flowers of an inflorescence. As such the botanists have grouped all the fruits into 3 main groups, namely: Simple fruits, Aggregate fruits and Composite fruits or Multiple fruits. This grouping is not relevant according to evolution.

Different Types of Fruits

Simple Fruits

These fruits can be either dry or fleshy, and these result from the ripening of a simple or compound ovary in a flower with only a single pistil. The former (dry fruits) may be either dehiscent (break open to discharge the seeds) or indehiscent (do not open for seed discharge).

The fruits in which a part or all of the pericarp is fleshy at maturity are called simple fleshy fruits. The fleshy fruits are classified into several types:

Fleshy Fruits

The fleshy fruits (the pericarp is fleshy at maturity) are of the following types:

Berries – These have one or more carpels and one or more seeds with a fleshy ovary wall (pericarp). These include Tomatoes, Egg plants and Grapes.

Hesperidia and ***Pepos*** – The former is a specialized berry with a leathery rind while the latter is a berry with a hard or brittle rind and the receptacle is partially or completely enclosing the ovary (an accessory fruit). Both of these are fleshy fruits and possess a leathery (or brittle) rind. The former includes *Citrus* fruits like Lemons and Oranges, while the latter include Cucumbers, Cantaloupes and Squashes.

Drupes – These are the fleshy fruits that have a single seed surrounded by a stony fruit (or bony or stone fruit) endocarp (the inner wall of the pericarp) which is sweet and juicy. These are derived from a single carpel with usually single seed. The exocarp is thin. These include plums, peaches and olives (all are pitted fruits).

Pomes – This is an accessory fruit that is derived from several carpels, receptacle and outer portion of pericarp is fleshy while the inner portion is papery or cartilaginous forming a core. These have many seeds with a fleshy tissue surrounding the pericarp that is sweet and juicy. These include Apples and Pears.

Hip - This is an accessory fruit in which several separate carpels are enclosed within a fleshy or semi-fleshy receptacle.

Dry Fruits

These fruits are dry because the pericarp is dry at maturity. These are classified into different categories as follows:

Dry Dehiscent Fruits

Follicles – These are composed of a single carpel and split along a single suture (ventral suture). These are pod-like and contain many comose seeds that disperse by the wind. These include pods of Milkweed (*Calotropis*) and *Magnolia*.

Legumes – these are pod-like fruits composed of a single carpel that open along both the sides (dorsal and ventral sutures) releasing several seeds. These include Peas, Beans, and Peanuts.

Capsules – These fruits are notable for opening along 3 or more lines at the tops of the fruit to release the seeds. The fruits develop from several carpels and open at maturity in one of the following ways. Lilies, *Hibiscus*, *Aristolochia*, *Datura*, *Argemone* and Poppies are the best examples. These may be: Loculicidal (along the middle of each carpel), Septicidal (along the line of union of the carpels, the septa), Septifragal (both loculicidally as well as septicidally), Poricidal (through the pores at the top of each carpel) or Circumscissile (along a circular horizontal line).

Siliqua **(or *Silique*)** – The fruit is composed of 2 carpels which separate at maturity leaving a persistent false partition (replum) between them to which the seeds are attached. These are seen in the members of Brassicaceae, such as mustard.

Silicule – This is a short and wide silique. This is best seen in Shepherd's purse and Pepper grass.

Dry Indehiscent Fruits

The fruits do not split open at maturity to liberate the seeds and are of the following types

Achenes **(or *Akenes*)** – These fruits have a single seed, fairly loosely held within, except for one small moorage (attachment at one point only), called the funiculus. The examples come from Sunflower family, Asteraceae.

Caryopsis **(or *Grain*)** – Caryopses are single-seeded fruits wherein the seed coat is adhered to the pericarp, at all possible points, as such they are called the grains. These occur in the plants of grass family (Poaceae) like Wheat, Corn, Rice and Oats *etc.*

Samaras – These are the modified achenes with one or two seeds that have flattened wing-like expansion of the pericarp to be carried away by wind. The examples are Ash and Elm trees.

Schizocarps – These fruits form from a syncarpous ovary with 2 carpels and do not really dehisce at maturity, but split into 1-seeded indehiscent segments, called mericarps along the midline. Examples are Maple tree, members of Parsley family, and Umbelliferae (Carrot and others).

Lomentum **(or *Loment*)** – This is a fruit with several seeds that breaks up into single-seeded bits (mericarps) at maturity. Examples are members of Mimosaceae, *Acacia arabica* and *Mimosa pudica*.

Nuts – These are similar to achenes except that their pericarps are hard, fibrous and composed of a compound ovary with a single seed. The pericarp is hard throughout. Acorns, Hazelnuts, and Hickory nuts are the examples.

Aggregate Fruits

These form from single flowers having multiple carpels not joined together (each pistil contains one carpel) and arranged over the surface of a single receptacle. Each pistil forms a fruitlet (the individual ovaries), and collectively all the fruitlets are called an etaerio. These are represented by 7 types, namely, a collection of Follicles (*Calotropis, Magnolia* and Peony), Capsules (Sweet Gum), Achenes (*Clematis, Ranunculus,* and *Sycamore*), Cypselas (Teasel), Druplets (*Rubus* species or Raspberry), Berries (*Annona* species), and Samaras (Tulip tree). In Blackberry, and Strawberry, the receptacle is elongated and part of the ripe fruit appears as an accessory-aggregate fruit.

Multiple Fruit (or Syconium) or Composite Fruits

This is formed from a cluster of flowers or an inflorescence and each flower produces a fruit, and all threes mature into a single mass or a clustered fruit. The mature ovaries of several to many flowers are more or less united to form a mass. The examples are: Pineapple, Fig, Mulberry, and Bread-fruit. The fruit of Pineapple, includes the tissue from the sepals as well as the pistils, and hence, it is a multiple fruit and an accessory fruit.

Syncarp

In Noni or Indian Mulberry (*Morinda citrifolia*), the different stages of flowering and the development of fruit can be observed on a single branch. In the beginning, an inflorescence of white flowers (head) is produced along the stem. Following fertilization, each flower develops into a drupe and as these expand, they become connate (merge with each other) into a multiple fleshy fruit (called a Syncarp). Some or all of the edible part of an accessory fruit is not generated by the ovary. The fruit can be simple, aggregate or multiple.

Parthenocarpic Fruits or Seedless Fruits

Fruit generally develops after fertilization, but in many plants fruits may develop without fertilization and seed development. This phenomenon is termed parthenocarpy. In some fruits of commerce, seedless nature is an important feature. The commercial cultivars of Bananas and Pineapples are the best examples of these fruits. Some varieties of Citrus (Grapefruit, Mandarin oranges, Navel oranges), Table grapes, Watermelons are valued for their seedless nature. Seedlessness is the result of parthenocarpy where the fruits are without fertilization. Majority of the seedless citrus fruits require a stimulus from pollination to produce a fruit. Bananas and Grapes without seeds are triploids. The seedlessness results from the abortion of the embryonic plant produced by fertilization (process is called as Stenospermocarpy) and this requires normal pollination and fertilization.

Anatomy of Fruit

On the basis of histology of the fruit wall, fruits may be distinguished into dry fruits and fleshy fruits.

Dry Fruit

A dehiscent fruit may develop from a single carpel (follicle, legume) or from more than one carpel (capsule, siliqua).

The method of dehiscence is highly variable. If the ovary is derived from a single carpel, the break may occur longitudinally through (1) the suture resulting from the union of carpel margins; (2) the back of the carpel; the suture and the back simultaneously (Figure 16.1 A-C). In syncarpous fruits with parietal placentation, dehiscence may occur through the suture between the carpels or through the backs of carpels (Figure 16.1 D-F). In syncarpous fruits with axile placentation (Figure

16.1 G), the separation along the line of union of contiguous carpels, that is through the septae (septicidal dehiscence), may be combined with breaking away from the central column of tissue (Figure 16.1 H). Dehiscence through the back of the carpels opens the individual locules (loculicidal dehiscence, Figure 16.1 I). In the fruit types mentioned above the longitudinal dehiscence may also occur without reference to the suture or the median vein. In some fruits, dehiscence occurs through horizontal circular region involving all carpels (circumscissile dehiscence) or through pores (poricidal dehiscence).

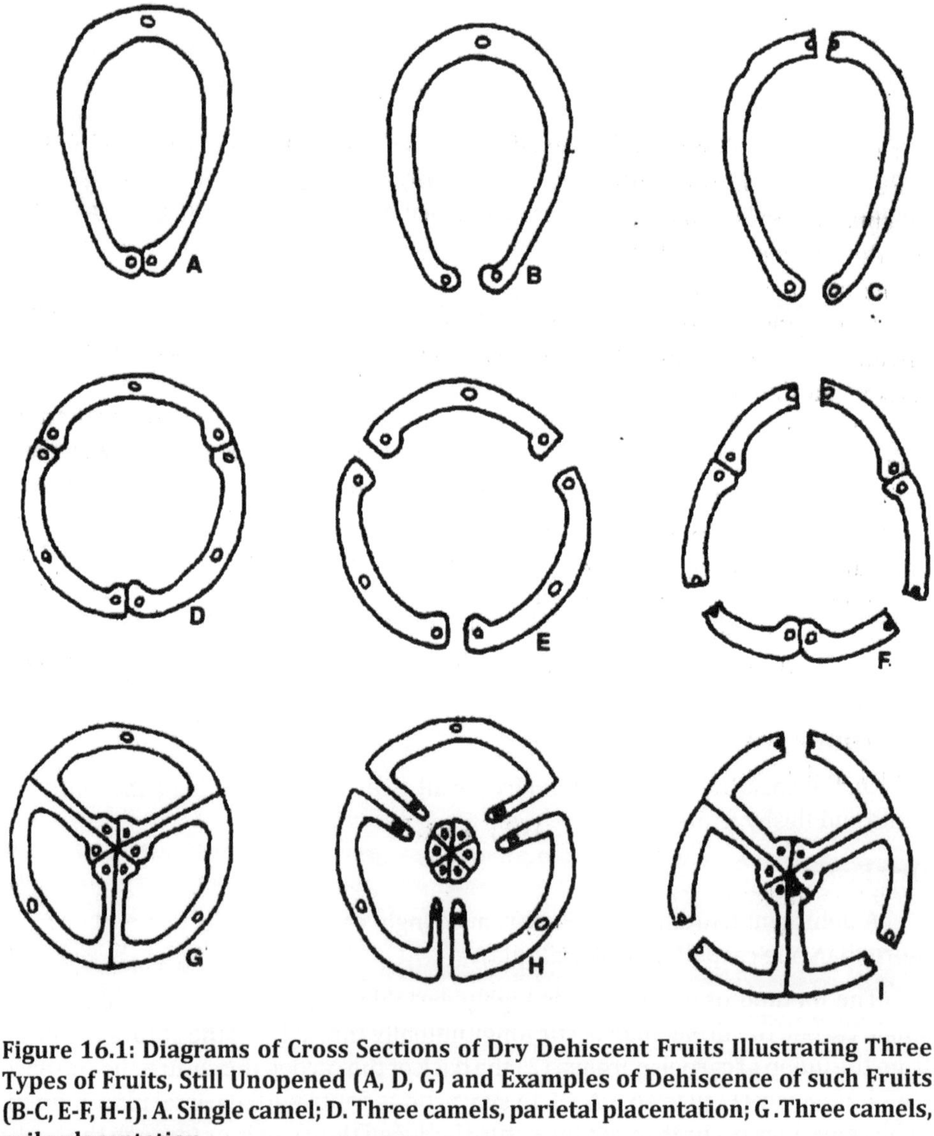

Figure 16.1: Diagrams of Cross Sections of Dry Dehiscent Fruits Illustrating Three Types of Fruits, Still Unopened (A, D, G) and Examples of Dehiscence of such Fruits (B-C, E-F, H-I). A. Single camel; D. Three camels, parietal placentation; G .Three camels, axile placentation.

Legume

Legume, found in many Fabaceae, is a well known example of a dry dehiscent fruit. The legume is derived from a superior ovary which embodies a single carpel. It dehisces along the suture of carpel margins and along the median vein (Figure 16.1 C). As seen in *Glycine max,* the exocarp is composed of outer epidermis and hypodermis, both with thickened cell walls. The mesocarp is composed of several layers of parenchyma while the endocarp is composed of sclerenchyma on the inside of which is usually a thin-walled epidermis. The vascular bundles are situated in the parenchyma of the mesocarp. The cells in the hypodermis and the sclerenchyma of the endocarp are elongated but the long axes of the two kinds of cells are oriented in opposite planes and the developing stresses promote the opening of the dehydrated mature fruit. Another feature found to be associated with the differential shrinkage of the pericarp in some legumes is the dissimilar orientation of cell wall microfibrils in different layers of sclerenchyma, in low-pitched helices in some, in steeply pitched helices in others. Tension developing during dehydration in some legumes is so great that the separated valves of the pod become twisted.

In *Phaseolus,* the inner epidermis undergoes numerous periclinal divisions and becomes a massive layer of usually non-photosynthetic inner parenchyma.

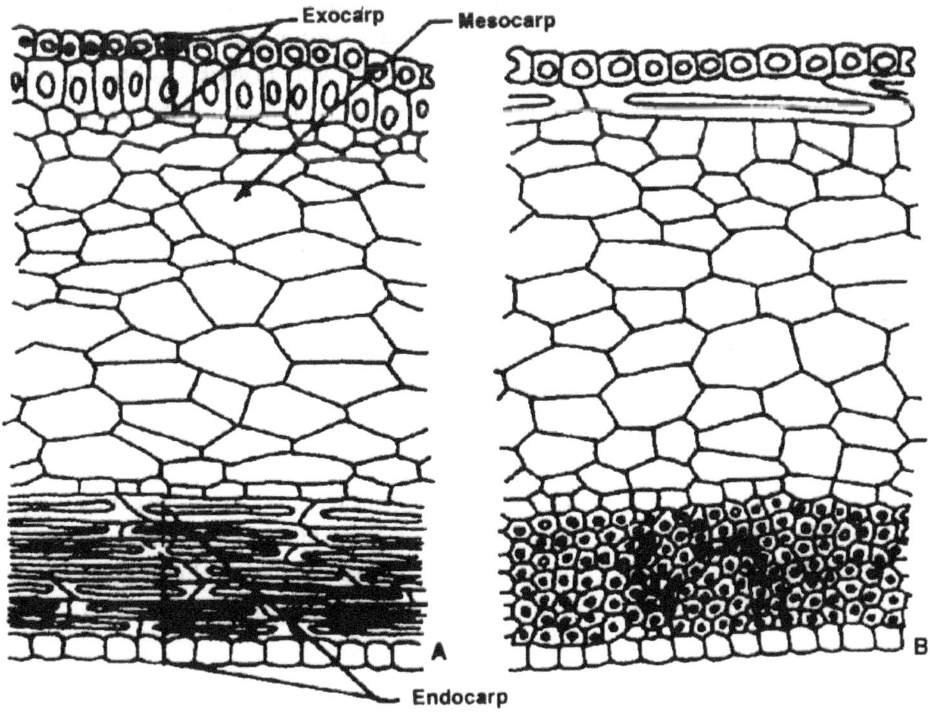

Figure 16.2: Pericarp of *Glycine max* (Soy bean). A. Cross section; B. Oblique longitudinal section.

Capsule

The outer part of the pericarp consists of rigid, highly lignified cells, the inner part of parenchyma cells. The inner epidermal cells may also be thick-walled. The dehiscence of capsule is brought out by the anisotropic swelling of thick walls of the outer or inner epidermis, and it is they that determine the direction of the bending of the teeth or valves.

Indehiscent Fruit

An indehiscent fruit usually originates from an ovary in which only one seed develops, though more than one ovule may be present. The pericarp of an indehiscent fruit often resembles seed coat in structure. The actual seed coat in such fruits may become obliterated to a considerable degree (Achene in Asteraceae) or fused with the pericarp (Caryopsis of the Poaceaee)

Cypsela

The anatomical structure and development of pericarp in *Adenostemma lavenia*, a member of Asteraceae has been described by Pullaiah (1979). At fully organized stage of the embryo sac, the integument consists of 6-10 layers in addition to the outer epidermis and integumentary tapetum. After fertilization, the cells around the integumentary tapetum become enlarged and deplete in appearance. This degeneration extends into the peripheral region and all the integumentary layers except the epidermis have become crushed, by the time a well-developed embryo has formed. Meanwhile, the integumentary tapetum also becomes crushed. Thus, in the fully mature seed, it is only the epidermis that forms the seed coat. The ovary wall at the organized embryo sac stage consists of an epidermis, a hypodermis and an inner layer of parenchymatous cells traversed by vascular elements. After fertilization, schizogenous cavities develop between the hypodermis and the inner layers of the ovary. A brown resinous substance exudes from the hypodermis and accumulates in the schizogenous cavities. At maturity, this resinous substance hardens into dark mass. Meanwhile, except for a few layers, the inner layers of the ovary wall, enlarge and then become obliterated. In the mature achene, the pericarp thus consists of epidermis, a hardened dark layer and a few inner layers.

Caryopsis

In grasses, the pericarp and the remains of the integuments of the single seed of the caryopsis are completely fused. The covering layers of wheat caryopsis are composed of the pericarp and the remains of seed coat. Five layers can be distinguished in the pericarp: the outer epidermis, the hypodermis, a zone of thin-walled cells, cross cells and tube cells. The outer epidermis and the hypodermis together form the exocarp. The cells of the exocarp are elongated in a direction parallel to the longitudinal axis of the caryopsis, and they become compressed and their walls thicken considerably. The cross cells are found below the parenchymatous layer and they have thick walls with pits which are elongated transversely to the cell. The longitudinal axis of these cells is at right angles to that of the exocarp cells. The tube cells constitute the inner epidermis of the pericarp and they occur on the inside of the cross cells. The longitudinal axis of the tube cells is parallel to that of the exocarp cells. The seed coat, which is united with the pericarp, is crushed in the mature caryopsis. In the development of seed coat, the outer integument disintegrates, and the inner becomes altered and compressed. The compressed inner integument contains pigment. Within the seed coat lies the endosperm which forms about 83 percent of the fruit.

In *Zea,* the outer pericarp consists of cells with thick pitted walls and is much compressed. The integuments disintegrate completely

Cremocarp

The fruit of *Apium graveolens* is described here as an example of a cremocarp. Each mericarp has five ribs in which the vascular bundles, accompanied by sclerenchyma are situated (Fg. 16.3 A-C). Between the ribs, one to three oil ducts may be present (Figure 16.3 B-D). The exocarp cells (*i.e.,* outer epidermis) are small and isodiametric. In surface view, they appear lobed. The mesocarp is parenchymatous and its cells have small papillae and finely striated walls. The oil ducts in the mesocarp are surrounded by polyhedral cells which become brown as the fruit ripens. The slightly elongated cells of the innermost layer of the mesocarp are wider than those of the endocarp. The endocarp (the inner epidermis) consists of narrow cells of which the longitudinal axis is parallel to the transverse axis of the mesocarp (Figure 16.3 C, D), or their arrangement is mosaic. Inside the pericarp, is the thin seed coat which surrounds the endosperm in which the embryo is completely enclosed. The maturing mericarps, each of which develops from a single carpel together with a small amount of tissue outside it (inferior ovary), separate from each other along the area of adnation between them, but they remain attached to the branched carpophores (Figure 16.3 E).

Fleshy Fruit

Hesperidium

The Citrus fruit, called hesperidium develops from a superior ovary and has about ten carpels (Figure 16.4 A, B). The placentation is axile. The exocarp (or flavedo - yellow tissue) consists of a cuticle-covered outer epidermis and compact sub-epidermal parenchyma with oil glands and crystal-containing cells. The mesocarp (or albedo - white tissue), consists of parenchyma with large intercellular spaces and cells with arm like extensions (aerenchyma). A vascular network occurs in this tissue. The endocarp is composed of the inner epidermis and a few layers of compact parenchyma. The partitions between the locules are extensions of endocarp and mesocarp tissues. The endocarp produces the juice sacs (pearls) that completely fill the carpel locules (Figure 16.4 D). The stalked, spindle-shaped juice vesicles, which fill the locules when the fruit ripens, develop from the cells of the inner epidermis and sub- epidermal layers (Figure 16.4 C). Each juice vesicle is covered externally by a layer of elongated epidermal cells which enclose very

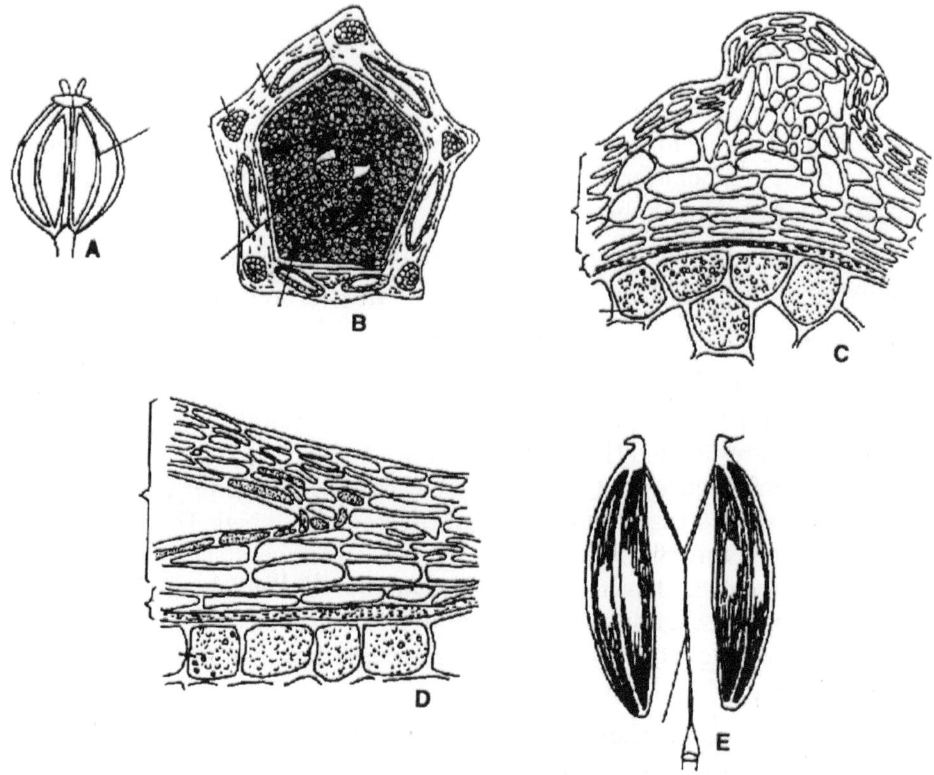

Figure 16.3: Fruit of *Apium graveolens*. A. Entire cremocarp; B. Cross section of a single mericarp; C. Portion of a cross section of a mericarp showing the structure of a rib; D. As in C, in the region of oil cavity; E. A mature cremocarp with separated mericarps.

large, extremely thin-walled juice cells. Development of juice sacs begins at the time when the flower opens and continues until the style withers. The chloroplasts, responsible for the green colour of immature fruits, occur in parenchyma of the exocarp. They become changed into chromoplasts as the fruit matures.

Berry

The fruit of tomato (*Lycopersicon esculentum*) consists of a pericarp and a placental tissue on which the seeds are borne. The exocarp consists of an epidermis and three or four layers of collenchymas cells. The epidermal cells are polyhedral and are covered by a thin cuticle. The glandular and other hairs that are usually present on the young fruit are shed as it matures. There are no stomata on the epidermis of the fruit. The mesocarp consists of a thick layer of large, thin-walled cells which enclose many intercellular spaces. In early stages of development, soon after pollination, the number of layers in the mesocarp increase rapidly but the main increase in thickness of the pericarp results from an enormous increase in cell volume in the later stages of development.

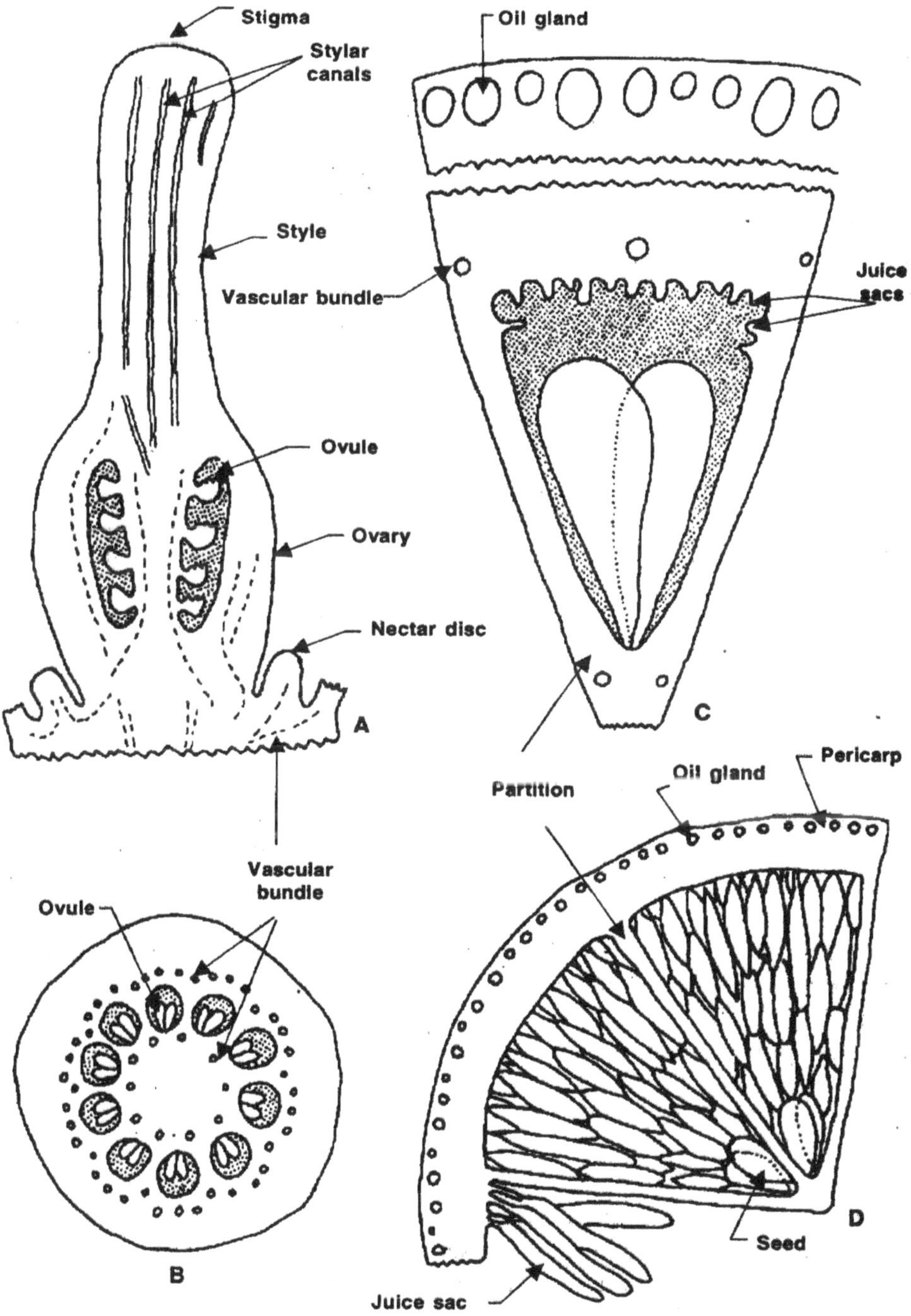

Figure 16.4: Citrus Fruit. (A, B) *Citrus ourantifolia* - young, ovary from flower in longitudinal; (A) and transverse; (B) sections; (C) Part of transection of young fruit of *Citrus sinensis* including one carpel; (b) *Citrus sinensis* fruit including two carpels and part of a third.

The change in the colour of tomato fruit during ripening occurs through transformation of chloroplasts into chromoplasts. Two kinds of chromoplasts develop: One, located in the gelatinous tissue that contains globular inclusions with β-carotene and the other, located in the pericarp which includes voluminous para-crystalline sheets of lycopene in addition to the globules with β-carotene. Towards maturity, the placental tissue becomes degraded and assumes a gelatinous consistency.

Banana

Musa paradisiaca var. *sapientum* is another good example of berry. It's wall consists of (i) an outer epidermis covered by cuticle, (ii) a multilayered parenchymatous hypodermis, (iii) a broad parenchymatous region in which vascular bundles are found scattered, (iv) another region of parenchyma with large number of well developed air spaces, (v) a zone in which the vascular bundles are surrounded by laticiferous cells, and (vi) a five to seven-layered parenchymatous region and an inner epidermis. The banana fruit has tannin containing cells in the pulp and skin which impart astringent taste to unripe fruits. In ripened fruits, the amount of tannin is gradually reduced.

Apple

The fruit of *Pyrus malus*, the pome, arises from an inferior ovary. The fruit here mainly develops from the hypanthium, *i.e.*, from the basal portions of the perianth and stamens and which are fused and adnated to carpels. The pericarp may be divided into a relatively thin exocarp composed of the epidermis and the sub-epidermal collenchyma, a fleshy mesocarp and a stony endocarp. The stomata and trichomes occur in young fruits. The lenticels later replace the stomata. The sub-epidermal tissue is compact and has thickened walls. It forms, together with epidermis, the relatively tough "*skin*" of the fruit. Anthocyanins occur in the skin of the red varieties of apple. The chloroplasts which change into chromoplasts with carotenoids during the process of ripening, are also located in the peripheral tissues of the fruit.

The outer parenchyma of the fruit develops from the hypanthium and it contains ten vascular bundles – five belonging to the sepals and five belonging to petals. These vascular bundles are branched and the branches penetrate into the parenchyma where they form a network. The bulk of the extra-carpellary tissue is parenchyma permeated with intercellular spaces. The endocarp is cartilaginous and it lines the locules. It consists of elongated sclereids with very thick walls which almost completely obliterate the cell lumen. In *Pyrus communis* (Pear), the groups of brachysclereids develop in the parenchyma. The succulent sugar-containing cells are arranged around such groups of brachysclereids in a radial pattern.

The structural changes occur in walls of the parenchyma cells of apple and pear fruits during the advanced stages of ripening. The middle lamella disintegrates almost completely and the adjoining outer region of the cell wall also undergoes partial dissolution of its cellulosic fibrils become dispersed.

Chapter 17

The Seed

Introduction

The term – seed – has a general meaning that antedates – "anything that can be sown – for example, seed potatoes – seeds of corn – or sunflower seeds *etc.* in the case of potato what is sown is a tuber, while in the other two, what is sown is the seed enclosed in the husk. Many structures referred to as the seeds are the dry fruits actually. The study of seeds and fruits is called `carpology`.

A seed can be defined as an embryonic plant enclosed in a protective outer covering, called the seed coat. The seed formation is a part of the process of reproduction in the seed plants (the spermatophytes) including the gymnosperms and the angiosperms. The seeds are the products of the ripened ovules, after the process of fertilization by the pollen grains and some growth within the mother plant. The embryo is developed from the zygote and the seed coat from the integuments of the ovule. The seeds have been an important development in the reproduction and success of both the gymnosperms and the angiosperms compared to more primitive plants (as ferns, mosses and liverworts) which do not have the seeds. At present, the seed plants dominate the biological niches on land – from the forests to grasslands both in the hot and cold climates.

Ovule – Components

The ovule after fertilization develops into a seed. In the mature seed, the following parts can be distinguished:

The funicle (funiculus or funiculi) – the seed stalk which attaches the ovule to the placenta and hence, the ovary or the fruit wall, at the pericarp, The funiculus, whole or in part, abscises from the ovule and leaves a scar, the hilum, through which water can penetrate with relative ease. In anatropous ovules, the part of funiculus that is adnate to the ovule remains recognizable as longitudinal ridge, the raphe, on one side of the seed.

The seed coat – the outer seed coat is called the testa, and the inner called the tegmen, that develop from one or the two integuments (the walls of the ovule),

The nucellus – the remnant of the megasporangium and the main region of the ovule where the megagametophyte develops,

The micropyle – a small pore or opening in the apex of the integument where the pollen tube usually enters during fertilization, This may remain visible or obliterated at maturity.

The chalaza – is the base of the ovule opposite to micropyle, where the integuments meet with the nucellus.

The endosperm – the reserve food material which may be present in small or large amount, due to complete or incomplete absorption by the developing embryo when a little amount is left at maturity (albuminous or endospermic) or no traces are left at maturity (exalbuminous or non-endospermic). A typical seed includes two basic parts – an embryo and a seed coat in addition to endosperm.

The perisperm – the nucellus may or may not remain to serve as the reserve material, as in the seeds of Centrospermae (either perispermic or non-perispermic).

Seed – Shape

The final shape of the seed is affected by the shape of the ovule. The general shape of the seeds is variable and the terms are self-explanatory. They may be reniform (bean shaped – resembling a kidney with lobed ends on either side of the hilum), square or oblong (angular with all sides more or less equal, or longer then wide), triangular (3-sided and broadest in the middle), elliptic or ovate or obovate (rounded at both the ends or egg shaped), discoid, globose, sub-globose, lenticular, sectoroid, striate (striped with longitudinal lines or parallel ridges), or suberose (with the consistency of cork) and crustaceous (hard and brittle).

Seed – Development

The first division of the zygote in majority of plants is transverse to the long axis which established the polarity of the embryo. The upper pole (chalazal pole) becomes the main area of growth while the lower pole (micropylar pole) produces the suspensor (stalk) that is attached to the micropyle absorbs and manufactures the nutrients from the endosperm during the growth of the embryo.

Embryo - Components

The main components of the dicot embryo are:

The cotyledons – these are the seed leaves (first leaves of the plant) attached to the embryonic axis. These may be one (monocots) or two (dicots), and are the source of the nutrients in non-endospermic dicots. These replace the endosperm and are thick and leathery. In endospermic seeds, these are thin and papery. In dicots, the cotyledons are attached opposite to each other on the axis.

The epicotyl – this refers to the embryonic axis above the point of attachment of the cotyledon(s).

The plumule – this is the tip of the epicotyl, has a feathery appearance (due to the presence of young leaf primordial at the apex) and this will become the shoot upon germination.

The hypocotyl – this is the embryonic axis below the point of attachment of the cotyledon(s). This connects the epicotyls and the radicle being the transition zone of the stem and the root.

The radicle – this is the basal tip of the hypocotyls which grows into the primary root.

In monocots, the embryo has two additional structures in the form of protective sheaths, the coleoptile and the coleorhiza respectively. The former covers the plumule and forms the first leaf, while the latter connects to the primary root and adventitious roots from the sides. The hypocotyl is rudimentary between the radicle and the plumule. The single large cotyledon is called the scutellum.

Seeds - Types

In his Comprehensive Seed Classification based on internal structure, Martin (1946) recognized 10 types of seeds based on the embryo and endosperm characteristics, in addition to 2 more additional types on the basis of the size of the seed – dwarf (0.3-2.0 mm); and micro (≤ 0.2 mm). He listed 17 plant families and 12 genera (in 5 other families) as having the dwarf seeds. In the revised key, all seeds are distinguished on the basis of embryo and endosperm characteristics. In his classification, he selected the embryo as the principal diagnostic feature and used its shape, size and position to divide the seeds (on the basis of the morphology of the embryo, and the amount of endosperm and the position of the embryo in relation to the endosperm).

The revised classification of Martin (1946) includes 12 seed types based on the morphology of the embryo, and the relative amount of the endosperm and the position of the embryo. The 12 types are: Broad, Capitate, Lateral, Periphereal, Rudimentary, Dwarf, Micro, Linear, Spatulate, Investing, Bent and Folded types. The differences between the different types are not clear. For example, the only difference consists in the relative size of the seed and the embryo between the rudimentary type, dwarf type and microtype; while the microtype seems type to be a reserved group to Orchidaceae family.

The Seed Coat

The seed coat in a mature seed can be a papery thin layer (groundnut) or thick and hard (coconut) or fleshy (Pomegranate) and protects the embryo from mechanical injury, predators or drying out. It is either bitegmic (two layers - testa and tegmen) or unitegmic (one layer) depending on the development. Seed coat develops either from one integument or from both the integuments. The seeds of angiosperms have mostly dry seed coats but some have fleshy appendages such as elaisome, or juicy layers such as the edible outer epidermis of the seed coat in pomegranate (*Punica granatum*).

All parts of one or both integuments may take part in the formation of the seed coat, *e.g., Viola tricolor*. However, in most seeds, much of the integumental tissue is destroyed and absorbed by the other developing tissues of the seed, and then the seed coat develops only from the remaining parts of the integuments. The parts that are destroyed are usually the innermost or intermediate middle layers of the integument.

In certain seeds, especially those of indehiscent fruits, only the two or three outermost layers of the integument persist. In the mature seed of the Apiaceae only the outer unspecialized epidermis of the single integument remains. In the seeds of certain genera of the Asteraceae, *e.g., Lactuca*, the integument is represented only by a thin layer of obliterated cells which persists beneath the fruit wall. The innermost layers of the fruit wall also disintegrate. In the seeds of certain monocotyledons, *e.g., Zea mays*, the integuments are completely destroyed. In seeds developing from ovules with two integuments, the two integuments may be present in the testa or only two or three of the outermost layers of the outer integument may constitute the major portion of the seed coat or the outer may be the better developed and be specially adapted for protection while the inner is non-specialized. The former type is characteristic of the Malvaceeae, Violaceae, Hypericaceae and Tiliaceae and the latter of the Brassicaceae, Papaveraceae, Berberidaceae and certain genera of the Liliaceae, Iridaceae and Araceae. In Onagraceae, Lythraceae, Aristolochiaceae and others, the protective layers develop from both integuments and the entire nucellus or its outermost layers only take part in the formation of the seed coat. In most of the Fabaceae (Leguminosae), Ranunculaceae and certain genera of the Liliaceae and Amaryllidaceae, the inner integument and the nucellus are completely obliterated. In the ovules with a single integument, only a few of the outermost layers and inner epidermis persists, *e.g.*, in the Plantaginaceae and Polemoniaceae.

Seed coats of pulses are impervious to water and gases due to deposition of thick layer of cuticle except in *Cajanus cajan* where it is very thin and smooth. The cuticle is smooth in *Cajanus cajan, Vicia faba, Dolichos biflorus* and *Phaseolus aconitifolius*, rough with dentate structures in *Lens culinaris* and *Lathyrus sativus,* and papillose in *Phaseolus aureus*. In pulses, seed's cuticle is followed by a layer of characteristic malpighian cells, called palisade layer. It consists of closely packed palisade-like cells that elongate at right angles to the surface and differentiate into macrosclereids with unevenly thickened walls. Two palisade layers occur in hilum region. The outer of these is derived from the funiculus. The cells of the sub-epidermal layer differentiate into the so-called columnar cells, also termed pillar cells, hour-glass cells, osteosclereids or lagenosclereids, depending on the distribution of wall thickenings and the shape of cells. The deeper tissue is a lacunate parenchyma with large tangentially elongated cells in the outer part and smaller much branched cells in the inner part.

The palisade layer attracts much attention because its structure in certain hard legume seeds is assumed to be causally connected with their high degree of impermeability. The so-called light line of the palisade cells is thought to be particularly the impermeable region. The light-line effect results from the intense refraction in a restricted region in epidermal walls. In sections of seeds, this region is oriented tangentially above the middle of a cell or close to its outer wall. The cell wall in the light-line region appears to be especially compact.

In members of Brassicaceae, the seed coat is partly derived from outer integument and partly from inner integument of the ovule. The bitegmic ovules have rather thick integuments. The outer integument has four to eight cell layers, the inner up to ten. The endotegmen is derived from the inner epidermis of the inner integument, and exotegmen from the outer surface of the inner integument. The endotesta is derived from the inner epidermis of the outer integument, and the outer layer of the testa from the outer surface of the outer integument is the exotesta. In many species, the epidermal cells of the outer integument become almost filled with mucilaginous material which appears in layers (Figure 17.1 A, B). The mucilage which consists of pectin and cellulose, swells when it comes in contact with water and, in some species, bursts the outer wall of the cell and forms a gelatinous film over the surface of the seed. If sub-epidermal parenchyma is present in the outer integument, it develops thickened walls (Figure 17.1 B) or becomes crushed (Figure 17.1 A). The inner epidermis of the outer integument, the palisade layer, is the strongest layer in most species, for its cells develop lignified thickenings on the radial and inner tangential walls (Figure 17.1 A, B). The inner integument dies and is compressed. In some species the inner epidermis of this integument becomes the pigment layer.

Corner (1976) in his book "*The seeds of Dicotyledones*" introduced an original classification of the seed coat constituents. The portion of the seed coat produced by the outer integument is termed testa and that produced by the inner integument tegmen. Seeds with characteristic testa are called testal and seeds with characteristic tegmen, are called tegmic, according to which tissue in the seed coat becomes sclerified. Corner suggested the following terms: Exotestal and exotegmic seeds depending on whether the mechanical layer develops from the outer epidermis in the testa or tegmen respectively, mesotestal and mesotegmic seeds in which the mechanical layer develops in the middle layer of one of the integuments, and endotestal and endotegmic seeds in which the mechanical layer develops from the inner epidermis of the appropriate integument.

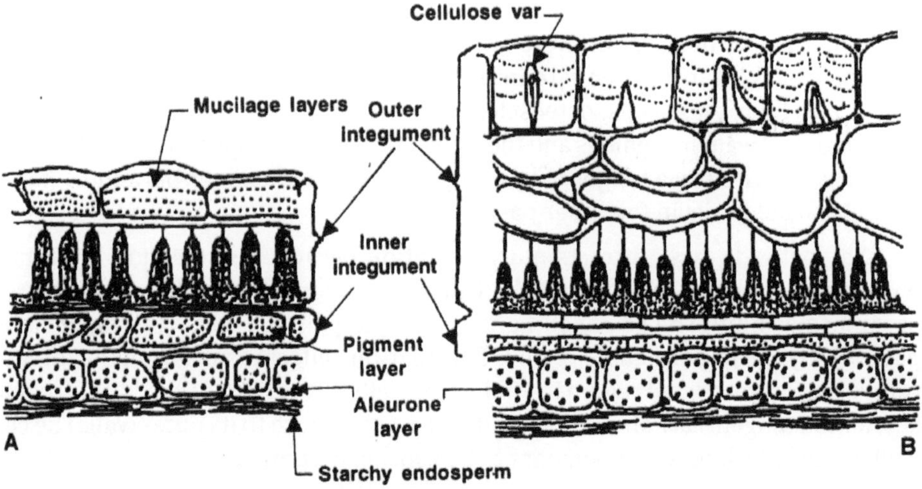

Figure 17.1: Cross Sections of Seed Coat and Aleurone Layer of (A) *Brassica*; (B) *Sinapsis.*

In general, the epidermal cells of seed coat are thick-walled and remain filled with colouring matter. But, there are certain variations, *e.g.*, in *Gossypium*, the epidermal cells elongate into hairs (the cotton fibres - coma) and in *Linum*, the epidermal walls are highly hygroscopic and become mucilaginous on contact with moisture. The epidermis of the seed coat of members of Cucurbitaceae consists of highly elongated cells with characteristic presence of branching ribs. In gymnosperms, which do not form ovaries, the ovules and hence the seeds are exposed. This is the basis for their nomenclature – the naked seeded plants. The seed is composed of the embryo and the tissue from the mother plant (which also form a cone around the seed (in conifers like pine and spruce).

Storage of the Nutrients

The form of the stored nutrients varies depending on the type of the plant. In angiosperms, the stored food is called the endosperm, derived from the mother plant through double fertilization (usually 3*x* and is rich in oil (fat), or starch and protein). In gymnosperms (conifers), the storage tissue (also called the endosperm) is a part of the female gametophyte (haploid). The endosperm is surrounded by the aleurone layer (peripheral endosperm) filled with proteinaceous aleurone grains.

The endosperm is considered to be farinaceous (mealy) in which the cells are filled with starch (farinaceous – cereal grains) or not (non-farinaceous). It may also be called as fleshy (or cartilaginous) with thick soft cells (Coconut) or oily (*Ricinus*, *Croton* and Poppy). The endosperm is also called as horny when the cell walls are thicker (Date and *Coffea*) or ruminated if mottled (Nutmeg, Palms and Annonaceae).

In most monocots (Grasses and Palms) and some dicots - endospermic or albuminous (Castor beans), the embryo is embedded in the endosperm (and nucellus) until germination. The endosperm is absorbed by the embryo (non-endospermic dicots) as it grows within the developing seed and the cotyledons become filled with the stored food (exalbuminous seeds - legumes like Beans and Peas - trees like Oak and Walnut - vegetables such as Squash, Radish and Sunflower). The uncommon storage is seen hypocotyls as in Brazilnut. The seeds of all gymnosperms are albuminous.

Seed - Size

The size of the seeds is variable. The seeds in orchids are the smallest and dust-like (about one million seeds per gram), are often embryonic with immature embryos and no energy reserves. The largest seed is the coco de mer (*Lodoicea maldivica*) weighing about over 20 kg. The plants that produce smaller seeds generate more number of seeds per flower while the plants with larger seeds produce fewer seeds with more amounts of resources. Smaller seeds are quicker to ripen and are dispersed sooner. The herbaceous perennials and woody plants have larger seeds produced over many years with more reserves to have a more established seedlings after germination.

Seed - Dispersal

The plants have evolved different ways to disperse their offspring by seed dispersal because of their limited ability to seek favourable conditions for growth and life. The fruits may be dehiscent (open up) or indehiscent (do not open) to release the seeds. The dispersal is: by wind (anemochory), by water (hydrochory), by animals (zoochory), by ants (myrmecochory), or by special mechanisms (autochory) *etc.*

Chapter 18

Dormancy in Plants

Introduction

Dormancy is the state in which a plant or a plant part exhibits little or no growth and in which most of the metabolic activity ceases for a period of time. Most of the plant life functions best when there is ample supply of water and required temperature (well above freezing) throughout the year. The majority of the plants are exposed to dry periods and temperatures below freezing for variable lengths of time during the year except those in moist, tropical regions of the globe. The plants do not have body insulation or locomotion (like animals), and hence, they cannot seek the shelter or use other ways to survive shortage of water and cold weather. So, many plants become dormant to avoid the unfavourable environmental conditions, during which their metabolic activity either ceases or is drastically reduced.

Dormancy – Definition and Terminology

Dormancy refers to a state in which a plant or a part of the plant exhibits little or no growth and in which most metabolic activity ceases for a period of time. The term is widely recognized as a behavioural and physiological state of both the animals and plants that generally involves inactivity and reduced metabolic rate. A similar term to dormancy is "torpor" meaning `inactivity` or `lethargy`. These two terms involve very different physiological states, in response to different stimuli, like low or high temperature, lack of water or food, and it can be a short-term event (less than 24 hours) or may continue for a few days or may last an entire season or even for many years. Dormancy can also involve a developmental arrest (hence, called as `diapauses`). A more extreme state than dormancy, with almost no detectable activity or metabolism is "cryptobiosis" literally meaning `hidden life`. This is most prevalent in lower vertebrates and is a seasonal survival strategy to cold or desiccation. "Quiescence" refers to the short periods of dormancy that are directly induced by adverse environmental conditions, in particular lower or higher temperatures.

Unlike the animals, many plants become dormant to avoid the unfavourable conditions of the environment. In dormancy, the metabolic activity either ceases or is drastically reduced. This evolved as a means of surviving the unfavourable conditions of the environment. The term 'dormancy' is generally used to indicate tissues such as buds, seeds, tubers, and corms that fail to grow even though they are provided with adequate moisture and appropriate oxygenation (*i.e.*, physiological temperature). Dormancy is a precisely a regulated phenomenon, cued by factors in the environment and maintained, and ultimately broken, by specific metabolic changes in the organism.

Dormancy - Main Types

The classification is based on its expression under different conditions. It is indicated as Thermodormancy when it is induced by elevated temperatures, and it is not known whether this type differs from Photodormancy in a mechanical sense. Hence, it was proposed to distinguish only two types of dormancy - Primary dormancy and Secondary dormancy (Karssen, 1982; Amen, 1986; and Hilhorst, 1995) based on timing. The former is associated with the development, while the latter occurs after dispersal and is basic to seasonal dormancy. Dormancy is the temporary inability of a viable seed to germinate under favourable environmental conditions (Simpson, 1990). Dormancy helps the seed to survive adverse environmental conditions and at the same time helps in the seed dispersal (Snape *et al.,* 2001).

Non-dormancy is not always a desirable trait because insufficient level of dormancy in some crops leads to pre-harvest sprouting. A higher level of dormancy may lead to poor establishment owing to slow and inconsistent germination which may adversely affect the malting of barley and also may lead to weed problems in the subsequent crop. Dormancy has a hereditary component which has been demonstrated for a large number of crop species including rice, soybean, and barley in addition to weeds like wild oats. The classical genetic analysis has demonstrated that 1-10 recessive genes are related to dormancy. The mutational and molecular analysis indicates that the number of genes involved in the regulation of dormancy is much higher, and also that the degree of it is hereditary. The environmental factors which affect the dormancy include: length of the day, quality of light, temperature, soil moisture, and mineral nutrition.

The buds normally form from spring to mid-summer in the temperate zones. There may be a little growth in the late summer, it virtually ceases in the fall in preparation for the winter. Entry into a dormant state protects the buds from the freezing temperatures. Many seeds contain a complete embryo along with a reserve food supply, and mature seeds are formed within the ripe fruit. If it were not for germination inhibitors present in the fruit or the seed, the seeds would begin to germinate while still in the fruit. In addition to this, the seeds in the temperate regions often reach the soil in late summer or early fall, when the plants often face low moisture and imminent cold weather. The survival would be unlikely, if the seeds were to germinate during this time of the year.

Dormancy tends to be closely associated with the conditions of the environment and they can synchronize their entry into a dormant phase with their environment through predictive or consequential means. The predictive dormancy occurs when an organism enters a dormancy phase before the onset of adverse conditions of the environment (for example, the photoperiod and decreasing temperature are used to predict the onset of winter by many plants). The consequential dormancy occurs when an organism enters a dormant phase after the adverse conditions have been onset. This is generally found in areas with an unpredictable climate. The sudden changes in the conditions may lead to a higher mortality rate (among the animals) and its use can be advantageous as the organisms remain active longer and can make a greater use of the resources available.

Patterns of Growth and Death in Plants

The type of response to dormancy depends on the pattern of growth and death of the plants. The perennials are the plants that live year after year, undergoing a special period of dormancy during the cold season. In the herbaceous species, the portions above the ground die while the plants survive as specialized underground stems. The woody shrubs and trees remain alive above the ground. The deciduous species shed their leaves in the winter, while the non-deciduous species (called the evergreen species) keep their leaves year-round but drastically reduce their metabolic rates. The biennials live for two years. In the first year, the vegetative growth and the formation of underground storage tissues are formed. The plant lies dormant through the winter and its second year of growth results in the stored food supply being used to produce the flowers and the seeds. The annuals complete their life cycle in one growing season (or one year). The plants die after producing the seed, which normally remain dormant until the following growing season. Most perennial trees and shrubs in the temperate regions produce the buds in the summer. These develop into leaves, stems or flowers, exhibit reduced metabolic activity even before the leaves begin to senesce (age). As the temperatures decrease in the fall, complete dormancy sets in. The bud scales (specialized leaves) cover and protect the dormant tissue. These block the oxygen diffusion into the bud and also prevent the loss of water from the tissue. All the flowering plants produce the seeds. The seeds develop as ovules within the ovary of the flower. As the ovary ripens to form the fruit, the ovules mature into seeds. Each seed is composed of a reserved food supply and a new plant with an embryonic root, leaf and the stem tissue. This along with the reserve food supply is surrounded by a tough seed coat. In many species, the seeds especially in the trees in the temperate zone, do not germinate immediately after maturation even under ideal conditions of moisture, temperature, and nutrients because there is a built-in period of dormancy. Though the seasonal dormancy is most often correlated with changes in the temperature, variation in precipitation is the primary factor in the regions where pronounced wet and dry seasons alternate. Some deciduous trees and shrubs drop their leaves and remain dormant during the dry season and grow new leaves after the rains return. The herbaceous perennials die back and go dormant at the beginning of the dry season and regrow only their above ground mass. Desert plants have the seeds that will only break the dormancy when the rains are sufficient, which in some regions may be only every few years. Some specialized seeds germinate only after they get tumbled in the waters of a flash flood which scrapes their seed coats.

Bud Dormancy

A bud is actually a meristem in which the internodes have failed to elongate and the leaves have not enlarged. The whole meristem is enclosed in a set of modified leaves, called bud scales. Bud scales serve a protective function, insulating the bud and preventing desiccation. Bud scales also contribute to reduced respiration since their removal usually results in a significant increase in respiratory rate.

Dormancy is induced by a number of environmental factors. In many plants, this occurs in response to low temperatures while in others, the short day photoperiod is responsible for initiating the dormancy. In still others, both the lower temperatures and short days are necessary to trigger the onset of dormancy. In addition, the bud dormancy has been shown to occur under conditions of limited nutrient supply or conditions of drought. Hence, dormancy is a survival mechanism. When the temperature, water or the nutrients are no longer favourable, the buds become dormant.

Many studies show that bud dormancy is induced in some varieties of certain species but not in other varieties of the same species. Such a difference suggests that there may be genetic variation in the control mechanism. But, the exact mechanism of genetic control is not clearly understood.

Seed Dormancy

Seed dormancy is the state or a condition in which the seeds are prevented from germinating even under the favourable environment for germination, such as temperature, water, light, gas, seed coats, and other mechanical restrictions. Seed dormancy can be defined as the failure of mature and viable seeds to germinate under favourable conditions for germination. Many seeds fail to germinate after falling from the plant even if favourable conditions of moisture, temperature and aeration are available. Dormancy occurs because some property of the seed prevents germination. In some dormant seeds morphological changes must take place before commencement of germination. For others, parts of seed must undergo physiological changes before germination can occur. Under natural conditions necessary changes take place gradually under varying combinations of aeration, moisture, temperature and light.

This can be caused by a number of factors, and the major causes for the seed dormancy are: light, temperature, hard and resistant seed coat, period after ripening, germination inhibitors, immaturity of the seed embryo, impermeability of the seed coat to water, impermeability of the seed coat to oxygen, mechanically resistant seed coat, and the presence of a high concentration of the solutes.

The presence of hard seed coat will often result in seed dormancy. In many cases, the seed coat is impermeable to water which is required for the germination; this will serve as an effective inducer of dormancy. In some cases, the seed coat may be impermeable to certain gases and prevent the diffusion of oxygen into the seed while others are impermeable to carbon dioxide. In a few species, the seed coat restricts the growth of the embryo physically as the growing embryo fails to develop sufficient thrust to break through the seed coat during germination. The seeds of almost all species have a minimal temperature below which they will not germinate. The exact mechanism by which the lower temperatures cause dormancy is poorly understood and it appears that temperature alters the membrane structure which prevents the seed to germinate. The light is also a factor in seed dormancy. In many species, the exposure to light will induce dormancy and some species exhibit sensitivity to the photoperiod. Some species are dormant during short-day cycles and germinate only during long-day cycles, while others remain dormant in long-day cycles and only germinate under short-day cycles. The light activates a plant regulator that blocks the metabolic reactions which are necessary for germination.

The studies on the chemical control of bud dormancy account for the fact that abscissic acid is associated with dormancy though the evidences are inconclusive. A number of studies have shown that when ABA is applied to the seed, it will block the enzyme activity that is necessary for germination. A number of germination inhibitors are present in the dormant seeds. The inhibitors like cyanide are produced in some seeds. The higher concentrations of different inorganic salts prevent germination in some species, while an assortment of phenolics are known to prevent the process in other species. Coumarin is particularly widespread in the seeds and is an effective inhibitor of germination. These are also present in both the fruit and the seed and these prevent the seed from germination while they are still within the fruit. Such seeds will remain dormant in nature until sufficient rain leaches the substances from the seeds.

The growth of deciduous perennial plants typical of temperate regions is cyclical – such plants normally undergo a cessation of active growth on an annual basis. Buds formed in the later part of the summer or early fall over winter in a dormant condition and do not normally renew their growth until more favourable temperature conditions return in the spring. Bud dormancy is thus a defence mechanism, helping to ensure that the plant is able to survive the adverse conditions of winter. To ensure survival, dormancy mechanisms must be in place before the arrival of unfavourable conditions. This means that the plant must be able to anticipate climate change. Mechanisms must also be in place to ensure that the buds do not break dormancy until such time that conditions in the environment are appropriate to sustain normal growth and development. Premature breaking of buds during an unseasonally warm period in the winter, for example, could have serious consequences. In other words, dormancy is more than simply a period of inactivity.

The onset of bud dormancy is coincident with leaf fall, decreased cambial activity and increased cold hardiness. Bud dormancy appears to be a typical activity, and increased cold hardiness. Bud dormancy appears to be a typical short day response, initiated by the short days of late summer. In most cases, phytochrome is involved and the leaf perceives the photoperiodic stimulus. However, in at least one case, that of birch (*Betula pubescens*), the response appears to be localized in the apical bud.

Temperature near or just above freezing appears to be most effective at breaking dormancy. The amount of chilling required varies with species, cultivar, and even location of the buds on the trees. Species such as apple (*Malus pumila*), pear (*Pyrus communis*) and cherry (*Prunus* sps.) require approximately 7 to 9 weeks of exposure to temperature below 7°C in order to overcome dormancy. Others may require up to 22 weeks (*Prunus americana*). Persimmons (*Diospyros kaki*) require only 4 days of low temperature and so can be grown successfully much further south than other fruit trees.

Types of Seed Dormancy

The dormancy is mainly of the following types:

1. Seed-coat Dormancy (External Dormancy)

In many plants the cause of dormancy is the presence of hard seed–coat, which may offer mechanical resistance to embryo growth (Malvaceae, Fabaceae), or it may be impenetrable to water (*Melilotus, Trigonella*) or oxygen (*Xanthium*). In nature, such seeds are rendered germinable by the weakening or breaking-down of the seed coat by microbial attacks, mechanical abrasion, passage through the digestive tract of animals or exposure to alternating low and high temperatures. Some of the treatments used to break seed-coat dormancy artificially are: (a) scarification, which means scratching the seeds with abrasives (*e.g., Cassia, Melilotus*), (b) soaking the seeds in concentrated sulphuric acid for 15-16 minutes (*e.g., Gossypium, Pistachia*), and washing the seeds in alcohol (*e.g., Cassia, Parkinsonia*).

2. Embryo Dormancy (Internal dormancy)

The most common type of seed dormancy is one in which the embryo requires certain period of time after harvesting to be able to germinate. During this period, which is called after-ripening period, the embryo may undergo structural maturation or physiological maturation. The most likely cause of embryo dormancy is the presence of germination inhibitors in the embryonic axis or in the food storage tissues or the seed. For germination to occur, these inhibitors must be metabolically inactivated or germination promoting substances must overcome their effect. There is considerable evidence that ABA is involved in regulating the onset of dormancy and in maintaining the dormant state. Germination inhibitors have been isolated and identified in a number of woody plant seeds with ABA, the most common inhibitor. ABA accumulation in developing seeds is low during early stages and is greatest during mid-development. So prevention of germination during development may be due to the endogenous ABA content of the seed, *e.g.*, *Acer platonoides, Malus pumila, Quercus rubra.*

3. Morphological Dormancy

This refers to an under-developed undifferentiated embryo. Morphological dormancy results from embryo not being completely developed when seeds are shed. At shedding, the seeds in some members of the Ranunculaceae contain an undifferentiated embryo. During after-ripening period the embryo undergoes further development and acquires the mature structure. In *Fraxinus excelsior*, a seed when shed contains a fully differentiated embryo but it undergoes considerable enlargement before germination. In Rosaceous plants, the embryo does not show any visible morphological or histological change during the after-ripening period. It is assumed that these seeds undergo physiological maturation during dormancy period. The physiological changes may include the disappearance of growth inhibitors, appearance of growth promoters *etc.*

4. Combined Dormancy

Combined dormancy is a condition in which two or more primary factors, such as seed coat dormancy and embryo dormancy, are present to the extent that each requires treatment to overcome. Some examples of combined dormancy are *Cercis canadensis* var. *mexicana, Rhus trilobata*. Seed coat scarification with acid for 10 to 40 minutes followed by moist scarification, is the recommended treatment to overcome dormancy.

5. Double Dormancy

Double dormancy is a condition in which there is dormancy in both the radical and the epicotyls of the embryo, but each has different conditions to overcome the dormancy, *e.g.*, Oats.

6. Secondary Dormancy

Secondary dormancy results from action, treatment or injury to seeds during collection, handling or sowing. Pine seeds can incur secondary dormancy if exposed to high temperatures and moisture at critical times. When stratified seeds are re-dried to storage levels, they are often said to have incurred secondary dormancy. Germination can certainly be delayed under these conditions.

7. Innate Dormancy

This refers to the condition of the seeds which is incapable of germination even if the conditions suitable for the growth of the seedling are supplied. This inability may be due to the embryo being immature at the time of dispersal.

8. Enforced Dormancy

This is the condition of the seeds which is incapable of germination due to an environmental restraint like amount of moisture, oxygen, light and a suitable temperature.

9. Induced Dormancy

This occurs when the seed has imbibed water, but has been placed under extremely unfavourable conditions for germination. As such, the seed fails to germinate even under more favourable conditions.

Methods of Breaking Seed Dormancy

Dormancy can be broken naturally when the embryo gets natural environment such as adaptive moisture and temperature. The hard seed coat becomes permeable by the action of natural agents like microorganisms, temperature and abrasion by the digestive tract of birds and animals that feed on them. Other methods include: completion of the over-ripening period, leaching of the inhibitors present in the seed coat, inactivation of the inhibitors by the supply of cold, heat and light, leaching of the excess and highly concentrated solutes from the seeds, and production of growth hormones which can neutralize the effect of inhibitors.

Dormancy can be broken artificially in order to obtain faster growth or increased plant production. The treatments to break dormancy can be divided into different categories: Action with hot water for the termination of waxes, surface inhibitors *etc.*, rupturing of the seed coats by filing, chipping, or thrashing through machine, exposure to heat, cold, or light depending upon the type of seed dormancy, by the application of hydraulic pressure for 5-20 minutes in order to weaken the seed coats, and seed coats are treated with concentrated sulphuric acid for the removal of all traces of mineral acids, in addition to the use of temperature, light, chemicals, or mechanical means, and the artificial exposure to lower temperatures, warmth, and altered photoperiods (singly or in combination) break dormancy.

The separate treatments to overcome the dormancy can be further divided into:

Seed Coat treatment

The treatment makes a hard seed coat permeable to water or gases either by softening or cracking (called scarification) and the treatment is either chemical or physical.

Embryo Treatments

Stratification: These include the incubation of seeds at an appropriate low temperature over a moist layer before transferring them to a temperature suitable for germination, called *stratification*. The mechanical breaking (scarification) of the hard seed coat has proven to be an effective method of overcoming dormancy in many seeds. Such broken seed coats allow the seeds to take up water and gases that are necessary for germination.

High Temperature Treatment: The incubation at 40-50°C for a few hours to a few days may have an effect in overcoming the dormancy in some species such as rice seeds treated with hot water at 40°C for at least 4 hours.

Chemical Treatments: Plant growth regulators or other chemicals can be used to induce the germination. A number of chemicals also break bud dormancy. Ethylene chlorohydrins (used for years to release the dormancy in many fruit trees), and gibberellins (the natural growth hormones) will break dormancy in most cold-requiring plants when they are directly applied to the buds. The natural hormones such as gibberellins, cytokinins and ethylene have been shown to be involved in breaking the dormancy. Thiourea and other substances are used to favour seed germination on a commercial basis.

Duration of Dormancy

In seeds the duration varies from species to species, and it has been noticed that the duration varies from few months to many years. The controlling factors that impose are many such as – the strength or the susceptibility of the seed coat – ability of the seed to absorb and utilize water and/or oxygen – the presence of inhibitors and environmental factors like temperature, water, day length *etc.*

Seed Viability

This refers to the period for which the embryo remains healthy within the seed coat and is capable of germination under permissible conditions (the longevity of the embryo). Beyond this period, the seeds fail to germinate even in favourable conditions. The half-life of the seeds vary from species to species. Some seeds have few days or months while some remain viable for a longer period as 100 – 1000 years. It was reported that the seeds of lotus obtained at an excavation site were as old as 900-1200 years and still they were viable (Horace Wester, 1971). The majority of crop plants are viable for 1-2 years.

Causes for Dormancy

Dormancy can be caused by various means such as:

Seed Coat

Many seeds with a hard seed coat are incapable of immediate germination as the developing embryo is incapable of breaking open the same. It also fails to imbibe water, and fail to diffuse oxygen which is necessary for germination. Hence, the seed coat acts as a protective structure and imposes dormancy (necessary evil) in the following ways:

Prevention of Uptake of water

Some species of plants produce the seeds with a hard seed coat which is also waxy. Hence, the seeds go impermeable to water. In *Pinus arborus*, water first enters through the hilum (made of hygroscopic tissue). As a result, the seed swells up and the micropyle gets closed which prevents the entry of water into the seed during germination.

Prevention of Oxygen

Some seeds though they are capable of taking up water, are incapable of taking up atmospheric air. For example, in *Xanthium* species, each fruit contains two seeds placed one above the other. The lower seed germinates under normal conditions, but the upper one does not until it is subjected to higher oxygen pressure. This is due to the presence of an inhibitor which will be oxidised only under higher concentration of oxygen; otherwise it inhibits the embryonic growth.

Prevention of the Growth of the Embryo

Species of *Amaranthus* and some others are capable of absorbing water and oxygen, to allow the embryonic growth, but the enlarging embryo cannot break open the hard and mechanically resistant seed coat. As such due to the prevention of germination, the seeds may remain for months to years until the seed coat gets loosened or cracked to favour the embryo to come out of it.

In the above cases, though the embryos are normal, and are without growth inhibitors, the growth is prevented by the presence of a hard and mechanically resistant seed coat which either prevents the entry of water or oxygen or the seed coat may fail to crack open thus preventing the emergence of the growing embryo. The only mechanism by which such seeds can be induced to germinate is to break the seed coat and make it weak so that the seeds can not only take up water and oxygen but also facilitate the embryo to emerge out of the seed coat. This process is called scarification which can be achieved by the following methods:

Mechanical Scarification

In this method, the seeds are shaken with abrasives or nicking the seed coat with sharp edged metals or chewing the seed coats without damaging the embryo makes the seed coat to crack open without damaging the embryo. This facilitates the emergence of the embryo out of the seed coat.

Chemical Scarification

The hard seed coats can be loosened by the strong acids or solvent treatments by which the seed coat becomes soft. The duration of the treatment has to be determined for each kind of seed. Otherwise, this may cause the damage to the young embryos contained within.

Immature Embryos

In many plants the seeds are shed before the embryos are fully mature, as a result of which germination fails to occur till the embryos reach the maturity. This can be achieved by storing the seeds under favourable conditions. But, in some plants, the seeds fail to germinate even under favourable conditions. In such cases, the storage is prolonged when the seeds undergo germination (after-ripening effect). This is due to changes in the metabolic activities during the period of storage. The maturation of such seeds can be accelerated by keeping the seeds in layers alternating with moist sand or moss at low temperatures. This can be explained due to leaching out of many inhibitors out of the seed.

Effect of Temperature (Chilling Treatment)

Though many seeds germinate under normal conditions of temperature, some fail to germinate if stored at room temperature. This is because they require chilling treatment for a period of time. Then only the seeds overcome dormancy and germinate. This is because the cold treatment destroys the inhibitors in the seed coat or in the embryo. This is identical with vernalization but the mechanism is different.

Effect of Light

The radiant energy has a profound influence on seed dormancy and germination in addition to initiating many biological processes. Many plants are sensitive to light radiation as such the seeds are divided into three types, namely: positive photoblastic seeds – negative photoblastic seeds – and non-photoblastic seeds.

The seeds which are positively photoblastic germinate when they are exposed to one or two cycles of an intense source of white light. The negatively photoblastic seeds do not respond to white light, but they germinate in complete darkness. The non-photoblastic seeds are insensitive to light and germinate irrespective of the presence or the absence of light. The red light (Pepper grass seeds germinate on exposure to red light) and far red light (inhibits the germination and the seeds go dormant) induces dormancy or germination in light sensitive seeds. These effects on the germination of the seeds can be reversed. For example, the number of alternate cycles of red and far red light treatment ending with red light, the percentage of germination gets enhanced, while with far red as the last of the cycle, inhibition percentage is more or less 90-93 per cent. The effectiveness of the red and far red light suggests the involvement of the phytochromes. The role of phytochromes was demonstrated by Ikuma and Thiman. This depends on other factors like water, temperature, pH and age of the seeds, duration of the light treatment and chemicals present in the seeds.

The phytochromes exist in two forms – PR and PfR which are inter-convertible in the presence of red and far red lights. PfR under the influence of red light undergoes a reverse reaction in the dark to produce PR form. In some cases, PR is converted to PfR form by specific enzyme mediated reactions that require the input of energy. It is also known that PfR requires an unknown factor called X to be in its active state (PfR-X). This pigment complex is known to be very effective in breaking the dormancy in red light of positive photoblastic seeds. The positively photoblastic seeds require a higher concentration of PfR-X complex than negatively photoblastic seeds, while the negatively photoblastic seeds require a minimum amount of PfR complex to be highly effective. The negatively photoblastic seeds do not germinate if the PfR concentration is more in them. The non-photoblastic varieties germinate without light treatment and less PfR has a promotive effect on them.

Induction of Dormancy by Chemicals

Many chemical compounds like coumarin, para-ascorbic acid, hydrogen cyanide, abscissic acid *etc.*, have been isolated from the seed coats, endosperm pulp and the juices of the fruits with embryos. During maturation of seeds and fruits, these compounds are actively synthesized and stored in different parts of the seeds and fruits. These compounds are non-toxic but induce the dormancy as such the compounds are to be either destroyed or leached out, otherwise the seeds remain dormant. In many fresh fruits, the ripening is also initiated by ethylene, and it can also induce ABA synthesis. The abscissins can bring about the induction of enzymes which enhance ripening of the fruits but also the permeability in the membranes and also transcriptional and translational activity. These slow down the metabolsim associated with the reduction of water content in the embryonic cells as such the seeds become dormant.

Chemicals that break Seed Dormancy

The compounds that are known to overcome dormancy are gibberellins, cytokinins, ethylene, chlorohydrins and thiourea *etc.*

Methods for Overcoming Dormancy

1. For Seed Coat Dormancy

Treatments are designed to breach the seed coat, or other covering structures, and remove barriers to moisture uptake, gas exchange, swelling of embryo, and radical emergence. Methods used to overcome seed coat dormancy are collectively known as scarification treatments, and there are risks to seed viability inherent in all of them. In selecting a scarification treatment, the most gentle methods should be tested first, then increasingly severe treatments until the desired effect is obtained.

Cold water soak: In some hard seeded species, the seed coats are not completely impermeable to water. Soaking such seeds in water at room temperature or 24 to 48 hours may be sufficient for full imbibitions and subsequent germination.

Hot water soak: Similar to cold water soak, seeds are put into very hot or boiling water and left there as the water cools for 12 to 24 hours. The hot water softens the seed coats or causes them to crack, and imbibitions occur as the water cools. The container used for this treatment should not be made of aluminium, as it may be toxic to the seeds. Numerous leguminous seeds can be treated in this manner.

Acid treatment: Treatment with concentrated sulphuric acid is the method of choice for many species. Seeds should be in contact with the acid for 15 to 60 minutes, depending on species or individual seed lot, and washed thoroughly in running water afterwards to remove any acid that remains on the seed coat.

***Mechanical treatment*:** Mechanical scarification is used extensively for large lots of seeds. Mechanical scarification can be done by rubbing seeds between two pieces of sand paper. Seeds may also be mixed with coarse sand, small rocks and shaken vigorously in a jar. Care must be taken not to injure the embryo. A mechanical device has also been developed to crack peach seed coats. For small samples, seed coats can be scarified by hand with knives, files, dippers, sand paper *etc.*

2. For Internal Dormancy

Treatments to overcome internal dormancy are expected to bring about physiological changes within the embryo that will enhance rapid germination.

Stratification (Pre-chilling): The usual procedure for stratification is to refrigerate fully to imbibe seeds at 1 to 5°C for 1 to 6 months. This procedure simulates the natural winter conditions of temperate-seeds that are lying on forest floor. During stratification (a) enzyme systems are activated (b) stored foods are changed to soluble forms; (c) the inhibitor/promoter balance change.

The optimum length of the stratification period varies greatly among species and among different seed lots of the same species. One benefit of stratification for nurseries is an increased uniformity of emergence. The low temperature used in stratification inhibit germination of the seeds that are no longer dormant while the remaining seeds are under-going the needed internal changes. When the seeds are finally sown in favourable temperatures, there is a flush of uniform germination and emergence, which is crucial to even seedling development.

Importance of Seed Dormancy

Seed dormancy is important because of the following reasons. They are:

1. In the stratification method uniform germination and seedling emergence can occur.
2. Dormancy is a great advantage when one wants to store seeds, but a disadvantage when prompt germination is desirable.
3. Plant enables the embryo to safely pass the unfavourable part of the year and germinate when the conditions are suitable. Hence, it is vital mechanism that connects one generation with the next.
4. Seeds are exhibiting primary dormancy when they are dispersed from the parent plant in a dormant stage. Prevention of germination during seed development may be due to endogenous ABA content of the seed. Absence of or insensibility to ABA during seed development, results in the production of viviparous, *e.g.*, maize viviparous (vp), tomato sitiens (sit) *etc.*

Dormancy - Advantages and Disadvantages

The seeds with a longer period of dormancy and viability are capable of surviving the worst hazards of the environment. When they get dispersed to longer distances, even then they are capable of germination under favourable conditions and propagate in distant areas. Some plant seeds with such a survival capacity (like weeds) create nuisance to humans and they cannot be destroyed easily. The seeds with longer periods of dormancy will help in storing the food grains.

Seed Dormancy - Importance

1. Dormancy allows the storage of seeds for a latter usage by the animals and man.
2. It helps in the seed dispersal through the unfavourable environment.
3. Dormancy induced by the inhibitors in the seed coats is highly useful to desert plants.
4. This allows the seeds to continue to be in suspended animation without any harm during cold or high summer temperature and even under the conditions of drought.
6. This helps seeds to remain alive in the soil for several years and provides a continuous source of new plants, even when all the mature plants of the area have died down due to natural disasters.

Chapter

19

Senescence and Abscission

SENESCENCE

Virtually, all the organisms after they age and do end their life with senescence followed by the death. The fundamental question in biology is - how and why the living organisms senesce and die? According to Lim *et al.* (2007), senescence is the age-dependent, programmed, degradation and degeneration process (of cells, organs or the entire organism) leading to death. The process of senescence occurs most distinctly at the organ and organismal levels (the outcome can be best observed in rice, corn, soybean, and wheat at the time of their harvest). At the organ level, senescence is manifested in the leaf colour changes and subsequent death of the leaves in autumn (post-mitotic senescence). Senescence in plants also defines the last stage of a developmental program which is degenerative, and occurs in a temporarily coordinated manner (Gan and Amasino, 1997; Lim *et al.,* 2007). How do the plants "know" when and how to "die"? - is a fundamental question yet unanswered. There are 6 major research themes which address this and constitute the basis of this issue are: changes in the photosynthetic apparatus during senescence - the interaction of endogenous and exogenous signals with the biochemical pathways - the integrative, multi-layered spatio-temporal networks – the evolutionary basis – extended applications of multi-omics to explore the genetic elements and their networks which control the process, and integration of the types of multi-omics, genetic and physiological data – and source-sink interactions for enhancing the yield.

Plant senescence is the process of aging and this is both stress-induced and age related developmental aging. The degradation of the chlorophyll reveals the carotenoids (like Anthocyanin and Xanthophyll) are the cause of leaf colour in autumn of the deciduous trees. This has the important function of recycling the nutrients (mostly nitrogen) to the growing and the storage organs. In plants, new organs are formed continuously while the older organs undergo a regulated senescence programme to maximize the export of the nutrients.

The process of senescence is simply defined as the changes which lead sooner or later to the death of an organism or a part of it. The term abscission is a process involved in shedding of plant structures, characterized by the degradation of cell walls at the point of weakening. The present evidence suggests that cells surrounding the fracture line secrete cell wall-degrading enzymes which hydrolyze the central region of the wall, allowing the cells to separate and fracture to occur. In contrast, the less common process of 'mechanical tearing' involves the generation of large forces which tear apart an inherently weak band of cells. The cells along the fracture line thus play a passive role in this process (the best example being the flaking of bark).

Distribution of Senescence Process in the Life Cycle of Plants

Senescence processes are encountered in all plants and at all stages of the life cycle. Figure 19.1 is a schematic diagram of the life cycle of a flowering plant in which senescence process associated with successive stages of life cycle are listed.

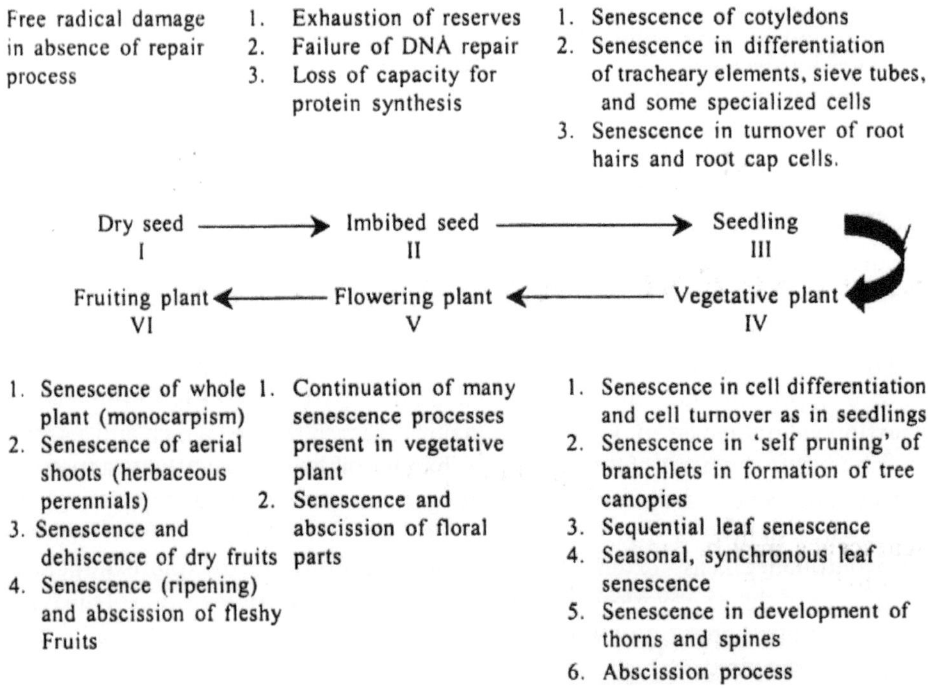

Figure 19.1: Summary of the Distribution of Senescence Process in the Life Cycle of Seed Plants. Most of the phenomena listed, except those which occur within the seed, are integral part of the development programme of the plant (after Sexton and Woolhouse, 1984).

A particular feature of this scheme is that the events which are listed are integral components of the developmental programme of the plant, with the notable exception of those that occur within the seeds.

Metabolic Changes associated with Senescence

Chlorophyll Degradation

One of the cardinal events by which senescence in leaves is generally recognized is the loss of chlorophyll, revealed in the change from green to yellow coloration.

The Synthesis and Degradation (Turnover) of Cellular Constituents

In many species there is a progressive decline in the amount of proteins and RNA in the leaves in the course of their maturation and senescence. The loss of these essential cellular constituents is considered a potential causal factor in senescence.

Influence of Hormones and Regulation of Senescence

The programmed senescence is influenced by plant hormones (abscissic acid, ethylene as a gaseous hormone, jasmonic acid and salicylic acid) which promote senescence, and other hormones involved are represented by gibberellins, brassinosteroids, and strigolactone. Cytokinins are also known to help maintain the plant cell and expression of cytokinin biosynthesis genes late in the development which prevent leaf senescence. The inability of the cell to perceive cytokinin may cause it to undergo apoptosis or senescence. The mutants who cannot perceive ethylene, delay the process. The genome-wide comparison of mRNAs expressed during dark-induced senescence v/s those expressed during the age-related developmental senescence account for the fact that jasmonic acid and ethylene are more important for stress-related (Dark-induced) senescence, while salicylic acid is important for the developmental senescence.

The experimental approach most frequently used to study the effects of growth substances on leaf senescence has been to place excised leaves with their cut petioles in solutions containing growth substances, or to float excised leaf disks on the appropriate solutions and observe the rate of senescence, usually in terms of loss of chlorophyll and protein. The results obtained varied greatly between species. In excised leaves of *Xanthium* and many other species, cytokinins delay senescence. Indole acetic acid is needed to delay senescence of leaves of *Prunus* species. Gibberellins will delay the senescence of excised leaves of *Taraxacum officinale* but cytokinins are without effect. In *Rumex* species and *Tropaeolum majus*, cytokinins or gibberellins will delay the senescence of excised leaves over similar ranges of concentrations. The reason for this diversity of responses is not known. In the case of *Tropaeolum* it was found that the gibberellins content of the leaves decreased resulting in their senescence.

Exogenously applied abscisic acid (ABA) accelerates senescence in leaves of many species. Treatment of leaves with ethylene will promote senescence. Auxin treatment of many plant tissues leads to an enhanced production of ethylene.

Theory of Hormonal Induction of Senescence

Though reasonably widely accepted that senescence is done hormonally, the botanists concentrate on ethylene and abscissic acid but neglect gibberellins and brassinosteroids which inhibit the root growth, because of the fact that the roots grow below the ground and harder to study. The important theories are as follows:

1. Hormonal Shoot Pruning Theory

This is a new simple theory according to which though ethylene is responsible for the final act of leaf shedding, ABA (abscissic acid) and strigolactones that induce the process in the leaves due to a runaway positive feedback mechanism. What happens is that both of these are released by the mature leaves under water or mineral shortage, because these two push out the minerals, sugar, gases, and cytokinin (in addition to jasmonic acid and salicylic acid). This causes more amount of ABA and strigolactones to be produced until the leaf is drained off all its nutrients. When the conditions get bad and the nutrients are emptied, the mature leaf will experience deficiency of sugar and oxygen which may lead to gibberellins and ethylene emanation. Then it is time excise the leaf.

2. Hormonal Root Pruning Theory

According to this theory, ethylene, GA, BA and Eth are induced by sugar (GA/BA) and oxygen (Eth) shortages (and the excess levels of carbon dioxide for Eth) in the roots, to push the sugar, oxygen and minerals in addition to water and growth hormones out of the root cell resulting in a positive feedback loop that results in the emptying and death of the root cells. The final death for a root might be strigolactone (probably ABA) as these are the indicators of the substances that should be abundant in the root. Hence, senescence is the result.

3. Parallels to Division

According to this theory, both auxin and cytokinin are needed before a cell divides, in the same way, ethylene and GA/BA (as well as ABA and strtigolactone) are required before a cell senesces.

Senescence of the Seed

A major determinant of crop yield is seed germination. Associated with the accumulation of damage of DNA the quality of the seed with age deteriorates. DNA damage occurs with the loss of viability of the embryos in dry, aging seeds of rye. The dry seeds of *Vicia faba* are seen to accumulate the damage of DNA with time in storage and they undergo the repair of DNA upon germination. A DNA ligase is employed in the repair of DNA single- and double-strand breaks during seed germination in *Arabidopsis*, and this is an important determinant of seed longevity. The cellular repair response to the damage of DNA in eukaryotes is orchestrated by the DNA damage checkpoint- kinase ATM, and this has a major role in controlling the germination of the aged seeds by integrating progression through germination with the repair response to DNA damages accumulated during the dry quiescent.

Senescence in Plants

Like all other living organisms, plants do have a specific life span during which they develop, grow and finally die. But, before their death, certain deteriorative processes occur in them which result in the termination of their functional life. Programmed cell death is the death of the plant (or parts of the plant) as a consequence of senescence. This process is not confined to the entire plant but to a particular plant organ (like leaf or flower). The senescence patterns in the plant are of 4 types:

a. Overall Senescence

This type occurs in general in the annuals where the entire plant is affected and dies.

b. Top Senescence

This is normally seen in perennials only in the parts of the plant that lie above the ground.

c. Deciduous Senescence

This occurs in the woody plants and is less drastic as it affects all the leaves of the plant.

d. Progressive Senescence

This refers to the gradual progression of the senescence of the leaves from the base upwards as the plant grows.

Conclusion

Senescence is a complex process wherein the mechanisms should exist that sense the age of the cells, organs and the whole plant as well as the various environmental signals that are integrated, transduced and the appropriate pathways are to be selected and the degeneration process is to be executed. The central questions also remain as to how the plants balance the environmental signals and developmental signals? Senescence constitutes a part of the overall developmental programme, in which multiple, internal and external signals are integrated into information about the developmental age through intricate regulatory pathways. The integrative analyses allow an assessment of the dynamic changes that occur in physiological, biochemical and molecular phenotypes required for an understanding of the systems and their principles. Different datasets - such as genomics, epigenomics, transcriptomics, proteomics, metabolomics, and phenomics are the integrated for systems biology approaches which provide more accurate picture of the networks that underlie senescence. One of the challenges to address such issues would be the establishment of a phenomic facility and integration of this data with the genomic, proteomic and metabolomic data to interlink the genetic and environmental inputs to plant growth and development with an understanding of the controlling mechanisms from birth to death.

ABSCISSION

Theophrastus, a Greek Philosopher, and a student of Aristotle, described many years ago the leaf abscission habit in higher plants, followed by a great deal of research in the area of abscission, and as on today, the manipulation of abscission is a common agricultural practice. The process may be defined as the separation of a part of the plant (like a leaf, a flower, a fruit, a seed, or a branch) from the parent. This occurs only in higher plants and is also very common in lower plants. The process is highly variable within the species and variables of a species.

The word Abscission comes from Latin, *ab* meaning "away" and *sciendere* meaning "to cut", refers to the shedding of various parts in a plant such as dropping of a branch, a leaf, bud scales, a fruit, a flower or a seed. It also refers to the separation of two daughter cells at the completion of cytokinesis. The process is a tightly controlled cellular progression which occurs at given developmental stages during the life span of a plant which also ensures that plants can shed their organs when they are no longer required (like the flowers after pollination and senescent leaves in the fall) – and also as a way of getting rid of the infected and the damaged organs. The process is also important for the reproductive success of the plants ensuring the seed and fruit dispersal.

The more common example involves the separation of the cells within specialized tissues in an AZ (abscission zone). The separation layer must be alive and must have the ability to generate the hydrolytic enzymes to help in abscission. The process can occur either slowly or very rapidly, and in some cases, the external forces like wind are necessary to complete abscission.

During the early stages of the development of the organ, many plants obtain one or several layers of morphologically distinct cells at the base of their organs, where later, a break occurs along a fracture plane called AZ (abscission zone) in response to the environmental and endogenous cues. The AZ can be distinguished by the morphology of the cells that comprise it, and they are small with dense cytoplasm interconnected by plasmodesmata. They generally differentiate simultaneously with the development of the organs formed from the shoot meristem (apical meristem). The secondary AZs can also develop in mature organs in a position where a zone would not normally form as a response to the injury of an organ.

The more conspicuous examples of abscission such as the shedding of fruits, leaves, bud scales, floral structures and branchlets will be familiar, but it is important to realize that virtually any aerial part of a plant can be shed in this way. Structures may range in size from the complete shoot system of tumble weeds down to the hairs shed from the developing leaves. Fracture is caused by wall breakdown in 'separation layers'. Plants do not have the ability to produce separation layers anywhere; they seem generally limited to specific locations called abscission zones and are characteristic of a given species.

The active separation of a leaf from a branch, without injury to the branch, is called leaf abscission. Leaves of gymnosperms and woody dicotyledons are usually shed as a result of changes that take place in the tissues of the leaf base prior to leaf death. In the base of mature deciduous leaves, a narrow zone, the so called abscission zone or region can be seen. The vascular system in this region is usually concentrated in the centre and the sclerenchyma and collenchymas are less well developed or absent. Two layers are discernible in the abscission zone, an abscission or separation layer, through which the detachment occurs and the protective layer, which at the leaf fall protects from desiccation and invasion by parasites to the exposed surface. In most leaves, flowers, fruits and some stems, the preparation of the abscission zone occurs during ontogeny, a short time before shedding of the organ in response to conditions causing abscission.

Abscission - Types

Abscission is mainly of 3 types in plants: Leaf abscission, Branch abscission and Flower abscission.

1. Leaf Abscission

The leaves which live from weeks to years earlier to senescence and abscission exhibit 3 patterns of abscission. 1. The seasonal abscission of leaves occurs in many deciduous trees and shrubs that are grown in the temperate zones within a very short period of time in response to different environmental factors like shortening photoperiod as well as cold. 2. In tropical and sub-tropical regions, many deciduous trees and shrubs shed their leaves in response to the changes in the growth and vigour of the plant. This is because, the stress due to drought conditions will promote the abscission, and when once the plant receives water, the growth of the leaves is promoted. Such a cycle of abscission and the formation of the leaves can occur many times over the year. 3. In the sub-tropical and the temperate regions, the vernal leaf abscission of the previous season's growth occurs during the new flush of growth in the spring in evergreen trees.

2. Branch Abscission

This type of abscission functions in plants as a type of self-pruning habit. The abscission of the branches occurring rapidly in some species suggests that the hydrolysing enzymes were produced to efficiently promote the process. There are special cases, where the partial separation occur which suggests that the hydrolytic process was not complete, and other factors like wind and other mechanical forces too assist in abscission.

3. Flower Abscission

In higher plants, the abscission of the flowers, fruits and the seeds commonly occurs to facilitate the process of reproduction. When some fruits are set and begin to develop, the excess flowers are abscised to avoid the production of more fruits which can develop to maturity. In some other species, a large number of young fruits are produced, and in that case, abscission of the excess fruits occurs as a safety mechanism. The seed dispersal is extremely important for any plant to survive. Without the separation of the seeds from the placenta and the dehiscence (a type of abscission), the dispersal of the seed fails to occur, as such the promotion or inhibition of abscission is important in agriculture.

Abscission - the Process

The process occurs in a series of 3 important events: 1. Resorption, 2. Formation of the protective layer and 3. Detachment. The last 2 steps may occur in either order depending on the species.

1. Resorption

This refers to the degradation of chlorophyll to extract majority of its nutrients, and nitrogen is found in chlorophyll and is often the limiting nutrient for the plants because the plants need large quantities to form the amino acids, nucleic acids, proteins and some plant hormones. After extraction, the nutrients will travel to other tissues of the plant. This causes the leaves to change their colours in the fall. The carotenoids are slower to degrade than the chlorophyll and hence, the autumn leaves appear yellow and orange.

2. The Formation of the Protective Layer

The cells under the abscission zone divide to form a layer of cork cells, and situated on both the sides of the zone are layers of parenchyma, which produce and inject suberin and lignin under the zone into the new layers of cork. These two substances help create durable and water-proof layer for the plant when the organ gets detached.

3. Detachment

This occurs in a variety of ways depending on the species but always at the abscission zone. This occurs when the parenchyma layers secrete cell wall enzymes to auto-digest the middle lamella, which holds the cell walls together at the abscission zone. This causes the cells of this zone to break apart and the leaf or any other plant part to fall off. A different way of detachment occurs through imbibition of water, when the cells at the abscission zone take in a large amount of water, swell and finally burst, making the organ fall off. Thus, the protective layer of cork gets exposed.

Abscission - Mechanisms

The mechanisms of abscission may be structural, lack of chlorophyll as a trigger, chemical, and hormonal.

Structural Mechanism

The abscission zone also called as a separation zone in deciduous trees is formed at the base of the petiole, and it is composed of a top layer that has cells with weak walls, and a bottom layer which expands in the autumn season, breaking the weak walls in the top cell layer which allows the leaves to be shed.

Lack of Chlorophyll as a Trigger

The reduction of the production of chlorophyll in the leaves due to decreased sun light in the autumn season explains why some leaves turn yellow to orange. This yellow colour can attract the aphids as a result of which some trees turn their leaves red by injecting a bright pigment. The loss of chlorophyll may also contribute to the process of abscission.

Chemical Mechanism

During the times of stress (either biotic or abiotic) including UV light, cool temperature, excessive light, pathogens, parasites and high salinity, a variety of ROS (reactive oxygen species) are generated by the plants. The presence and continuous production of these species causes disruption of in the homeostasis of the cellular components which leads to metabolic dysfunction and expression of the WDEs (cell wall degrading enzymes).

Hormonal Mechanism

Originally it was believed the abscissic acid to be the hormone that simulates abscission, it was later proved that it does not play a primary role. But, auxin (a plant hormone) and ethylene (a gaseous hormone) have been implicated as important regulators of abscission signalling. These two work in a synergistic fashion – and as the level of auxin decreases, the flux of the auxin to the zone of abscission is reduced, and exhaustion of auxin makes the zone sensitive to ethylene. The plants are then exposed to ethylene, the gene expression of cell wall-degrading enzymes (cellulose and polygalacturonase) are activated. The decreasing levels of the auxin are also capable of influencing the colour change in the autumn leaves.

Abscission – Anatomy

Hugo Von Mohl (1860) was the first to describe the 2 important parts of the process of abscission – separation and protection (wound healing), and both these occur simultaneously during leaf abscission, although there are cases where they occur separately. Several histological processes were observed during abscission, not all of them in the same species. Cell division may precede the actual separation, the new cell walls being specifically affected by the subsequent degeneration phenomena.

In the abscission zone lignification of parenchyma cells, tylose formation in the tracheary elements, and callose deposition in sieve elements was also observed. The final event in abscission involves the enzymatic degradation of cell walls. This starts with the removal of calcium and pectin from the middle lamellae and continues with the hydrolysis of the cellulose walls. At the end, the sclerified tracheary elements rupture. The middle lamellae between the cells of the vascular tissue also disintegrate. Xylem vessels and interstitial parenchyma of vascular bundles may also become weakened by hydrolysis of cell wall material prior to separation. Golgi vesicles may be involved in secretion of abscission-specific enzymes. In most monocotyledons and herbaceous dicotyledons no dissolution takes place and abscission is apparently affected by physical stresses.

The protective layer, or cicatrice, is formed through deposition of protective substances such as suberin and wound gum, in the cell walls and the intercellular spaces. In woody species the protective layer is sooner or later replaced by periderm that develops beneath the protective layer in continuity with the periderm elsewhere in the stem.

It has been shown that auxin applied to leaf blades inhibits the abscission of the leaf. Ethylene, on the other hand, is regarded as responsible for the induction of the enzymes active cell wall degradation. Peroxidase brings about an increased synthesis of ethylene, and phosphatase is associated with the aging phenomena.

Genetic Regulation of AZ Differentiation

The studies on the mutants with defects in abscission shed light on the process of limited regulation of AZ differentiation. In a fruit market, you can observe the clusters of tomatoes that stick firmly to their stems, and these refer to joint-less mutants. JOINTLESS is a MADS box transcription factor (TF) is essential for the formation of an AZ in tomato stalks. In the model system of *Arabidopsis thaliana*, the proteins responsible for AZ differentiation are not MADS box TFs but belong to NONEXPRESSOR OF PR GENES1 (NPR1), which is a key regulator of the systemic acquired resistance response in plants, that ensures the induction of defence genes upon the accumulation of salicylic acid. Two NPR1-like genes, BLADE-ON-PETIOLE (BOP1) and BOP2 acquired a different regulatory function than that of plant-pathogenic interaction and play a role in plant morphogenesis and development. Under normal conditions, *Arabidopsis* only displays floral organ and seed abscission, the 2 proteins – BOP1 and BOP2 function redundantly to control AZ cell differentiation, not only at the base of the floral organs, but also at the base of the cauline leaves and the pedicel of the siliqua; where abscission in vestigial AZs can be activated in response to abiotic and biotic stresses.

Function of Abscission

The leaves of most plants as they cannot withstand the cold weather face death during the winter. If the leaves did not prepare for the onset of cold weather by senescence, the first freeze would kill the leaves before the materials within them are salvaged. A plant will cut off a part to discard a member that is no longer necessary (leaves during autumn season, a flower following fertilization, or for reproduction). Majority of the deciduous plants drop their leaves by abscission before winter, while the evergreen plants do it continuously. Similarly, the plants abscise the fruits while they are still immature in order to conserve the resources necessary for the remaining fruits to come to maturity. A plant may abscise the damaged leaves to conserve water or photosynthetic efficiency. The parts that fall off are greenish-gray in their colour. This can also occur as a means of plant defense in premature leaves, in response to the infestation by the gall aphids. The abscission is selective, and the chance of dropping leaves increases as the number of galls increase. Without this process to remove the dead leaves, the senescent tissues would shade the new spring growth that appears the following season. Hence, both senescence and abscission provide a means by which the perennials can recycle a major portion of the leaf material as the plants prepare for both the cold weather and the following growing season.

Chapter 20

Anatomy in Relation to Taxonomy

The anatomical characters of both the vegetative and floral pars of the flowering plants have been successfully employed to solve the taxonomical problems and also for the elucidation of the phylogenetic relationships. The very first attempt to use the characters of anatomy in plant classification was made by Bureau (1864) for the delimitation of the taxa of various levels within the family Bignoniaceae. The anatomical data have been used extensively as a taxonomical tool only after the 19th century. This data has not only been useful at the higher levels but in certain instances, have been successfully employed even at the specific level. One of the pioneer taxonomists was Auguste Mathiew, who used the features of wood anatomy in the description of the forest plants in his *Florae Forestiere*. Later, Solereder (1899) in his classic book, *Systematische Anatomic der Dicotyledonae* discussed the systematic value of the anatomical structures in dicots and emphasized the importance of xylem structure in plant classification, the English translation of which was later published in a modified form in the 2-volume book, *Anatomy of the Dicotyledons* by Metcalfe and Chalk. Bailey (1951) and Metcalf and Chalk (1950) have listed numerous anatomical features that can be made use in taxonomy. The *Anatomy of Dicotyledons* (Metcalfe and Chalk, 1950) and *Anatomy of Monocotyledons* (Metcalfe, 1960 onwards) are some of the books in this direction.

Anatomical Evidences in Systematics

The evidences which can be useful in taxonomy in several ways:

1. A piece of wood, can be well exploited taxonomically in the identification of the fragmentary material.
2. The anatomical study may prove useful when the characters of morphology are of no help in the primary identification of the herbarium material.

3 The anatomical data is very useful in discerning the evolutionary trends and the interrelationships of the taxa at and above the level of the species as well as at higher taxonomic categories. These are useful in determining the relationship between the different genera, families, orders and other categories.

While studying this data, it is advised to study the ranges of variability of the characters within the same individual and also between different individuals of the same species, and not rely on the data from a single sample of an organ or tissue, as the similarities in the structural organization may not necessarily reflect the close relationship but may be the result of parallel or/and convergent evolution.

Vegetative Anatomy

a. Leaf Anatomy

This provides various characters of taxonomic importance as was stated by Carlquist – that the leaf is anatomically the most varied organ of angiosperms and its anatomical variations often concur closely with the generic and specific and family lines. The anatomy of the leaf has been used widely in several taxonomically different groups like – Euphorbiaceae, Cyperaceae, and Poaceae of Angiosperms and Coniferae of Gymnosperms.

The leaf anatomy has been one of the most reliable characters in the systematics of grasses. Koyama and Govindarajulu studied the leaf anatomy of several species of Cyperaceae and formulated the keys to identify various species of *Cyperus*. Similarly, Brown surveyed 72 genera of grasses on the basis of their tissue arrangement to recognize 6 main types. Vidakovie used several characters of leaf anatomy to differentiate the species of *Pinus*. The taxonomic implication of leaf anatomy have also been established by several workers for several genera of Musaceae, Zingiberaceae, Xanthorrhoeaceae and Ericaceae.

The scope of leaf anatomy was to support the generic delimitation by contributing to the identification of the species and constructing a dichotomous identification key that includes the anatomical characters. The taxonomic characters which were considered to be the most important and used in the identification for the studies in the species of Malphigiaceae were – the presence and location of the glands, presence of phloem in the medullary region of the mid rib, the type of the mesophyll, the presence and the type of the trichomes, and the presence, quantity and disposition of the accessory bundles in the petiole. It was also possible to indicate the promising characters for the future taxonomic and phylogenetic studies in Malphigiaceae for certain genera (such as: *Banisteriopsis, Byrsonima,* and *Heteropterys*).

Important Characters of Taxonomic Significance in Leaf Anatomy

A. Epidermis – Nature and Thickness

The size and shape of the epidermal cells is of great value in the taxonomy of several taxa, and characters of the cuticle of the epidermis and stomata have also proved to be of great value. Conde studied 5 species of the genus *Opuntia* with reference to thickness of the cuticle, papillosity of the epidermis, size and frequency of stomata, thickness of the hypodermis, number of vessels *etc.*, to find the each species was distinct with reference to degree of papillosity of the cells of epidermis, thickness of hypodermis and width of the vessels. In addition, the information from the anatomy of trichome has also proved to be useful in certain taxa like *Veronica* species.

B. Structure and Types of Mesophyll, Storage Parenchyma, Mid Vein Structure, Bundle Sheath, Secretory Apparatus *etc.*

Anderson and Crech suggested precise grouping of *Solidago* and other species of Asteraceae on their study of leaf anatomy, including both qualitative and quantitative differences in the mesophyll, storage parenchyma, bundle sheath extension and the structure of mid vein.

C. Pattern of Sclerenchyma

In *Carex* and *Festuca*, the patterns of distribution of sclerenchyma have been used in distinguishing of species.

D. Silica Bodies and Chloroplast

The silica bodies in the epidermal cells of Zingiberaceae, Musaceae and Palmae among the monocotyledons, and Rosaceae in dicotyledons have been used as diagnostic character in the systematic at generic as well as specific levels. Similarly, the structure of the chloroplast also proved to be of taxonomic significance.

Epidermal Characters - Stomata

Leaf characters especially epidermal characters and stomatal characters are very much useful in taxonomy. The importance of epidermal characters was emphasized by Stace (1965) while the micromorphology (study of morphological characters with scanning electron microscope) has been emphasized by Dehgan (1980) and Barthlott (1981).

Van Cotthem (1970) divided the stomata into 8 types. They are:

Anomocytic (Ranunculaceous type)

(Anomocytic means irregular celled). Stoma surrounded by a limited number of subsidiary cells that are indistinguishable in size, shape or form to those of remainder of epidermis (Figure 20.1a). *e.g.*, Ranunculaceae, Malvaceae, Papaveraceae, *Tridax procumbens, Boerhavia diffusa, Cleome* spp.

Anisocytic (Cruciferous Type)

(Anisocytic means unequal celled). Stoma remains surrounded by three accessory or subsidiary cells of which one is distinctly smaller than the other two (Figure 20.1 b), *e.g.*, Cruciferae, Solanaceae.

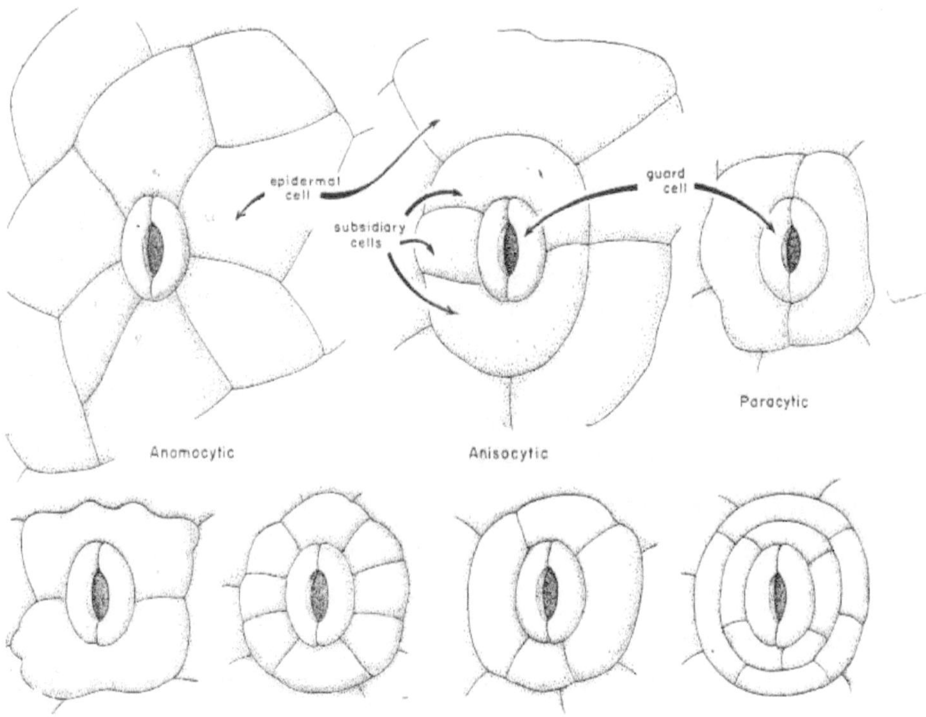

Figure 20.1: Types of Stomata. a: Anomocytic; b: Anisocytic; c: Paracytic; d: Diacytic; e: Actinocytic; f: Tetracytic; g: Cyclocytic.

Paracytic (Rubiaceous type)

(Paracytic means parallel celled) Stoma remains surrounded by two subsidiary cells on either side which are parallel to the long axis of the pore and guard cells (Figure 20.1c). *e.g.*, Rubiaceae, Magnoliaceae.

Diacytic (Caryophyllaceous type)

(Diacytic means cross celled). Stoma remains surrounded by a pair of subsidiary or accessory cells whose common wall is at right angles to the guard cells (Figure 20.1d), *e.g.*, Caryophyllaceae, Acanthaceae.

Tetracytic

Four subsidiary cells are present, two lateral and terminal (Figure 20.1f), *e.g.*, Most families of Monocotyledons.

Actinocytic

Stoma surrounded by a circle of radially elongate subsidiary cells (Figure 20.1e), *e.g.*, Commelinaceae, Musaceae.

Cyclocytic

Stoma surrounded by four or more subsidiary cells which form a ring around each stomata (Figure 20.1g), *e.g.*, Tribe Laguncularieae of the family Combretaceae.

Hexacytic

Stoma accompanied by six subsidiary cells consisting of two lateral pairs parallel to the long axis of the pore and two polar (terminal) cells; the second lateral pair as long as the stomatal complex.

Besides the above, several other characters, the presence of stomata on the upper side (epistomatic) or the lower side (hypostomatic) or on both the sides (amphistomatic) of the leaf blade, can be used and are useful in taxonomy. For example the presence or absence of stomata on the upper surface of a leaf is a good diagnostic feature in the separation of *Alnus subcordata* and *A. orientalis* of Betulaceae, which are otherwise difficult to distinguish. In the former, epistomatic condition prevails while in the latter, the hypostomatic. Similarly, shape of the guard cells is also useful in plant classification. For example, stoma with bean- shaped guard cells is met with in Dicotyledons while stoma with dumbbell-shaped guard cells is seen in Poaceae. In Cyperaceae, rectangular guard cells are seen.

Development of stomata has been divided by Pant (1965) into three types.

1. Perigenous

Stoma in which the subsidiary cells do not have a common origin with the guard cells, but are formed by cells lying around the meristemoid that divides to form the guard cells, are referred to as perigenous.

2. Mesogenous

Stoma in which the subsidiary cells are generated from the same meristemoid (initial) of the guard cells are called mesogenous.

3. Mesoperigenous

Stoma in which at least one of the subsidiary cells has a common origin with the guard cells, but others do not, are referred to as mesoperigenous.

Development of stomata is useful in plant classification. Ramayya (1977) has emphasized the importance of stomatal characters in taxonomy. Ramayya and Rajagopal (1968, 1971) gave a key to genera of Portulacaceae and Aizoaceae based on epidermal characters.

In the family Acanthaceae, the stomata are diacytic, whereas in the closely related Scrophulariaceae, the stomata are anomocytic. In the family Combretaceae, the stomata are paracytic in the subfamily Strephonematoideae and anomocytic in the subfamily Combretoideae except the tribe Lagunculaireae where they are cyclocytic. Stomatal characters are not, however always so reliable and the different types have obviously each originated on many different occasions and they often exist together on one plant. In the genera *Streptocarpus* and *Saintpaulia* of the Gesneriaceae; for example, the cotyledons bear anomocytic stomata and the mature organs anisocytic stomata (Saharsrabudhe and Stace, 1974). In *Lippia nodiflora* of Verbenaceae, anomocytic, anisocytic, diacytic and paracytic stomata occur on the same leaf (Pant and Kidwai, 1964). In *Alternanathera pungens* of family Amaranthaceae also all the above four types of stomata are found.

Inamdar (1968) who studied the stomata of *Nyctanthus* expressed the opinion that this genus should be kept in Oleaceae. In *Nyctanthus* and Oleaceae the stomata are Tetra-Mesoperigenous (sometimes aperigenous), paracytic while Verbenaceae shows Dia-Mesogenous stomata. So, Inamdar (1968) said the *Nyctanthus* should be kept in Oleaceae instead of Verbenaceae.

Bremekamp (1953) said that the genus *Elytraria* should be removed from Acanthaceae and should be included in Scrophulariaceae. Paliwal (1966) studied the phylogeny of stomata to ascertain this. He found that in *Elytraria* and Acanthaceae, the stomata are diacytic and development is Syndetocheilic, but in Scrophulariaceae, stomata are anomocytic and development is Haplocheilic. Hence, Paliwal (1966) retained the genus *Elytraria* in the family Acanthaceae.

Trichomes

Trichomes are epidermal outgrowths. They exhibit great variation in structure and function and thus are of much taxonomic value. Glandular and non-glandular hairs, papillae, setae and prickles are all considered as hairs. These trichomes are useful in plant classification and understanding the relationship between groups. Schulz (1936) used the type of hair as the major criterion in the subdivision of the family Cruciferae into tribes and genera. Ramayya (1962, 1969) studied the trichomes of Compositae and gave key to the genera based on trichome characters. Based on trichome characters he considered Heliantheae as the primitive tribe. Rollins (1944, 1945, 1946, 1949) studied the trichomes of *Parthenium argentatum* and *P. incanum* and their natural hybrids. He has shown that in *P. argentatum* the trichomes are T-shaped while in *P.incanum* they are whip-like with a long thread. Intermediate types of trichomes are observed in the hybrids between these species. In the family Combretaceae, Stace (1965, 1969a, b, 1973) found that trichome anatomy is of immense significance in classification at all levels, from the circumscription of the family down to the separation of species and even varieties. In particular, it has led to an improved tribal classification within the family and an improved subgeneric and sectional classification within the largest genus *Combretum*.

Epidermal Cells

The epidermal cells, other than those modified by their relationship to trichomes, stomata, the venation system or other special structures, provide many characters of taxonomic value. These include their size, shape, orientation, anticlinal wall undulation and periclinal wall curvature. The epidermal cell modification occurs also on the leaf margin and when silicified, calcified, suberized, gelatinized or containing various types of crystals. Silicified and suberized epidermal cells are of major taxonomic importance in the Poaceae.

Other Epidermal features are also useful in plant classification. The genus *Morus* is divided into species based on Idioblasts in epidermis. Tomlinson (1956, 1959a, b) and Metcalfe (1968) said that silica bodies in the leaves are useful in classification of the families Zingiberaceae, Musaceae, Arecaceae and Rosaceae up to genera and even species.

Internal Structure

Carlquist (1961) said that leaves in Angiosperms show great diversity and this character is useful in classification. The nature of epidermis, its thickness, type of mesophyll, sclerenchyma patterns, sclereids, crystals, venation patterns, *etc.*, may be useful in classification. The importance of foliar anatomy in the systematics of grasses was realized quite early by Duval Jouve (1875). According to Avdulov (1931), Poaceae have two types of internal structure in leaves. In the Festucoid (Pooideaea) grasses the mestome around the stele is layered and mesophyll cells are straight while in Panicoid grasses mestome may or may not be present and the mesophyll cells are arranged radially.

Similarly, Ayensu (1974) found that sclerenchyma patterns in the Velloziaceae can be used to separate the genera *Vellozia* and *Barbasenia*. The genera in Cyperaceae can be segregated based on internal structure of the leaves (Govindarajulu, 1969).

Crystals, especially calcium oxalate crystals, are found to be useful in the classification of Angiosperms. Foliar sclereids have, of late, been demonstrated to be of taxonomic importance in many taxa.

Laticifers are cells or a series of fused cells containing latex and are common features of many succulent plants and other plants of arid regions. They vary widely in their structure and latex in their composition. Their presence or absence and when present, their structure has been of some taxonomic value. In Aroideae (De Bary, 1884), for example, certain species lack laticifers or any related structures while others have longitudinal rows of elongated, cylindrical sac-like cells. Some authors consider that Fumariaceae, which lack laticifers (Sperlich, 1939), should be considered distinct from the related family Papaveraceae having a well developed latex system. In the former family, laticifers are replaced by idioblasts. It is one of reasons for considering these two families as distinct from the rest of the families in Engler's Rhoedales.

The taxonomic usefulness of petiole vascularization patterns has become increasingly apparent in recent years (Radford *et al.,* 1974). Sinnot (1914) gave a classification of nodal types in megaphyllous plants. In general there are three major types of node — unilacunar, trilacunar and multilacunar (Figure 12.2). Trilacunar nodes occur in the majority of Dicotyledons, mutilacunar nodes in the primitive orders such as Magnoliales, Piperales, Trichodendrales and advanced orders such as Umbellales and Asterales and unilacunar nodes in Laurales, Caryophyllales,

Ericales, Diapensiales, Ebenales, Primulales, Myrtales and a majority of families of Asteridae (Paliwal and Anand, 1978).

Stem anatomy has been studied in the herbaceous members where it has proved to be of some diagnostic value. In the stem of *Ranunculus repens* there is relatively little strengthening tissue and is correlated with the creeping habit of the plant, while in the closely related, species, *R. acris*, interfascicular tissue is sclerified and therefore an erect habit is possible.

Stace (1970) has shown that in the subgenus *Genuini* of *Juncus*, anatomical characters of stem and foliar bract can be used to distinguish most British species. Further, it is also possible to identify parents of several hybrids on anatomical grounds.

There is considerable variety in the anatomy of stems of *Dioscorea* and can be successfully employed in the delimitation of species that are not easily separable on exomorphic grounds. Ayensu (1970) found anatomy useful in separating *Dioscorea rotunda* and *D. cayennensis*, which are closely related and difficult to distinguish.

The classical case of anatomy in relation to taxonomy is the distribution of vascular bundles. In Dicotyledons vascular bundles are arranged in a ring while in Monocotyledons vascular bundles are scattered.

Bentham and Hooker (1862-83) included Basellaceae as a subfamily in Chenopodiaceae. Anatomically, Basellaceae are characterized by a single ring of vascular strands with intraxylary phloem in the larger bundles. On the other hand, Chenopodiaceae are characterised by the development of polycyclic, anomalous secondary vascular bundles.

Bentham and Hooker (1873-1876) split the Gentianaceae into two sub- families — Menyanthoideae and Gentianoideae. But, Hutchinson (1969), Cronquist (1981) and Takhtajan (1980) raised them to independent families. Vascular bundles in Menyanthaceae are collateral while Gentianaceae show bicollateral vascular bundles.

The most important anatomical features that have been used in taxonomy and phylogeny are those of the secondary wood. This is because of their conservative nature. Wood anatomy can be used at all taxonomic levels. At and above the family level it has been successfully used along with other lines of evidence to establish the systematic position of the primitive vesselless Angiospermic families Winteraceae (all seven genera-*Drimys, Pseudowintera, Bubbia, Belliolum, Exospermum, Zygogynum* and *Tetrathalamus*), Trochodendraceae, Tetracentraceae, Amborellaceae and Chloranthaceae. Similarly it has been agreed by all phylogenists that the Englerian group of primitive Angiosperms, namely Amentiferae cannot be considered primitive since they have specialized wood (Tippo, 1938). Anatomical evidence also supports the case for considering the remarkable genera *Paeonia* and *Austrobaileya* as the type of independent families (Bailey and Swamy, 1949).

At and below the generic level, the most useful contribution of wood anatomy is to provide evidence for placing of taxa of uncertain affinity. In *Quercus* (Fagaceae), different subgenera and sections can be delimited with the help of anatomical evidence. The placement of the genus *Myristica* and the related genera close to Lauraceae is supported by the evidence of wood structure (Eames, 1961), while on the same basis Myristicaceae are considered unrelated to Annonaceae and Eupomaticeae. Similarly, placing of Calycanthaceae in or near Rosales or Myrtales is not supported by wood structure.

Sieve cells also provide taxonomical evidence. For example, unspecialized sieve cells occur in lower vascular plants — Pteridophytes and Gymnosperms. Only one Angiosperm, *Austrobaileya scandens*, shows unspecialized sieve cells in phloem.

Stem Anatomy

This has also been long relied as a taxonomic tool and the 2-volume work by Metcalfe and Chalk reveals the taxonomical significance of the anatomical characters in plant classification to be used at various levels of Dicotyledon - Monocotyledon distinction to separate various species of the same genus. This has also proved to be of diagnostic value in the herbaceous plants, and for example, the stem anatomy has successfully employed in the delimitation of the species of *Dioscorea*. Carlquist has used the anatomical features of the genus *Fitchia* (family Asteraceae) in the classification of various species. The anatomical features that can be taken into account to be useful include: the degree of elevation of the stem ridge - the distribution and the abundance of collenchyma - the pattern of thickenings of collenchyma - the transformation of the cells of the ground tissue of the cortex into transfusion tissue (as in *Casuarina*) - and the distribution of the fibres (as in *Genista*) - variation in cell structure of the endodermis of the stem (in families like Piperaceae, Asteraceae and Lamiaceae) - stem pith features (as in the species of *Dubantia* and *Fitchia*) - the shape and the size of sclerenchyma girders (as in the subgenus *Genuini* of *Juncus*) - and the arrangement as well as the type of vascular bundles of 2 species of *Dioscorea* (such as *D. cayenensis* and *D. rotundata*).

Anatomy of the Petiole

Metcalfe and Chalk and Howard have suggested that the anatomy of the petiole might also be of taxonomic significance, and according to Howard, the families, the genera, and even the species may be identified. These characters include: the position of the petiole on the stem - the presence or the absence of the stipules (stipulate or exstipulate) - the outline of the petiole - the number of layers of cortical parenchyma - the type of vascularisation and the distribution of the perivascular fibres - and the number of traces. The anatomy of the petiole of 64 species of *Baphia* of Leguminosae have been studied by Soladave, and some species of *Phlomis* and *Eremostachys* of Lamiaceae by Azizian and Cutler, provide clear support of the use in the taxonomy of the genera.

Nodal Anatomy

This has gained much importance in the taxonomy and phylogeny (of angiosperms) in recent years, and the correlations with some other families might help significantly in tracing the phylogeny. A comparative study may also show the important relationships of genera or even species. In general, there are 3 major types of nodes (Figure 12.2): 1. *Unilacunar* – these occur in Laurales, Caryophyllales, Ericales, Ebenales, Primulales, Myrtales and some families of Tubiflorae and a majority of Asteridae. 2. *Trilacunar* – these occur in the majority of Dicotyledons. This is primitive according to Sinnott. 3. *Multilacunar* – these occur in the primitive orders like Magnoliales, Piperitales, Trochodendrales and a few advanced orders like Umbellales and Asterales. The unilacunar and multilacunar nodes are advanced according to Sinnott. 4. *Unilacunar Two trace* – This is the 4th type which was discovered by Marsden and Bailey, and is now considered as the basic type found in angiosperms. The mode of nodal vasculature is uniform in the family but exceptions have also been reported when different types of nodes occur in the same individual plant. The best example is the sub-family Icacinoideae of the family Icacinaceae which is divided into 2 distinct groups – one section characterized by trilacunar nodes while the other section is characterized by unilacunar nodes.

Wood Anatomy

This has been used at all taxonomic levels because of the conservative nature; anatomical features of the secondary wood have been useful in both taxonomy and phylogeny. This has been successfully used in deciding the systematic position of the primitive vessel-less families (like Amborellaceae, Tetracentraceae, Trochodendraceae, and Winteraceae) included under Magnoliales. It has also been agreed by all the phylogenists that the Englerian group of primitive angiosperms (like Amentiferae which include the Families like Salicaceae, Betulaceae, Fagaceae, Juglandaceae *etc.*) cannot be considered as primitive. Some of the important features of wood anatomy of taxonomical significance are: the Vessel elements, Axial parenchyma, Vascular rays, Storied wood, Sclereids, Cellular contents, *etc.*

a. Vessel Elements

These have been derived from the tracheids and their advancement (evolution) is considered to have occurred along the following lines: decrease in the length of the vessel – transition from angular outline to circular outline – loss of borders and decrease of bars on the perforation plates – alteration from oblique to nearly transverse angle of the end wall – and pitting of the lateral walls from scalariform to transitional to opposite to alternate (evolution). In addition to this, the abundance of the vessels, distribution of the vessels and sculpturing on the walls are also of taxonomic significance. The solitary vessels are primitive to aggregate, (like pore clusters, pore multiples and pore chains), and diffuse porous wood is primitive to ring porous wood.

b. *Axial Parenchyma*

The distribution and characteristics of the cells and the length and thickness and lignification on the walls are also useful in taxonomic considerations. The evolution of axial parenchyma occurred along the following lines: the absence this tissue is primitive at least in some cases (as in Winteraceae) – and the diffuse arrangement is primitive. On the basis of the distribution of cells, this is of 2 types – *Apotracheal* (parenchyma distributed without any specific relation to vessels) - and *Paratracheal* (parenchyma distributed in close association with the vessels).

c. *Vascular Rays*

The characteristics of these rays that can prove useful as taxonomic criteria include: the ray abundance – the cellular composition of the rays – dimensions of the rays in tangential section – the degree of wall thickness, and the pitting of the cells. The rays with the cells radially elongated (homogeneous rays are considered advanced to the heterogeneous rays), that is rays with both vertically and radially elongated cells.

d. *Storied Wood*

This type of wood refers to the planes of divisions of the cambial initials. The wood with non-stratified cells is primitive to storied wood.

e. *Other Characters*

Finally, the presence or the absence of the latex vessels, resins, gums, crystals *etc.* in the wood are also of taxonomic importance.

f. *Sclereids*

These are the cells with very thick lignified walls and widely distributed in the plant have been used as diagnostic tools in many taxa like Connaraceae, Nymphaeaceae, Oleaceae, Theaceae, and a few genera of Aracaeae, Acanthaceae, Ericaceae and Melastomaceae. In dicots, sclereids are more common in woody trees, but are extremely rare in monocots, except certain genera of Araceae, Agavaceae, Arecaceae and other families. As these exhibit different shapes, sizes and characteristics of their walls, these have been of some taxonomical significance. These are mainly of two types – isomorphic and polymorphic types, and are characteristic of a particular species and are of taxonomic value.

G. Cellular Contents

Many types of microscopic cell contents such as – chemical deposits can serve as diagnostic tools and sometimes prove useful in delineating species, genera as well as families. The crystals and crystalliferous cells are of systematic importance in several families of angiosperms like Euphorbiaceae, Leguminosae and Verbenaceae *etc.* Some of the important deposits of taxonomic significance are:

1. **Albuminoids** – in *Laportea* systematic significance.

2. **Starch Grains** - The diversity in the starch grains is of a good taxonomic character as in *Solanum tuberosum* (Potato).
3. **Protein Bodies** – The deposition of solid protein deposits have systemic use in some members of Cactaceae.
4. **Large Silica Bodies** – These occur in the epidermal cells of many families such as Arecaceae, Musaceae, Zingiberaceae and Rosaceae are of use at generic and specific levels.
5. **Calcium Oxalate Crystals** – These are widely distributed in plants and are of many types like prismatic, styloid and idioblasts. The distribution is specific for an individual taxon and of taxonomic importance (as in *Eichhornia* and *Allium*).
6. **Cystoliths (Calcium Carbonate Crystals)** – These occur in families like Cannabinaceae, Moraceae and Urticaceae.
7. **Tanniferous Cells** – The presence or the absence of the tanniferous cells in the cortex of the root of Rapataceae and Xyridaceae can be used as a systematic criterion in classification.
8. **Laticifers** – In addition to the above, the presence or the absence of laticifers (the cells or a series of cells that contain the latex and their structure) has also been of some taxonomic value, and these are a common feature of many succulents and plants of arid regions which vary widely in structure and the composition of the latex. Certain species of Aroideae lack laticifers while others have longitudinal rows of elongated, cylindrical, sac-like cells.

Floral Anatomy and Taxonomy

The reproductive organs show a high degree of conservation and as such they have been widely used in the classification. It is also likely that the vascular supply to the floral organs is conservative and is more reliable in taxonomic and phylogenetic interpretation. The role played by the floral anatomy in the solution of morphological problems has been emphasized by Puri. The floral anatomical characters of the characters of the families and genera are well marked and useful in solving the fundamental questions such as the nature of the flower, carpel, inferior ovary and several others related with homologies, phylogeny and taxonomy. The following are some of the examples in resolving the taxonomic position of some disputed taxa.

1. The confirmation of the origin of the families of Annonaceae, Calycanthaceae, Menispermaceae from Ranunculaceae.
2. The separation of *Paeonia* from Ranunculaceae and inclusion under a separate family Paeoniaceae.
3. The derivation of Polemoniaceae and Caryophyllaceae from Caryophyllaceous stock.

4. The inclusion of Solanaceae and Scrophulariaceae under Scrophulariales due to uniformity in floral vasculature.
5. The confirmation of the transfer of *Hydrocotyle asiatica* L. to the genus *Centella* as *C. asiatica* L.

The taxonomists have to look forward to workers in different disciplines of botany for the features and correlations that may be of interest to them in making the classification the epitome of our knowledge of plants. This has been our endeavour to indicate the way in which the studies in floral anatomy have helped the taxonomist in achieving his object.

The variations in the vascular plan of the flower are numerous and they have been classified under different headings, such as – reduction, amplification, cohesion and adnation. All these modifications helped the taxonomist by providing him the additional data in one way or another. Some specific examples from the families Ranunulaceae, Umbelliferae (Apiaceae), Polygalales, Gentianaceae, Papilionaceae, Rutaceae and Apocynaceae are given to show how the floral anatomy has been helpful in strengthening or sometimes rejecting his inferences. Besides, it is inferred that the floral anatomy may also be contributed valuable data in determining the correlations in larger groups like orders.

Floral anatomy has a wide application in determination of status of high ranking groups such as genera and species. On the basis of similarity of floral anatomical pattern, affinities of families Papaveraceae, Capparaceae, Brassiaceae and Magnoliaceae could be established. These families have a common parietal placentation and inverted placental bundles with regard to floral axis on inner side of secondary marginal strands with normal orientation. This fact also supports placement of Moringaceae in Rhoedales. Puri (1947, 1949) suggested affinities between Passifloraceae, Moringaceae and Cucurbitaceae on floral anatomical basis.

Burtt (1965) opined that *Cyrtandromoea* has no close affinity with other members of Gesneriaceae and advocated its transfer to Scrophulariaceae with which it exhibits more resemblance. Singh and Jain (1978) have supported this treatment on the basis of floral anatomy. Kale and Pai (1979) have supported the removal of the genus *Trichopus*, on the basis of floral anatomy, from the family Dioscoreaceae of Bentham and Hooker and erection of a unigeneric family Trichopodiaceae as has previously been suggested on anatomical evidence by Ayensu (1966, 1972). In conclusion, it is also pleaded that the floral anatomists should make their studies more comprehensive than what they hitherto have been. In addition, they should also realize that, like any other branch, floral anatomy also has certain limitations and ignoring them will be as harmful to the cause of floral anatomy as exaggerating them.

Fruit Wall and Seed Coat Anatomy

Fruit wall and seed-coat anatomy have also come to the help of taxonomists in many cases. The distinction between the two genera *Ammannia* and *Rotala*

(Lythraceae) has been puzzling taxonomists for quite some time. The diagnostic feature was in the fruit-irregularly rupturing fruits in *Ammannia* and valvular capsules in *Rotala*. But, this difference was very difficult to appreciate, especially in herbarium specimens. Van Leemjen (1971) found the fruit-wall character to distinguish the two taxa. The orientation of the cells of the two main layers of pericarp are different: in *Ammannia* the cells of both layers are very similar, whereas in *Rotala* the cells of the inner layer are linear, strongly lignified and transverse to those of the outer.

Two different views were expressed regarding the systematic position of the Asteraceous genus *Lagascea*. Cassini (1815), Lessing (1832), De Candolle (1836) and Merxmuller (1954) included the genus *Lagascea* in the tribe Vernonieae. Bentham (1873), Hoffmann (1894) and others place *Lagascea* in the tribe Heliantheae. Stuessy (1976) on the basis of morphological and chromosomal studies, supports the assignment of the genus *Lagascea* to Heliantheae. Pullaiah (1981) on the basis of pericarp structure favours retention of *Lagascea* in the tribe Heliantheae. In Heliantheae, three different zones are distinguished in the pericarp. Between the middle and outer zones characteristic schizogenous cavities develop which get filled with a brown resinous and hardening substance. Such schizogenous cavities are absent in Vernonieae.

Seed coat morphology has yielded valuable taxonomic information. Seed coat anatomy has also been used for systematic purposes.

Ultra-structure in Relation to Taxonomy

The Electron Microscopy (EM) has brought revolution in all the fields of biology including Systmatic botany. Heywood and Dakhshini have demonstrated the benefits of SEM (Scanning Electron Microscopy) and suggested that this device represents one of the most powerful taxonomic tools now available for research in taxonomy, in addition to TEM (Transmission Election Microscopy).

However, only a few ultrastructural characters are of wide application in plant classification.

'Dilated Cisternae' (DC)

These structures were of common occurrence in Brassicaceae and of the order of Capparales. DC in the endoplasmic reticulum was first noted and reported by Bonnet and Newcomb (1965) in the root cells of *Raphanus sativus* (radish) and later by Favali and Gerola (1968) in the phloem parenchyma of foliar veins in *Brassica chinensis* and in *Capparis cynaophallophora* (Jorgensen *et al.*, - 1977). Iverson and Flood (1970) found that such structures are common occurrence in the Brassicaceae. According to Behnke (1977) these organelles are typical of Brassicaceae and Capparaceae. The significance of ER in sieve elements has been focused by Spanner andf Moattari.

Sieve Element Plastids

The potential uses of ultra-structural features of sieve element plastids of systematic value were noted one decade ago. Behnke (1972, 1977, 1981) in a series of observations since about three decades ago, has demonstrated ultra-structural details of sieve element plastids, which can be used in plant taxonomy. Behnke studied more than 1500 species from 380 families with regard to this character. There are broadly 3 types of such plastid elements, *S – type, P-type,* and *S-0 type.*

1. S-type

These mainly accumulate starch and about 65 per cent of the flowering plants have these plastids in their sieve tube elements. These are present in sub-classes like Dilleniidae, Hamamelidae, Ranunculidae and majority of the orders in Rosidae, half of the members of Magnoliidae, a few orders of Caryphyllidae and Rhamnaceae.

2. P-type

The elements of this type commonly accumulate protein, and are differentiated on their number, shape of the crystalloids and the nature of the filaments around them. These are seen only in Pinaceae (in Gymnosperms) and in all the 21 families of Monocotyledons and a few groups of Dicotyledons (like Vitaceae, Leeaceae and half of Magnoliidae) in Angiosperms.

3. S-0 type

These are neither the elements which have starch nor the protein accumulated within (such as Crassulaceae, Rafflesiaceae, and some species of Moraceae, Ulmaceae and Urticaceae). According to Behnke, P-type is considered to be ancestral while the S-type is considered have to be derived from the P-type by loss of protein. The presence of P-type plastids in all groups of Monocotyledons, and S-type in the basal groups of Dictoyledons have been interpreted as an additional evidence for the independent origin of Monocots and Dicots.

P-type plastids are further differentiated on the number and shape of the crystalloids and the nature of the filaments surrounding them.

(i) **P I subtype:** The plastids contain single crystalloids of different sizes and shapes and/or irregularly arranged filaments. PI subtype sieve element plastids are present in Magnoliales, Laurales and Aristolochiales.

(ii) **P II subtype:** The plastids contain several cuneate crystalloids oriented towards the centre of the plastid. PII subtype sieve element plastids are present in Monocotyledons.

(iii) **PIII subtype:** The plastids contain a ring-shaped bundle of filaments. These plastids are seen only in Centrospermae

(iv) **P IV subtype:** The plastids contain a few polygonal crystalloids of various sizes. This type of plastids are seen only in Fabales.

(v) **PV subtype:** The plastids contain many crystalloids of different sizes and shapes. It is seen in the order Ericales and family Rhizophoraceae.

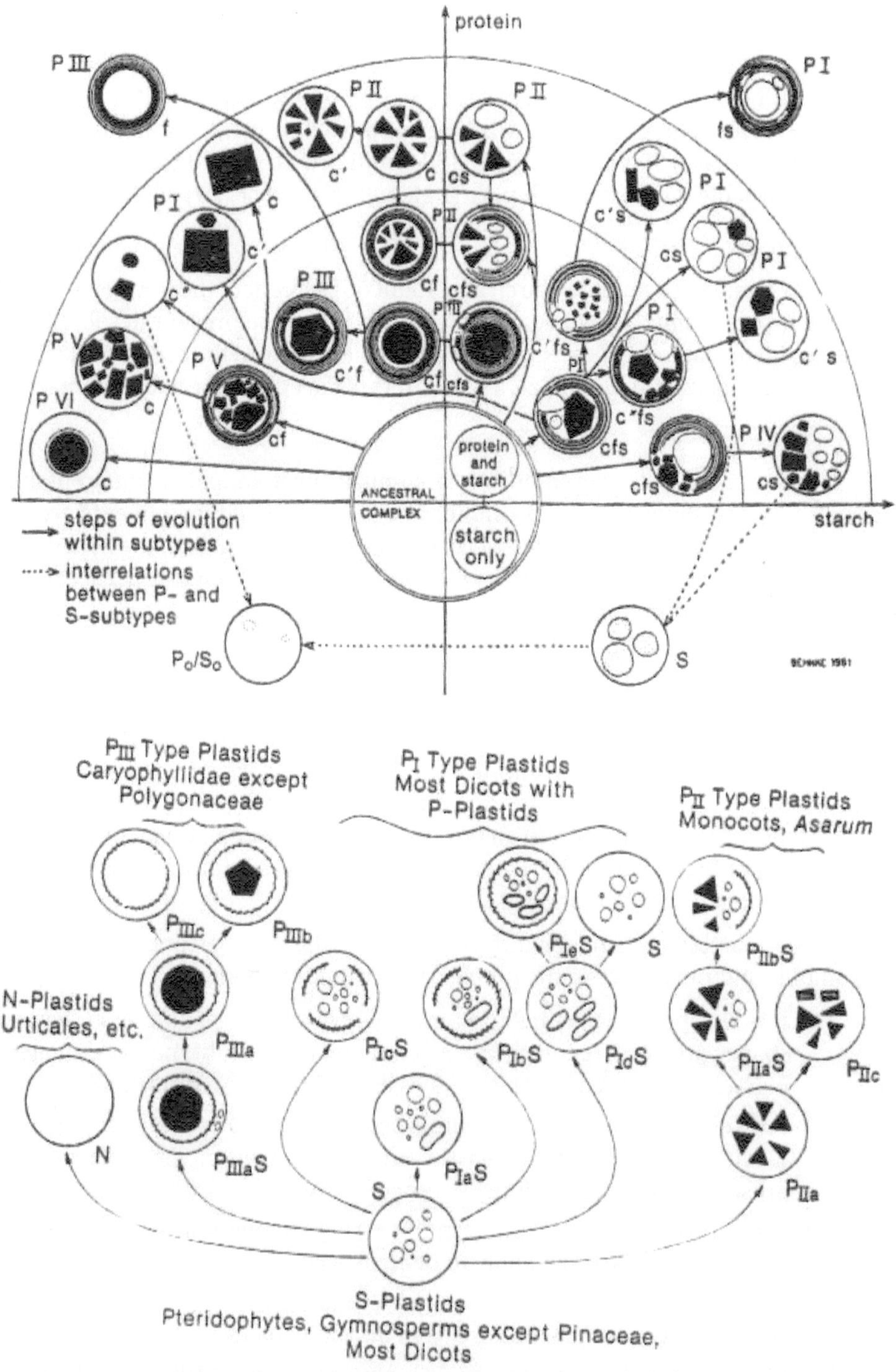

Figure 20.3: Sieve Element Plastids in Angiosperms.

(vi) **PVI subtype:** The plastid contains a single circular crystalloid. It is found in family Buxaceae.

Behnke (1974) investigated the order Rhamnales and found that the family Rhamanaceae had S-type while Vitaceae and Leeaceae had P-type sieve element plastids. The heterogeneity of the Magnoliidae as to its plastid types is attributed to its basal position in the evolution of flowering plants (Behnke, 1975). P-type are considered to be ancestral and S-type has been thought to be derived by loss of protein. P I subtype is held to be the most primitive among P-subtypes of flowering plants.

The best use of plastids has been made in elucidating the interrelationships and circumscribing the order Caryophyllales (Centrospermae) (Behnke, 1976). All the families kept in this order show P III subtype sieve element plastids. These families are : Aizoaceae, Nyctaginaceae, Phytolaccaceae, Portulacaceae, Caryophyllaceae and Molluginaceae. In earlier classification, other families like Polygonaceae, Vivianiaceae, Bataceae, Gyrostemonaceae, Plumbaginaceae, Fouquieriaceae, Frankeniaceae, Rhabdodendraceae and Theligoniaceae have been kept in Centrospermae but in recent classifications these have been removed from Centrospermae because they show S-type sieve element plastids.

Chapter 21

Programmed Cell Death (PCD)

Introduction

Though death may be compared in opposition to life, the death appears as life's un-dissociable companion. Hence, a sophisticated system for programming death is a common trait for higher organisms, at the level of the species and at the level of pluri-cellular individual. The limited life spans of the individuals are determined genetically and allowing the renewal of the generations. The somatic cells also continuously die to guarantee the survival of the individual. This process is referred to as PCD (Programmed Cell Death). It is believed that the cells activate an intrinsic death programme and actively contribute to their own death, and this is referred to as "apoptosis" (refers to an expression that defined as a particular morphological appearance of cell death currently used as a synonym for PCD). The term – Apoptosis – was first used in a classic paper by Kerr, Wyllie and Currie (1972) to describe a morphologically distinct form of cell death. This is defined as cellular suicide involving a specialized initiation and execution mechanism within the cell. The term literally denoted the drop of leaves from the tree in the fall season, reflecting the sporadic and gradual loss of cells in the tissues. Both these are opposing to necrosis – which is caused by physical, chemical or osmotic damage with disruption of the internal and external membranes.

PCD (Programmed Cell Death – sometimes referred to as Cellular Suicide) is the death of a cell as a result of the events that occur inside of a cell (such as Apoptosis or Autophagy). This is carried out in a biological process in which confers has an advantage during the life cycle of an organism (*e.g.* differentiation of fingers and toes in a developing human embryo because the cells between the fingers apoptose, as such the digits separate). PCD serves many fundamental functions during plant and animal development. Apoptosis and Autophagy are two forms of PCD, while Necrosis is the death of a cell caused by some external factors (like trauma or infection) and occurs in several different forms (a non-physiological process that occurs as a result of infection or an injury). In 2000s, necroptosis (a form of programmed necrosis) was recognized as an alternative form of PCD, which can serve as a cell-death backup to apoptosis when the signalling is blocked by an endogenous or an exogenous factor (like viruses or mutations). Other types of regulated necrosis have been discovered more recently that share several signalling events with necroptosis and apoptosis.

PCD involves the active participation of the endogenous cellular enzymes in the mediation of death before the membranes lose their integrity. Lytic necrosis entails the denatured proteins and DNA fragments into the intercellular space, and also the inflammatory responses. The cells involved in PCD exhibit near-to-intact external membranes entail the late phase of metabolic suicide, and allow the adjacent cells to engulf the dying cell and also to eliminate it before its contents are released, and before the inflammatory reactions are triggered. PCD is associated with different types of higher eukaryotes (all plants, slime-molds, nematodes, insects and vertebrates). The failure to undergo apoptosis during organogenesis causes malformations in *Coenorhabditis elgans* (a nematode) as well as some insects and mammals. The bio-chemical mechanisms involved in the execution of cell death are poorly understood. A second messenger system can either prevent or induce PCD and many different signal transduction pathways may participate in the induction phase of apoptosis.

Programmed Cell Death - History

The concept and the usage of PCD (Programmed cell death) is used by Lockshin (Lockshin and Williams, 1964) in relation to insect development 8 years before "apoptosis" was coined. This is a physiological process that leads to the selective elimination of the unwanted cells in multi-cellular animals. The term has been a source of confusion and Durand and Ramsen have developed a concept and provided mechanistic and evolutionary definitions. It has become the general term to all the types of cell death that have a genetic component. The first insight into the mechanism came from BCL2 studies, the product of a putative oncogene that gets activated by the translocations of chromosomes found in follicular lymphoma. It was found that BCL2 promoted cancer stops lymphoma cells from being able to kill themselves. The trend of increasing attention and research efforts on PCD was highlighted with the 2002 Nobel Prize in Physiology and Medicine to Sydney Brenner (UK), H. Robert Horvitz (USA) and John E. Sulston (UK).

PCD is characterized by the invariable involvement of the dying cells' own machinery as an 'executioner', *i.e.*, it is a case of cell suicide. It is reported to play a major role in controlling cell number and turnover especially in the skin, gut and reproductive tract, tissue homeostasis and specialization, tissue sculpturing and pattern formation, oncogenesis and other disease incidents, defence reactions, and above all, in maintaining the overall integrity of the animal. PCD is distinct from necrosis. While the former is an energy-dependent physiological process that is genetically programmed with the involvement of regulatory genes, stimulatory events, signalling pathways and morphologically expressed distinctive features, the latter is a non-physiological process dissociated from developmental and morphogenetic events and involving cell swelling, lysis and leakage of cell contents with no obvious changes in DNA (*i.e.*, no chromatin condensation and DNA fragmentation). PCD has also been reported in plants although plant PCD studies have been initiated with greater vigour only in very recent years. PCD is as important in plants as in animals and that it can occur on a local or systematic scale. Krishnamurthy *et al.* (2000) in an excellent review gave details of Programmed Cell Death in plants.

PCD - Types

The two important signal transduction pathways involved in apoptosis are: Apoptosis (Type I cell-death) and Autophagy (Type-II cell death - cytoplasmic - being characterized by the formation of large vacuoles which eat away the organelles in a specific sequence prior to the destruction of the nucleus in the cell).

According to Lockshin and Zakeri (2004) and Bras *et al.* (2005), the cell death in animal systems has been sub-divided into 3 categories, namely: Apoptosis (Type I) - Autopohagic cell *death* (Type II) - and Necrosis (Type III), and this classification provides an important framework for the study of PCD. Type II cell death appears distinct from the other two, according to Schweichel and Merker (1973). The term – autophagy – is derived from Latin (meaning "self-eating", and it describes the process in which the cytoplasmic material is degraded through the lysosomal machinery of the cell, and is characterized by the formation of vacuoles and dilation of mitochondria and ER as well as a slight enlargement of Golgi material (Bras *et al.*, 2005).

Categories of Cells that underog Death in Plants

A number of categories of cells can be recognized within those that undergo death in plants.

Category (i)

Cells that have served their functions – A number of plant cell types belong to this category: root cap cells, cells undergoing senescence, cells involved in abscission of plant parts, vegetative cell of pollen, synergids, antipodals, anther tapetal and wall cells, transmitting cells of stigma, style and the ovary wall, endosperm cells, including haustoria, aleurone cells of grass endosperm, suspensor cells or developing embryo, *etc.* The cells of this category lose their earlier identity, morphology and functions and their remnants (or corpses) may persist. Available evidence strongly indicates that these cells undergo physiological PCD; cells at the line of abscission may undergo PCD while the other cells of the organ may die indirectly because of not receiving nutrient, water or growth regulators.

Category (ii)

Cells that are unwanted from the beginning – The stamen and carpel primordial cells respectively in male and female flowers, non-functional megaspores in monosporic and bisporic embryo sacs, non-functional microspores in the members of Cyperaceae and Epacridaceae, non-functional cells formed during embryoid initiation in cultured cells, *etc.* are examples of plant cells belonging to this category.

Category (iii)

Cells that die during redifferentiation into specialized cell types – xylem tracheary elements, cells of certain trichomes, thorns and spines, sclerenchyma (sclereids and fibres), cork cells, *etc.* are some of the plant cells belonging to this category. The most important characteristic of these cells is that they all start functioning only after their death. The other characteristic feature of these cells is that significant changes (morphological and chemical) take place in their cell walls. In other words, cell death leads to terminal differentiation in this category of cells.

A sub-category under this includes those cells which undergo '***partial***' death during differentiation. The sieve elements of the phloem belong to this category (The nuclei persist in mature sieve elements, albeit with a lot of structural changes and loss of transcriptional activity in species of *Selaginella* and *Isoetes* and in some ferns). Here, the nucleus is lost, but not the cytoplasm and its integrity. The functional life span of the cell is, however fixed; a sieve tube element functions from a few days to a few years depending upon the nature of the plant.

Category (iv)

Cells that are subjected to hypersensitive response – Cells subjected to hypersensitive response (HR) due to pathogenesis and other environmental and abiotic stresses such as osmotic, oxidative (H_2O_2 and salicylic acid), temperature, salt, water, heavy metals, UV light, nutrient deprivation, toxins, chemicals *etc.*, often undergo PCD. Here, cell death is used as defence against the stress or as a means of emanating signals for the other nearby cells to build up defence/immune reactions. PCD is reported to be employed in these instances not only to kill the host cells (to result in hypersensitive resistance response – HR lesions) to prevent further infection, but also the invading pathogens. In many instances, cells experiencing lethal stress doses often react by activating their PCD mechanisms to commit suicide before they are killed. A critical survey of the available literature indicates that not all cells subjected to stress undergo PCD; there are cells which may undergo necrosis, especially if the dosage of stress is beyond a particular level.

Category (v)

Cells that are damaged and unable to function further – In plants, cell death due to injury and other reasons may occur, but it is not known whether such deaths are programmed? It is often difficult to distinguish this category of plant cells from the previous category.

Category (vi)

Cells that are present in wrong places – Cells that die to result in lobe formation in lobed leaves and leaflets in compound leaves or to result in holes in the lamina of taxa like *Monstera* and in some *Croton* varieties are typical examples of this category.

Process Associated with PCD in Plants

PCD in animal system is reported to result in the disassembly of cells involving the condensation, shrinkage and fragmentation of the cytoplasm and nuclei into several sealed packets (often called the apoptotic bodies), which are then fagocytosed by the neighbouring cells or macrophages. Thus, there are no remnants of cells corpses left. Nuclear fragmentation is preceded by chromatin condensation and marginalization in the nucleus. Fragmentation of DNA at the nucleosome linker sites then takes place. When all these events are combined to result in a distinct morphological expression then PCD is termed as apoptosis, a word coined by Kerr *et al.,* in the year 1972. In other words, apoptosis is a distinct form of PCD. However, there are others who consider apoptosis and PCD as one and the same. The apoptotic PCD involves a syndrome of mechanisms with effectors, adaptors, regulators and signals as well as the activity of endonucleases.

Although several plant systems have been studied to understand the mechanism of PCD, very little attention has been paid to the details of cytoplasmic changes, if any, during cell death. Condensation and shrinkage of cytoplasm were reported in dying aleurone cells, root cap cells, sister cell of the embryonal initial and HR lesion cells of *Arabidopsis*. In the embryoid sister cell, the break-down products of the cytoplasm are reported to form several small membrane-sealed packets. TEs of intact plants is reported to become lobed, condensed, shrunken and finally gets broken into small packets. A critical perusal of hundreds of published micrographs of dying plant cells belonging to megaspore dyads and tetrads, nucellus, endosperm, embryo, suspensor, anther tapetum, root cap, *etc.* indicates the occurrence of cytoplasmic condensation and shrinkage but not its breakage into bits; there are evidences of vacuolar autophagy in most cases which accounts for the elimination of the cytoplasm.

A different mechanism of programmed autolysis was first suggested by Nagl with reference to the death of embryo suspensor cells. The filiform leucoplasts of the suspensor cells are transformed into cytosome-like structures displaying high acid phosphatise activity. during transformation, polar swelling of the plastids takes place incorporating into it the invaginations of the degenerating cytoplasm to form spoon-like structures, which may later become 'ball-like'. These structures were termed 'plasto-lysomes'. They form the characteristic elements of the lytic compartment. Plastolysomes include various organelles and cytoplasmic components, particularly ER, polysomes and lipid bodies. At a later stage, the stroma of these plastolysomes disappears completely and finally the bodies are transformed into autophasic vacuoles surrounded by a single membrane; all the included cytoplasmic components also degenerate. Whether it be by vacuolar autophagy or by plastolysomy, the plant cells that died are always evident as corpses and these are invariably represented by the remnants of the cell wall.

Effect of Caspases

Animal PCD is due to a syndrome of well organized and executed mechanisms involving effectors, adaptors, regulators and signals. The chief effector molecules are caspases. No caspase has yet been cloned from plants. But there is a growing information on the family of caspase-like proteinase and their involvement in plant PCD. The expression of cysteine proteinases (CyPs) is reported during cell death in *Solanum melongena* (Xu, 1999). Several types of cysteine proteinases such as actinidin, oryzain, barley aleurain *etc.* are known from the plant system. Phytepsin D, a barley vacuolar aspartate proteinase, is reported to be expressed very highly during autolysis and development of TEs and sieve elements. It is however, interesting to note that these proteinases are expressed not only during cell death but also during other cell events. There is still only a speculation that proteinases reported in plant PCD are involved in a function analogous to that of cysteine proteases of the interleukin l^2 converting enzyme (ICE) family reported in the animal system.

The substrates for almost all these plant proteinases are yet to be identified. It is quite possible that the plant proteinases are involved in the clearance of cytoplasmic and nuclear proteins from dying/already dead cells such as TEs, *i.e.*, they may merely be doing a scavenging job, thereby helping the recycling of nitrogenous compounds. Moreover, the results of Xu and Chye (1999) in brinjal on the isolation and characterization of cDNA encoding CyPs and their expression indicate their coincidence with several rather than specific events in cell death in many organs/tissues such as ovules, xylem, placenta, receptacle, anther, epidermis, endothecium, nucellus, *etc.* Although CyPs of brinjal has 90-92 per cent amino acid identity to the other CyPs from Solanaceae family (tobacco CyP 8 and tomato CyP2), they show only 39 per cent homology to cathepsin H, an animal protease. Caspase inhibitors, such as the chloromethylketone Ac-YVAD-CMK or aldehyde Ac-DEVD-CHO, are reported to completely abolish Hypersensitive resistant response (HR) cell death in tobacco. This is cited as an evidence for the presence of caspase-like plant proteases that participate in HR cell death. But, it is not yet clear whether these inhibitors are targeting the proteases that are evolutionarily related to metazoan caspases.

There are reports of a gene implicated in 'caspase' production and function in plants. *SAG12*, an *Arabidopsis* gene, is reported to encode a putative cysteine protease involved in senescence. *HSR 23 J*, a tobacco gene, is activated during HR specifically, while H1N1, a tobacco gene is activated during HR as well as lately in senescence, but the last two genes are reported to be activated only after the switching of 'cysteine proteases'.

Adaptor Molecules and Regulator Molecules

Neither Adaptor molecules nor Regulator molecules have been reported in plant PCD.

Signal Molecules

Oxidative stress, as a signal event, is shown to trigger cell death in plants; the involvement of O_2 in Hypersensitive Resistance (HR) responses and of H_2O_2 in cell death mediated by pathogens is reported. However, we do not have a clear idea about the role of these two ROS in cell death process in both animals and plants, although H_2O_2 is likely to induce DNA fragmentation and an influx of extracellular calcium. In plants H_2O_2 has been reported not only in cells infected with pathogens and subjected to stress but is also regularly produced when there is an activity of peroxidase and polyaminic oxidase, lignifications and cross-lining of extensions in the cell wall. It is true that H_2O_2 is present not only in living cells but also in cells that undergo death due to wounding, pathogen entry and xylogenesis, but available evidence is not sufficient enough to unequivocally implicate ROS species in PCD of these cells; they are more likely to be involved in necrotic cell death which is found hand in hand with PCD systems subjected to biotic and abiotic stresses, especially at a level more than threshold.

Like in animal PCD, in plant cells undergoing death an increase in cytosolic calcium is reported to be an important signalling event. The probable role of calcium is the activation of endonucleases in both PCDs. However in plants, Zn^{2+}-dependent endonuclease is known in addition to Ca^{2+}-dependent endonuclease. The former is known in cells of wheat leaf undergoing senescence, while the latter is known in differentiating TEs, aleurone cells and in HR lesion cells.

Plant cells subjected to oxidative stress (*i.e.*, treatment of H_2O_2) showed an increase in intracellular reduced glutathione. Plasma membrane blebbing has not so far been reported in plant cells undergoing death, although PS exposure has been demonstrated through annexin V binding assay in the protoplasts of tobacco subjected to abiotic stress. Whether this has any relation to the presence of the cell wall in the plant system is not clear.

Chromatin Condensation, Endonucleases and DNA Fragmentation

The programmed cell death of aleurone cells in barley during germination displays several characteristics associated with apoptosis. TUNEL staining of aleurone cells that had been triggered to die indicated that nuclear DNA in the cells undergoes fragmentation starting near the embryo. The fragmentation proceeds outward in a time-dependent manner *in vivo* and under osmotic stress in isolated cells *in vitro*. The visualization of DNA fragmentation precedes visible signs of death. Abscisic acid inhibits the aleurone cell death and DNA fragmentation both *in vivo* and *in vitro*. Okadaic acid, a protein phosphatase inhibitor that suppresses apoptosis in some animal systems, also inhibits the death of aleurone cells.

The root cap consists of living parenchyma cells derived continuously from the apical meristem. As new cells are produced in the interior, those on the periphery of the root cap are shed in a very orderly manner, reminiscent of Kerr's original connection of the term apoptosis to cell death as a natural programmed process. Cap cells of fresh onion and tomato root tips, double stained by TUNEL assays and a DNA stain, revealed that nuclear DNA in cells throughout the root sections remained intact until the cap cells reached the surface of the root. The region of TUNEL-positive cells was restricted to the outermost 1-3 cells of the root cap. The appearance of nuclear DNA changed from a compact circular form to elongated TUNEL-positive structures that eventually took on the form of numerous membrane-bound circular bodies that migrated to the periphery of the dying cell. The formation of these bodies appeared to be the final step reached before a cell is shed from the root cap. The morphology of PCD in aleurone and root cap cells confirmed the conservation of at least the terminal phases of the apoptotic process in plants.

Stress Proteins

One interesting feature about the death of some types of plant cells is the active production of special categories of structural protein rich in hydroxyproline, proline, glycine, arabinogalactan and occasionally threonine. These are actively synthesized in the ER packed in the Golgi and transported to the cell wall where they become integral components. In sclerenchyma cells and xylem elements such proteins have been repeatedly reported during the process of cell death.

Apoptosis

This process of PCD occurs in multi-cellular organisms, and the biochemical events lead to characteristic morphological changes in the cells and finally death. The changes include: blebbing, shrinkage of the cell, fragmentation of the nuclear material, condensation of the chromatin and fragmentation of the chromosomal DNA. The cells are induced to positively commit suicide (in a developmental context), while the absence of survival factors provide the impetus for suicide. Some evidence suggests that endonuclease activation is induced to presume that apoptosis and PCD must be genetically mediated.

Autophagy (Macroautophagy)

This is a catabolic process resulting in the autophagosomic - lysosomal degradation of bulk cytoplasmic contents, abnormal protein aggregates and excess or damaged organelles. This process is generally activated by nutrient deprivation but is also associated with physiological and pathological processes (like development, differentiation, neuro-degenerative diseases, stress, infection and cancer).

Other Types of PCD

Other pathways that have been discovered include: Necroptosis (also called as - Non-apoptotic PCD or Caspase-independent PCD - a type of necrosis or an inflammatory cell death, associated with un-programmed cell death due to cellular damage or infiltration by the pathogens); Nemosis (necrosis that takes place in fibroblasts); Eryptosis (a type of a suicidal RBC's death); Anoikis (identical with apoptosis except in its induction); Cornification (a type exclusive to the eyes); Excitotoxicity; Ferroptosis (iron-dependent form of cell death); and Wallerian degeneration. In addition to these, Aponecrosis (refers to an incomplete apoptosis that is completed by necrosis); NETosis (a process of cell death generated by NETs); Paraptosis (a non-apoptopoic cell death mediated by MAPK through the activation of IGF-1, and characterized by the intra-cellular formation of the vacuoles and swelling of mitochondria); and Pyroptosis (unique in being mediated by caspase-1, an enzyme that is not involved in apoptosis, in response to an infection by certain microbes). The plant cells undergo the processes similar to autophagy, but some common features are highly conserved in both plants and metazoans. The term – Phenotopsis (derived from "pheno" meaning 'showing' and 'ptosis' meaning 'programmed death' or 'falling off') was proposed by V. P. Skulachev (1999). It is a process of PCD that an organism's genes include the features that will cause the organism to rapidly degenerate and die off. This can also be referred to as "fast phenoptosis" as aging is being explored as "slow phenoptosis".

PCD – in Plant Tissue

PCD in plants has a number of molecular similarities to apoptosis in animals and it has also differences, the main one being the presence of a cell wall and the lack of an immune system that removes the pieces of the dead cells. Instead, the dying cell synthesizes the substances to break itself down and place them in a vacuole that ruptures as the cell dies. According to Martin Bonke and his colleagues, in the vascular plants, xylem consists of several cell types, the differentiation of which involves the deposition of the elaborate cell wall thickenings and PCD. The basic morphological and biochemical features of PCD have been conserved in plant and animal kingdoms, and specific types of plant cells carry out unique cell death programmes. These have common features with apoptosis of animals, such as degradation of nuclear DNA, and this is triggered by the collapse of the vacuole in the tracheary elements of xylem. J. Balk and C. J. Leaver, Dept of Plant Sciences, University of Oxford, carried out research on the mutations in mitochondrial genome of cells of Sunflower, to suggest that mitochondria play the same role in vascular plant PCD like in all the eukaryotic cells.

PCD in plants is a facet of a wide range of developmental programmes that vary from the beginning of the life cycle through xylogenesis (essential development) until senescence (the end of life cycle) according to Beerts (1997) and Rogers (2005). Compared with the animals, the cell death is important to plants as cell division, and PCD is used to describe all the instances of cell death observed in plants that lead to confusion.

PCD in Pollen

Plants enforce self-incompatibility (SI) during pollination as an important means to prevent self fertilization. Research carried out on *Papaver rhoeas* (Corn Poppy) has revealed that the proteins in the pistil on which the pollen lands, interact with the pollen and trigger PCE in incompatible pollen. S. G. Thomas and V. E. Franklin-Tong found that the response involves rapid inhibition of pollen tube growth, followed by PCD.

Conclusion

PCD in plants is a crucial component of development and defence mechanisms, and this is widely used to describe more instances of observed death in plants. The PCD is mediated by the proteolytic enzymes, caspases which trigger the cell death by cleaving the specific proteins in the cytoplasm and the nucleus. These enzymes exist in the cell as procaspases (the inactive precursors) which are activated by other caspases, producing a proteolytic caspase cascade. The activation is initiated by either extra-cellular or intra-cellular death signals which cause intra-cellular adaptor molecules to aggregate and activate the procaspases. The activation of caspases is regulated by the members of the Bcl-2 and IAP protein families.

PCD is important to the growth and development in plants as well as finding different ways to manipulate the pathways that have agronomic benefits. The elucidation of the molecular pathways of caspase-dependent apoptotic cell death has led to use this term interchangeably with PCD. Having reviewed plant PCD, it is apparent that 3 forms of cell death can be described – autophagy, AL-PCD and necrosis. The second type is the most common form of induced PCD in cell culture systems, with necrosis. The first one appears to be associated with the starvation response in the cultured cells occurring at higher levels.

References

Gilchrist, D.G. 1998. Programmed cell death in plant disease – The purpose and promise of cellular suicide. *Ann. Rev. Phytopathol.* **36:** 393-414.

Krishnamurthy, K.V., Krishna,R., Chochavendan,R. and Samuel, F. 2000. The programme of cell death in plants and animals – A comparison. *Curr. Sci.* **79:** 1169-1181.

Index

www.ingramcontent.com/pod-product-compliance
Ingram Content Group UK Ltd.
Pitfield, Milton Keynes, MK11 3LW, UK
UKHW041306190726
13845UKWH00001B/6

9 789354 616372